EARTH OBSERVATION SYSTEMS FOR RESOURCE MANAGEMENT AND ENVIRONMENTAL CONTROL

NATO CONFERENCE SERIES

I   Ecology
II   Systems Science
III Human Factors
IV Marine Sciences
V   Air—Sea Interactions

II SYSTEMS SCIENCE

EARTH OBSERVATION SYSTEMS FOR RESOURCE MANAGEMENT AND ENVIRONMENTAL CONTROL

Edited by
Donald J. Clough
University of Waterloo
Waterloo, Ontario, Canada

and

Lawrence W. Morley
Canada Centre for Remote Sensing
Ottawa, Ontario, Canada

Published in coordination with NATO Scientific Affairs Division

PLENUM PRESS · NEW YORK AND LONDON

Library of Congress Cataloging in Publication Data

Main entry under title:

Earth observation systems for resource management and environmental control.

(NATO conference series: II, Systems science; v. 4)
"Proceedings of a conference . . . held in Bermuda, November 15-19, 1976, spon-
sored by the NATO Special Program Panel on Systems Science."
1. Remote sensing—Congresses. I. Clough, Donald J. II. Morley, Lawrence W. III.
Nato Special Program Panel on Systems Science. IV. Series.
G70.E23   621.36'7   77-13989
ISBN 0-306-32844-5

Proceedings of a conference on Earth Observation and Information Systems held in
Bermuda, November 15-19, 1976, sponsored by the NATO Special Program Panel
on Systems Science

Acknowledgments

The Co-Directors would like to thank His Excellency the Governor of Bermuda, Sir Edwin Leather, K.C.M.G., K.C.V.O., L.L.D., for his good offices and for his resounding opening address to the ARI. His Excellency upheld the concept of the NATO peacetime alliance as a force for the advancement of science and technology to serve peaceful international purposes.

We would also like to thank the NATO Deputy Island Commander, Commander David I. Aldrich, R.N., Resident Naval Officer for the West Indies, H.M.S. Malabar, Bermuda. His exceeding good humour as a host and raconteur provided a nice change of pace from the heavy scientific deliberations of the ARI.

We would also like to thank Captain Raymond Smith, Aide-de-Camp to the Governor of Bermuda, for his kind assistance and services on behalf of the Governor.

The friendship and support of many Bermudians was remarkable. In particular, staff of the Board of Education and a number of government branches and commercial establishments made special efforts on our behalf.

Dr. Edward Pestel represented the sponsoring NATO Science Committee, and Dr. Bulent Bayraktar represented the Special Programme Panel on Systems Science, as NATO observers.

Mary-Lou Clough and Beverley Morley assisted in arranging social activities, and provided support during the workshop and plenary sessions.

Finally, the Co-Directors owe a special debt of thanks to Mrs. Ingeborg K. Moerth, Research Administrative Assistant to Professor Clough at the University of Waterloo, Canada. Mrs. Moerth served as Administrative Chief of Staff at the ARI and provided staff support in the production of these Proceedings.

Preface

The NATO Science Committee and its subsidiary Programme Panels provide support for Advanced Research Institutes (ARI) in various fields. The idea is to bring together scientists of a chosen field with the hope that they will achieve a consensus on research directions for the future, and make recommendations for the benefit of a wider scientific community. Attendance is therefore limited to those whose experience and expertise make the conclusions significant and acceptable to the wider community. Participants are selected on the basis of substantial track records in research or in the synthesis of research results to serve mankind.

The proposal for a one-week ARI on Earth Observation and Information Systems was initiated by the NATO Special Programme Panel on Systems Science (SPPOSS). In approving the ARI, the senior NATO Science Committee identified the subject as one of universal importance, requiring a broad perspective on the development of operational systems based on successful experimental systems.

The general purpose of this ARI was to address the critical problems of integrating the relatively new science and technology of remote sensing into operational earth observation and management information systems. The main problems of concern were those related to systems design, organization, development of infrastructure, and use of information in decision processes. The main emphasis was on problems of transferring technologies and methods from *experimental* to *operational* systems.

Thirty-two scientists and science administrators participated in the ARI, including two observers from the NATO Science Committee and SPPOSS. Twenty-six background papers were prepared in advance as a basis for initiating discussions. Eight working groups were formed: four to deal with *systems* aspects and four to deal with *applications* aspects. Each participant served on one systems group and one applications group, to provide cross-linking and eventual integration of ideas. The schedule of working group meetings was designed to accommodate this cross-linking scheme.

Short plenary reporting sessions were scheduled at the end of
each day, to provide further opportunities for cross-fertilization
of ideas. Final conclusions and recommendations were discussed in
a major plenary session on the last afternoon, and consensus was
achieved on major points.

This publication of the ARI Proceedings may be read in three
parts, from less detailed to more detailed, as follows:

1. An overview summary report of conclusions
    and recommendations, by the Co-Directors.

2. Eight summary working group reports by
    the working group Chairmen, based on notes
    taken at working group and plenary sessions.

3. A set of twenty-six background working papers
    prepared by the individual participants.

Editing by the Co-Directors was minimal and preserved all the original basic contributions of the participants, individually and in
groups.

The lead-off background paper by the Co-Directors, entitled
"Earth Observation Systems", provides a fairly broad introduction,
statement of purpose, structure of working groups, background notes
on systems and background notes on applications. It is recommended
as a starting point for readers who are not familiar with the array
of technologies and issues involved.

Donald J. Clough
Lawrence W. Morley

Contents

PART 3:   BACKGROUND PAPERS

OVERVIEW SUMMARY REPORT:

CONCLUSIONS AND RECOMMENDATIONS

Donald J. Clough
Lawrence W. Morley

EARTH OBSERVATION AND INFORMATION SYSTEMS

- Great progress has been made in recent years toward the development of operational earth observation systems and related information systems for resource management and environmental control.

- As the world's population increases, both national and international observation and information systems will become
more important for the management of food, fuel and other
scarce material resources and for the control of environmental pollution.

- In this ARI, the NATO Science Committee and its subsidiary
Special Programme Panel on Systems Science have provided a
timely forum for discussion of a variety of scientific,
technological and socio-economic issues related to implementation of an operational earth observation system. This
is one step toward elucidation of the issues. Hopefully it
will lead to further steps at political levels under the
auspices of a multi-nation organization such as NATO or the
international umbrella of the United Nations - steps toward
the development of an operational earth observation system.

- The ARI was organized into eight workshop groups:

Systems

A.   Platforms and Sensors
B.   Data Handling Systems
C.   Management Information Systems
D.   Socio-Economic Systems Aspects

Applications

1.   Atmospheric Applications
2.   Land Applications
3.   Ocean Applications
4.   Evaluation Methodology

- Main conclusions and recommendations of the working group reports, ratified in plenary session discussions, are summarized in the following sections. (See the individual working group reports for further amplification.)

PLATFORMS AND SENSORS

- Earth observation systems require both *remote* and *in-situ* platform/sensor systems to monitor both environmental parameters and human activities (e.g., land use, offshore engineering).

- No single platform or sensor type can satisfy all requirements. Integration of various platforms and sensors is required to support specific missions.

- There is a need for a central repository of information concerning all state-of-the-art sensors, platforms and their uses. *Recommendation:* an international agency, such as the International Astronautical Federation, should promote an annual publication describing advances in sensors, platforms and their uses.

- There is a need for international cooperation in research and development of advanced operational remote sensing platform/sensor systems which may exceed the economic capabilities and interests of individual nations.

- The greatest single improvement to the real-time global weather data base, for weather forecasts beyond 18 hours, would come from development of accurate vertical temperature and humidity profiling sensor systems. Present systems have not lived up to original expectations.

- The first Geostationary Operational Environmental Satellite (GOES), operating since October, 1975, has been surprisingly successful as an aid to field meteorologists, and is changing the approach to forecasting hazardous weather events. Success is attributed to the satellite's *on-demand* availability of imagery and computer-generated image grid.

- The success of GOES could be exploited further by developing an operational system of multiple polar orbiting TIROS-N satellites to provide *on-demand* imagery of high-latitude regions in conjunction with GOES imagery of mid-latitude regions.

- There is a need for higher resolution infrared imagery for atmospheric and ocean applications. *Recommendation:* the design objective of 1-kilometer resolution at the sub-satellite point, from geosynchronous orbit altitudes, should be pursued.

- Current experimental LANDSAT Multi-Spectral Scanner (MSS) systems are capable of 80-metre ground resolution in the visible and near-I.R. bands. Planned follow-on systems may be capable of 30 or 40 metre resolution. *Recommendation:* MSS sensor research should pursue the long-range objective of providing 10 or 20 metres resolution, for operational systems providing 9-day repeat coverage of the globe, over selected areas.

- The Synchronous Earth Observatory Satellite (SEOS), proposed for possible launch in the mid-1980s, could be as surprisingly successful as GOES because of the *on-demand* coverage. The trade-off of ground resolution against more timely coverage may be favorable. A SEOS would be naturally complementary to a LANDSAT-type satellite. *Recommendation:* uses and requirements for a SEOS for land sciences should be surveyed, to define technology requirements and economic applications.

- Satellites such as GOES and NIMBUS have been used successfully for retransmission of weather and ocean data from ocean data buoys, and tracking data from drifting buoys. *Recommendation:* integration of remote sensing satellites, surface platforms, and data retransmission requirements, should be planned for all future operational systems.

- There is a need for an operational system of spaceborne high-resolution imaging radars of the SEASAT-A Synthetic Aperture Radar (SAR) type. Because of its all-weather, day and night monitoring capabilities, the introduction of the SEASAT *SAR* (1978 launch) may be as momentous as the introduction of the LANDSAT *MSS* (launch 1972).

- As a prelude to an operational radar satellite system, there is a need for international cooperation in R and D on spaceborne SAR and SAR data processing. Otherwise the cost to individual nations may be too great.

- *Recommendation:* basic research on remote sensors for the following parameters should be intensively supported:

  (1) high resolution rainfall intensity

  (2) soil moisture content

  (3) sea ice thickness

  (4) air temperature over oceans

  (5) sea level pressure measurement

  (6) air motion (winds) in the free atmosphere independently of cloud motion

  (7) aerosol measurement.

- *Recommendation:* corrected images obtained by different remote sensing systems, above a specified level of resolution,
  should be provided in the same standardized format, with
  standardized geo-referencing and registration. Superposition should be sufficiently accurate for visual interpretation purposes.

DATA HANDLING SYSTEMS

- Multi-disciplinary activities, including data handling specialist activities, are especially important in the formulation stage of remote sensing systems design projects and
  application projects. Under current practice, data handling
  specialists are seldom involved in project formulation stages.

- A complete systems design approach should be adopted in the
  development of any operational data handling system, relating to the following components:

  (1)   sensors and platforms

  (2)   on-board data handling and processing

  (3)   telemetry

  (4)   data relay systems

  (5)   ground receiving facilities

  (6)   data-base architecture

  (7)   ground data handling and processing
        facilities

  (8)   information extraction methods

  (9)   user-oriented models

  (10) information services and distribution

  (11) final user requirements.

- With few exceptions, computer processing power and data storage systems now available are adequate to handle most applications up to 1980. One exception is digital processing for SEASAT-A synthetic aperture radar data; near real-time processing is not feasible with the existing state-of-the-art. (But we are optimistic about breakthrough developments by the early 1980's.)

- Economies of scale will accompany future operational systems. Unit costs of components and subsystems such as in-situ sensors, decentralized computing, and receiving stations, will probably decrease dramatically as operational systems evolve.

- The following potential problem areas require extensive study:

  (1)   data ownership and confidentiality

  (2)   technologies for data storage and retrieval

  (3)   bandwidth selection for high bit rates

  (4)   combination of multiple-source data

  (5)   dedication to reducing costs.

MANAGEMENT INFORMATION SYSTEMS

- Because of transmission channel capacity constraints, the amount of data processed on board spacecraft and other platforms such as ocean data buoys may increase. There is presently little coordination in the specification of management information requirements on which to base decisions concerning *on-board* versus *ground* data processing.

- Raw or partly processed data are acceptable to users in experimental systems. However, in an operational system the data must be transformed into specialized *information-products* which are readily useable for decision-making purposes. The design of an effective management information system (MIS) requires close cooperation between data processing specialists and users concerning (1) operations of the data base, and (2) production of tailored information products.

- At present, *centralization* of data bank operations and production of some information products seems to be most cost-effective. However, rapid changes in technology and methods, and rapid adoption by users, may lead to more cost-effective *decentralization*. General conclusions about the relative merits of centralization versus decentralization in future systems would be premature.

- *Standardization* is of vital importance to the M.I.S. component of an operational remote sensing system, including standardization of radiometric and geometric scaling, geo-referencing, and formatting for superposition of images from different systems and different times. Development of standards should be accelerated prior to development of an operational system.

- User instruction, training and education are essential to the effective utilization of an M.I.S. Regular user education programs are presently lacking, and should be strongly promoted as a necessary prelude to an operational earth observation system.

- *Reliability* is an important consideration in M.I.S. operations. Relationships between reliability and cost, during parallel maintenance and subsequent merging of old and new M.I.S. operations, should be taken into account in the design of an operational system.

- *Learning time* must be taken into consideration by the designers of management information systems. Users' expectations should be based on a realistic evaluation of the learning curve in each application.

- *Data lifetime* is an important consideration in the design of a data bank. Cooperation between designers and users is essential for the development of rules for storage time and purging of data from storage.

- At present our *knowledge* of requirements for data and derived information is limited. Consequently our ability to design effective operational systems is limited. The process of design is evolutionary, and the need for system flexibility and adaptability is obvious.

SOCIO-ECONOMIC ASPECTS

- *Continuity* of remote sensing information is becoming essential for the detection and prediction of hazardous events and for the detection of subtle temporal changes in the environment. The guaranteed continuity of the LANDSAT program, for example, is a matter of great international concern.

- There now exists a substantial international demand for early transition from the experimental LANDSAT system to a fully operational system having guaranteed reliability and continuity.

- Both the technical feasibility and the high probability of large economic benefits of operational earth observation satellite systems have been demonstrated by LANDSAT.

- The technology embodied in the experimental LANDSAT satellites provides an adequate basis for the design of an operational system, and could be extended and improved by providing :-

  (1)   the improvements planned for LANDSAT-C,
        with the addition of a Thematic Mapper

  (2)   high reliability through multiple satellites
        and ground stations

  (3)   nine-day repetitive coverage, and

  (4)   more rapid availability of data and
        information products.

- The United States would probably be looked upon as a leader in the development of either a national or an international operational system based on LANDSAT technology.

- An organized international effort is now required by the community of remote sensing scientists and users to ensure steady progress toward an operational system.

- Existing international organizations such as the U.N. World Meteorological Organization (W.M.O.) and the U.N. Food and Agriculture Organization (F.A.O.) could expand their areas of concern to include participation in the development and operation of an international earth resources remote sensing system. However, new international mechanisms may be required.

- *Recommendations:* The term "system" below includes both technologies and organizations. Separate parts of a cooperative "system" (e.g., satellites, ground stations) may be owned or operated by individual nations. The Working Group recommended that:-

  (1) an international operational Earth remote sensing system be established in which participating States obtain certain rights and accept certain responsibilities

  (2) each State or region be encouraged to establish an administrative focal point for coordinating earth observation activities

  (3) each State or region be encouraged to establish one or more centres for the collection, evaluation and dissemination of remotely sensed data of interest to it, to aid all States and regions in making effective use of the data

  (4) States or regions, through their academic, scientific and technical organizations, develop the individual skills required for the use and application of remotely sensed data

  (5) international mechanisms be established to:

    (a) assist in the world-wide collection of remotely sensed data

    (b) disseminate remotely sensed data throughout the world

    (c) analyse world-wide user needs

    (d) provide assistance and training in the use of remotely sensed data

  (6) existing international organizations should be considered with reference to recommendation number (5), and

  (7) an international conference be convened to consider the establishment of a global operational earth remote sensing system.

- The timetable for achieving the goal of recommendation (1) above would be subject to national and international political judgements. The international conference of recommendation (7) would have to be at a *political* rather than simply a *technical* level, perhaps under the auspices of the United Nations. If planning were to start immediately, the recommended conference could be held about 1980.

- If the development of an international operational earth observation system is not fostered through a United Nations mechanism, some individual nation or group of nations may proceed alone to develop a system having global monitoring capabilities.

- It may be timely for a multi-nation group or alliance such as NATO to play a vigorous role, not only in the establishment of an operational system but also in the devising of safeguards that would ensure its use for the benefit of all mankind - now and in the future.

ATMOSPHERIC APPLICATIONS

- Improvements in monitoring the atmosphere are required in both remote sensing and *in-situ* modes. The most important remote sensing improvements required in the near future are noted in the PLATFORMS AND SENSORS report. The most important *in-situ* improvements include low-cost ocean surface platforms for air and sea temperature and sea level pressure measurements, and low-cost automated agrometeorological observing stations.

- There exists an urgent need for better global communication of earth observation data. A communication satellite system is required that could relay data over the globe within two hours of real-time data acquisition.

- In many parts of the world there is a lack of knowledge of present meteorological satellite systems, particularly the recent application of GOES to short-range weather forecasting. Countries should be made aware of existing education and training opportunities, such as might be supported through the WMO Voluntary Assistance Program and NATO Science Committee Programs.

- It is of concern to members of the community of atmospheric scientists, and users of data, that there has been a steady diminution of meteorological data from *conventional* sources in recent years, particularly from the oceans and tropical land areas. This trend should be stopped, and conventional *in-situ* data and *remote sensing* data balanced to improve weather, climate and agrometeorological programs.

- An operational remote sensing system should be planned to function for a predetermined number of years without major changes. Subsequent changes should only be made if improvements are clearly indicated. Stability is necessary to facilitate user learning and to encourage user investments.

- The convergence of satellite observing techniques and computer solutions of atmospheric equations points to an impending scientific/technological breakthrough to provide *accurate one or two day predictions of the atmosphere-ocean system in the next five to ten years.*

LAND APPLICATIONS

- Early candidate applications for an operational system include (a) crop forecasting, (b) integrated resource surveys for land use, land capability, and water inventories, forest and rangeland inventories, mineral and fuel exploration, pollution monitoring, and (c) topographic and thematic mapping and map revisions up to scales of 1:50,000.

- MSS spectral bands within the region from 0.53 µm to 2.60 µm, with band widths varying from 0.04 µm to 0.60 µm, have proven to be most useful for remote sensing applications in forestry and agriculture. Improved interpretation and classification is possible with bands as narrow as 0.04 µm.

- Current LANDSAT MSS spatial resolution of about 80 metres is not adequate for many economic applications. Resolution of about 30 metres is required. Because of limitations in trade-offs among resolution, swath width, frequency of coverage, and data transmission and processing rates, no consideration should be given at present to global MSS coverage with resolution *less* than 30 metres. However, research should aim at eventual 20 metre resolution with 9-day repeat coverage.

- For most land applications of economic interest, the current LANDSAT 18-day cycle of repeat coverage is inadequate, particularly because of the frequency of cloud cover in many areas. A nine-day cycle is a desirable design objective for an operational system.

- All-weather, day or night imaging systems should be developed for several important land applications. A combination of passive infrared and microwave imagers and synthetic aperture radar (SAR) is desirable. Use of the SEASAT-A SAR will provide the first opportunity to evaluate such a spaceborne system, and experimental SAR time should be allocated to a representative selection of land applications.

- Full use should be made of all possible methods of obtaining stereoscopic images from satellites and from the space shuttle. Consideration should be given to developing a system to provide global stereoscopic mapping with a resolution of about 20 metres and a base to height ratio of 0.6 to 1.0.

- Future operational satellite systems should have the capability to provide data availability globally within five days of data acquisition in all cases, and a special capability to provide data within hours for some critical applications.

- Further research should be stimulated on improving the geometric accuracy of matching data from different remote sensing sources and from the same source at different times. Multi-temporal matching of data is particularly important in agriculture and forestry applications.

- Remotely piloted vehicles (RPV's) and high-altitude rockets equipped with cameras and other sensors may be cost-effective in some special applications requiring on-demand remote sensing of remote land and coastal ocean areas under circumstances that militate against the use of manned aircraft and other platforms. Experimental remote sensing from such platforms should be continued.

- Special aid is required to help developing countries to build remote sensing capabilities and to prepare for participation in an international earth observation system.

- Special public relations, promotional and educational programs are required to increase public understanding of remote sensing and the value of an operational earth observation system.

- Research on the applications of remote sensing in natural resource management and environmental sciences should be increased in conjunction with the development of an operational earth observation system.

OCEAN APPLICATIONS

- Satellite remote sensing methods are most promising for provision of ocean surface information, including real-time information on waves, sea ice, surface temperature, current boundaries, ocean colour, pollution, and near-surface winds and other weather parameters required in economic applications such as transportation, fishing and offshore engineering. Satellite systems are also required to provide long-term wave and ice climatology statistics for vessel and offshore engineering design purposes. Existing satellites such as NIMBUS, NOAA, GOES and LANDSAT have proven to be useful in oceanology. SEASAT-A, planned for launch in 1978, represents a remarkable step forward in satellite technology for ocean remote sensing applications.

- Desired frequency of observation, resolution, precision, accuracy and range of required ocean parameter measurements have been specified by user groups, and many requirements are within the state-of-the-art of satellite remote sensing (e.g., SEASAT-A).

- The optimum system for ocean observation will be a complementary combination of various platforms, sensors, and data transmission/data handling technologies. Satellite sensors can provide unique synoptic data, telemetry and position-fixing capabilities, but cannot satisfy all surface, subsurface and near-surface atmospheric data requirements (e.g., surface pressure, currents, etc., provided by ocean data buoys).

- Ocean management information systems can be made more effective by the use of shipboard receivers for data retransmitted from various platforms and data processing stations via
satellite. Development of low-cost shipboard equipment, in
conjunction with communication satellite systems, is important to facilitate adoption by users.

- An international community of scientists and users of remote sensing data have expressed keen interest in the SEA-
SAT-A sensor suite, particularly the synthetic aperture radar (SAR). Further development of these sensors and their
applications would benefit greatly from the availability of
SEASAT-A for experimental use by research groups outside the
U.S.A. This would require the organization of international
cooperative efforts. To avoid wasteful duplication, the development of future spaceborne microwave sensor systems, including radar, should be a multi-national effort.

EVALUATION METHODOLOGY

- While the scientific community and part of the potential user
community are convinced of the present and future utility of
information produced by satellites, national policy and decision-making communities are not similarly persuaded.

- Economic rationales such as cost-benefit analysis have been
important in evaluation processes for such programs as LAND-
SAT and SEASAT. It seems important to emphasize the need
for economic justification of any future operational system.

- Non-economic rationales have also been important, including
(1) the "progress imperative", based on the assumption that
new technology will automatically produce unexpected as well
as expected benefits, (2) the "sovereignty imperative",
based on the expectation that some other nation will develop
an operational system alone and may obtain better information,
and (3) the "knowledge imperative", based on the premise that
knowing everything about everything is itself a positive social good. It seems important that these rationales be understood by those involved in promoting an international operational system.

- *Recommendations:* The Working Group recommended that:-

(1)   Future economic studies contain precise descriptions of objectives of the technology, the decision environment in which new information will be used, and criteria for the measurement of improvements of the new system compared with existing systems

(2)   applications scientists should be motivated to report the way the results of their research fit into decision making processes of resource managers

(3)   economic field tests should be designed into experimental remote sensing projects, as a basis for validation of socio-economic assumptions as well as technical assumptions in supporting analyses

(4)   case studies of successful and unsuccessful technology transfer should be made, with emphasis on how successful experimental applications have lead to successful operational applications

(5)   more effort should go into modelling decision processes in the user community. Examples of formal decision models exist in such fields as water resource and electric power management.

SUMMARY REPORT A
PLATFORMS AND SENSORS

B.P. Miller, Group Chairman
Vincent J. Cardone
Donald J. Clough
J. D'Hoore
Hervé Guichard

1.   Introduction

Earth observation and information systems are intended to provide information on the natural environment of the planet Earth, and
on the interaction of human activities with this environment. The
natural environment consists of the atmosphere, and the surface and
subsurface regimes of both the land masses and the oceans. Human
activities occur in all of these environments. The interaction of
these environments with each other, with human activities, and with
radiation from the sun to produce climate and weather is not explicitly understood. However, it is accepted that changes in one regime of the environment will often cause change in several of the
others, and that changes in solar or human activity can significantly modify the environment.

Since the beginnings of history man has sought to understand
and predict the behavior of the environment. With increasing population and intensity of organized human activity in areas such as
agriculture, commerce, resource development, and transportation, the
need for an improved capability to understand and predict the behavior of the environment is increasingly apparent. The ability to
measure accurately the state of the natural environment, and the
activities of man in this environment, is an important step in the
development of models to predict the behavior of the environment,
or to predict the levels at which man can hope to extract food and
fiber products as well as other resources from this environment. The
range of parameters to be measured and phenomena to be observed is
large. Some phenomena are global and require observation at that
scale while others are regional or local in nature. With the present
state-of-the-art of sensors some phenomena can be observed remotely
while others require in-situ measurements.

17

The observation of the environment requires a platform or station from which the phenomena of interest can be observed and sensors to record the measurable attributes that describe the phenomena. Platforms or stations in operational use today include aircraft, balloons, land stations, ocean stations and data buoys, satellites and ships. These platforms provide the capabilities for observations ranging from in-situ measurements to remote sensing, and from local to global observation. The development of sensors for in-situ measurement of environmental phenomena is reasonably mature. On the other hand, instrumentation for remote sensing of the environment is a newer field and much work remains to be done in the development of remote sensors.

The operational systems and the sciences that are the consumers of the data produced by earth observation systems are primarily dependent upon observations. As data accumulate the physical processes which occur in the environment are explained partly by empirical or inductive methods, and partly by using the data to test theoretical ideas. The range of problems to be solved in climatology, land sciences, meteorology and oceanography requires data from all regimes of the environment. With the present state-of-the-art of platforms and sensors this data need cannot be fulfilled by a single type of platform or sensor. There is a need to integrate the capabilities of different types of platforms and sensors in order to support functionally oriented missions such as crop production forecasts or offshore resource exploration.

Early remote sensor developments employed instruments operating in the visual and infrared parts of the spectrum. The value of the information produced by these sensors has been demonstrated by the several environmental and earth observational programs such as TIROS, Geostationary Operational Environmental Satellite (GOES), and LAND-SAT. Experience with these systems has shown that their usefulness could be greatly enhanced by improved instrumentation. Improved vertical temperature and humidity sounders are needed for use in global numerical weather forecasting. All-weather sensors such as synthetic aperture radar are needed to provide high resolution coverage of areas that are frequently obscured by clouds.

The operational use of geostationary satellites for environmental observation is a recent development. In a short period of time, field meteorologists have achieved a high degree of success in the use of data from GOES for the prediction of short-lived, local, environmental phenomena. The ability of this system to produce "on-demand" information on a nearly continuous, nearly real-time basis appears to be the factor which has led to its success. The success of this system in environmental observation should be exploited by providing higher resolution sensors, particularly in the infrared part of the spectrum. The possibility of providing similar "on-demand" information for land science applications should also be explored.

2.   Issues in the Development of Operational Remote Sensor
      and Platform Systems

2.1 Systems

      2.1.1 The Integration of Systems to Perform
             Functional Missions

        Existing systems appear to have been developed along the lines
of educational disciplines such as the land sciences, meteorology
and oceanography. On the other hand, it is believed that operational remote sensing systems will provide information for use in monitoring or management functions that will draw upon the services of
several disciplines. Examples of these interdisciplinary missions
include crop production estimates (involving land sciences and meteorology) and offshore oil and gas exploration (involving oceanography and meteorology). Existing systems now make some use of the
unique capabilities of sensor platforms that operate at different
altitudes such as spacecraft (geostationary and low altitude), aircraft, and ground-based sensors.

        There is a further need for the systems designers to explore
optimum combinations of sensors and platforms, particularly as a
function of the specific resource monitoring or management mission.
The need for surface truth information for sensor calibration must
also receive careful consideration in the design of operational systems. For these reasons, it is the belief of this working group
that the design of operational systems to perform functional missions
will best be achieved through the *integration* of systems that operate
in the environmental regimes of interest, with the capabilities to
provide data over the range of scales desired. For example, an operational crop forecasting system involving both land sciences and
meteorology will probably require spacecraft for global monitoring
of crops and aircraft for local monitoring. The meteorological inputs to the crop forecasting system will probably be obtained from
a mix of spacecraft, balloons, land and ocean stations, ships and
ocean data buoys.

        In the development of operational meteorological and oceanogra-
phic systems, the working group also has observed the requirement to
consider the need for improved *integration* of data from various sources. At the present, this data integration is largely done by the
user. This group recommends that consideration be given to the feasibility of performing this integration of data at the collection
stage, thus providing the end-user with a fully integrated data set.
An example of this integration could be the combination of surface
radar data on precipitation with the visual or infrared imagery from
geostationary meteorological satellites. The working group also notes that there may be a relationship between the degree of integration required for a specific mission and the time between required

successive observations, or the time between cause and effect. It
is the recommendation of this working group that the requirements
and feasibility of data integration at the point of collection receive further study.

2.1.2 Dissemination of Information Concerning Sensors, Platforms and their Uses

At the present time there is no central repository of information concerning sensors, platforms and their uses. Although several professional societies and national agencies now publish in
this field, the publications are mainly organized along discipline
lines (e.g., meteorology, oceanography, agriculture, photogrammetry), or describe the results of specific projects.* The interested
applications scientist or educator in the field of remote sensing
is often handicapped in his work by a lack of information concerning
the state-of-the-art of sensors or platforms, or by a lack of information concerning applications outside of his discipline. It is the
recommendation of this working group that it would be desirable to
promote the improvement and expansion of the use of existing public
and private data banks with sensor, platform and applications content. It is further recommended that an international agency such
as the International Astronautical Federation consider the desirability of promoting an annual publication that will describe advances made in the development and applications of sensors and platforms
in the remote sensing field.

2.1.3 The Need for International Cooperation in the Research and Development of Advanced Operational Remote Sensing Systems

The costs of research and development of advanced operational
space-based remote sensing sensors and platforms may exceed national
economic capabilities or interests. It is recommended that the interested nations investigate the desirability and feasibility of cooperation in this field, and if such cooperation is deemed to be desirable and feasible that the mechanisms for such cooperation be developed. This working group further suggests that cooperation in
this manner may lead to cooperative efforts aimed at providing continuity of data from operational systems.

---

* - For examples see the Space Systems Summaries published by
      the American Institute of Aeronautics and Astronautics in
      Astronautics and Aeronautics, and the U.S. NASA Report "Sa-
      tellite Capability Handbook and Data Sheets", 624-3, pub-
      lished by the Jet Propulsion Laboratory, July 1976.

2.1.4 On-Demand High Resolution Coverage

On-demand high resolution coverage may be required for the observation of transient phenomena that occur in the environment, or phenomena such as forest fires or disasters on the surface that may require urgent response. In this instance there may be a need for on-demand high resolution coverage. Depending upon specific requirements, it may be possible to provide on-demand coverage from aircraft, remotely piloted vehicles, rockets, pointable imagers in low altitude satellites, or from a geostationary earth observation satellite. The requirements for on-demand high resolution coverage in disciplines other than meteorology are not sufficiently understood at the present and will benefit from further study. If sufficient requirements are found to exist for on-demand high resolution coverage, sensor and platform development should be undertaken to demonstrate the capabilities that could then lead to operational systems that meet these needs.

2.2 Platforms

2.2.1 Geostationary Land Sciences Satellites

The first Geostationary Operational Environmental Satellite (GOES), launched in October 1975, has been surprisingly successful as an aid to field meteorologists, and is changing the entire approach to field meteorology. This success may be attributed to the satellite's "on-demand" capability.

The Synchronous Earth Observatory Satellite (SEOS) is proposed to be launched in the mid 1980s. Like GOES, the SEOS satellite may turn out to be surprisingly successful because of an on-demand capability, providing continuous rather than intermittent coverage. The trade-off of ground resolution from a SEOS against more timely coverage may be favorable.

The working group recommends that the uses and requirements for a SEOS for land science uses be investigated in order to define the economic applications and technology needs for this platform.

2.2.2 Use of Satellites for Data Retransmission

Earth observation satellites such as GOES and NIMBUS have been used successfully for the retransmission of weather data from ocean data buoys and tracking data from drifting buoys. The integration of remote sensing, surface platform location, and data retransmission should be examined comprehensively in the planning of all future systems.

2.3 Sensors

2.3.1 Microwave Sensors

There is a need for an operational system of spaceborne high-resolution (25 meter) imaging radars of the SEASAT-A Synthetic Aperture Radar (SAR) type. This is particularly important for northern countries because of its potential for monitoring sea ice along shipping lanes and in offshore exploration and production areas, in foggy and cloudy conditions. It may also have a surprising variety of land mapping and monitoring applications (e.g., hydrology) that have not yet been fully explored by potential users. The introduction of spaceborne SAR, with resolution of about 25 meters, may be as important to the development of remote sensing as the introduction of the LANDSAT-A multi-spectral scanner.

There presently exist technological problems to be overcome before an operational system can be developed, including:

- Very large data acquisition rates that lead to problems of onboard storage and processing, telemetry, and ground storage and processing

- Choice of radar frequencies that lead to problems of ionospheric signal distortion, sea scatter (signal-noise ratio) and satellite power requirements

- Limitations of state-of-art digital and optical correlators for SAR signal processing.

As a prelude to an operational system there is a need for international cooperation in research and development on spaceborne SAR, particularly for developing shared operational systems as well as shared scientific and commerical opportunities. Otherwise, the cost to individual nations may be too high and the perceived imbalances of opportunities may have an undesirable political impact.

It is already evident that microwave scatterometer systems, such as the system to be deployed on SEASAT-A, can be improved in specific ways to increase their usefulness as ocean wind stress sensors. For example, the sensor could be improved to provide multiple or simultaneous looks at specific footprints in different polarizations, azmith angles, nadir angles and frequencies.

2.3.2 Improved Resolution for Land Sciences

Current LANDSAT type systems are capable of providing resolution of 100 meters in the visible, and planned systems may be capable of providing resolution of 30-40 meters. This may be useful in nations with large homogeneous land surfaces and large agricultural fields, but may not be adequate in many densely populated areas or in regions with small agricultural fields, such as Europe or East Asia. It is recommended that sensor research be continued towards the objective of providing 10-20 meters resolution in the visible and near infrared parts of the spectrum.

2.3.3 Improved Vertical Temperature and Humidity Sounders

The greatest single improvement to the real-time global weather data base for application to numerical weather forecasts for 18 hours and beyond would come from the development of more accurate vertical temperature and humidity profiling devices. Present systems have not lived up to original expectations. Errors in the temperature are at least 2.5° C in a random sense in present systems and systematic errors are introduced by cloud and precipitation interference effects. Humidity profiling devices measure broad vertical averages and are hardly useful today. The errors need to be halved before a significant improvement can be expected and before the satellite-borne sounder can serve as a substitute for the conventional radio-sound.

Present sounders already exploit infrared and microwave portions of the electromagnetic spectrum. Thus, no breakthrough appears possible, although system refinements through better mathematical modelling, design, and joint use of different portions of the electromagnetic spectrum, could achieve sufficient system improvement.

The sounder design process should more closely involve and exploit the experience and views of those who have attempted to use the present systems in an operational or quasi-operational mode.*

Sounder coverage and resolution appear to be adequately provided by the next generation of polar orbiting satellites, TIROS N. The possibility of deploying sounders from geosynchronous orbits should be explored as an R&D effort.

2.3.4 Improved Sensors for Short Range
Meteorological Forecasting

GOES cannot cover high-latitude regions because of its equatorial geostationary orbit. The success of GOES in very short range forecasting can be exploited further by combining GOES with a system of, say, four polar orbiting satellites to provide improved temporal resolution for "on-demand" use by field meteorologists. However, the possibility of other orbits, such as indirect geosynchronous orbits, should be explored.

---

* - For example, see Halem and Chow, 1976, *Journal of Applied Meteorology*.

There is a need for higher resolution infrared imagery. Infrared sensors should be developed that provide 1 kilometer resolution from geosynchronous orbit altitudes. Based upon current experience, the ultimate useful resolution for atmospheric applications appears to be about 0.5 kilometers.

Basic research in the development of remote sensors for the following parameters should be intensively supported:

- High resolution (20 kilometers) rainfall intensity

- Soil moisture content

- Sea ice thickness

- Air temperature over the oceans

- Sea level pressure measurement

- Air motion (winds) in the free atmosphere independently of cloud motion

- Aerosol measurement.

2.4 Data Registration and Formating

The monitoring of the interplay of climate, water, vegetation and soil, which appears to be the key to some of the most effective applications of remote sensing, is still to a great extent dependent on second order deductions and requires the input of information from many different platforms and sensors.

Some of these inputs must be observed simultaneously and in near real-time, while time delays are permitted for others. In some cases the timelags between causes and effects provide significant information of their own. Some data must still be provided by ground observation, but an increasingly greater quantity of information can be obtained from multispectral, multitemporal, synoptic observations.

Much of the interpretation is still effected by visual means. Differences in formats of the user products provided by different remote sensing systems make it difficult for the interpreter to combine this wealth of information to achieve full synergistic benefits.

For this reason, it is recommended that the format of corrected images obtained by different remote sensing systems above a certain level of resolution, and which contain information on climate, water, vegetation and soils, should be provided in the same standardized format. This will allow a superposition that is sufficiently precise for visual interpretation purposes.

Although it would be desirable to prepare computer-compatible tapes to the same standarized format, the cost of achieving this degree of standardization at the present time may be prohibitive for the purposes of routine analysis.

SUMMARY REPORT B
DATA HANDLING SYSTEMS

Charles Sheffield, Group Chairman
A.D. Kirwan   Archibald B. Park
Philip A. Lapp   Robert E. Stevenson
Other Contributors to Case Study on Crop Production
Wolfgang Baier   J.A. Howard
Eric C. Barrett   R.A.G. Savigear
Robert N. Colwell   Paul M. Wolff

1. Introduction

The data-handling problem arises in all disciplines, and with all observing instruments. In fact, the manipulation of the observational data forms the natural bridge between the remote sensing instrumentation on the one hand and the user community on the other. The chain linking instruments and disciplinary use is shown in Figure 1. It will be observed that the data-handling activity overlaps largely with sensors and platforms, at one end of the chain, and with management information systems at the other. (See the background paper by Sheffield.)

In Figure 1, the thrust runs from bottom to top. In other words, the final user needs are seen as the driving factor which ultimately decides the sensors and platforms to be used, and the amount of data analysis that is needed. In practice, the sensors and platforms must often be regarded as pre-decided, in that the user may have little impact on the design of the original hardware. Nonetheless, the logical point of departure is always user requirements, whether these are involved explicitly or only implicitly in hardware design.

With this general philosophy, the task of the data-handling group was perceived as that of examining the links between user requirements and basic instruments, and determining which were the critical links - critical, in the sense that they impose the heaviest load on the data-handling technology, both hardware and software. Since applications differ greatly in their data needs, the activities of the group, in addition to defining the overall approach, were strongly directed to the more detailed analysis of represen-

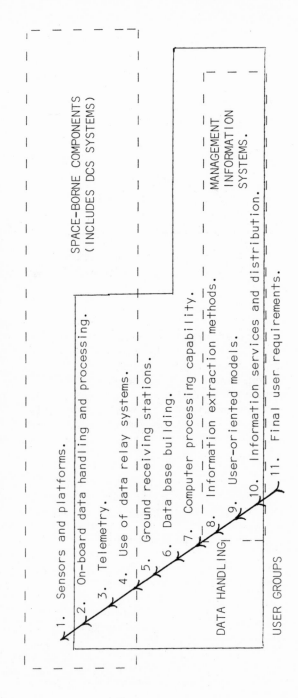

Figure 1 - Overall Methodological Framework.

1. Sensors and platforms.
2. On-board data handling and processing.
3. Telemetry.
4. Use of data relay systems.
5. Ground receiving stations.
6. Data base building.
7. Computer processing capability.
8. Information extraction methods.
9. User-oriented models.
10. Information services and distribution.
11. Final user requirements.

SPACE-BORNE COMPONENTS
(INCLUDES DCS SYSTEMS)

MANAGEMENT
INFORMATION
SYSTEMS.

DATA HANDLING

USER GROUPS

NOTES: The above system operates 'bottom-to-top' - i.e., it is driven by user group requirements. It is re-entrant, in that data relay systems, for example, can be used for raw telemetry, or for transmission of processed information.

Note the strong overlap between data-handling, management information systems, and sensors and platforms.

tative applications, carrying these through the complete logical
sequence from needs to observing instrumentation.

In addition to the assigned members of the data-handling group,
listed on the title page of this report, representatives were drawn
from other working groups to review work in the area of crop production, as a case study in data handling. The background papers
by Baier, Colwell, Howard and Park deal with various aspects of
crop forecasting.

2. Summary of main issues

Almost every attendee at this NATO ARI, in both their prepared
background papers and in their discussion of issues at the meeting,
referred to problems of data-handling. However, their issues of
concern varied over a wide range, from the highly specific to the
very general.

The major explicit questions of concern in data handling are
seen to be the following:

(1) What institutional frameworks exist, or should exist, for
the processing and timely distribution of remotely sensed data? In
particular, what is the appropriate way to handle the problem on an
international basis - recognizing that various international bodies
already have this responsibility for particular kinds of data (e.g.
weather satellite data)?

(2) As a general rule, should data storage and data manipulation systems be conceived and implemented in centralized or de-centralized modes?

(3) Is it possible to provide effective service for both
short-lived data and data with a long half-life, in a single service
facility? If so, how should this be done?

(4) What parameters can be effectively measured by remote
sensing, and what must rely on in-situ sensors? For the latter class,
what data handling methods should be used to transmit them to a suitable processing facility?

(5) How can data that are now being acquired be distributed
more rapidly and more reliably to the users?

These explicit questions are related to a series of implicit
concerns, which a few conference attendees identified in specific
terms, though most did not. The implicit concerns include:

(1)   Questions of data ownership and data confidentiality. What data should be or will be available to all users? If there are to be restrictions on data release, what are these restrictions and how can they be implemented in a safe and impartial way?

(2)   Technologies for data storage and data handling. What are the available technologies that can be employed to store and process data, particularly in comparing the virtues of centralized and de-centralized facilities?

(3)   Bandwidths. Are the available bandwidths adequate to carry projected data loads for remote sensing systems now in the planning and design stages? What are the actual bit rates - in user demand terms - that the new systems will generate?

(3)   Multiple source data. Almost all users recognize the potential value of multiple source data (time, wavelength, polarization, resolution, and other variables). What computer techniques are available for the combination of multiple source data?

(5)   Cost is the single most important factor affecting the widespread use of remotely sensed data. What will the controlling factors be that decide the cost of the data? Will it be data collection (platforms and instruments), data transmission, data storage, data combination, or data processing (information extraction)?

Answers to these questions, because of their widely differing degrees of institutional, technical and social elements, make sense only when looked at in terms of specific applications. Few general statements on data handling are useful here, other than the general statements that can be made on the price and technical state-of-the-art of data processing hardware and software. Again, analyses here must be specific to applications.

The most general questions raised above need attention by policy groups. Answers to the more specific ones evolve from the analysis of each application. This point becomes clearer when two applications are addressed in detail, later in this report.

3.   Recommendations

Specific recommendations that were developed during the discussions of the data-handling group are as follows:

(1)   Multi-disciplinary activity, including the use of specialists in data-handling and data methods, is especially important in the formulation stage of projects involving remote sensing inputs. Such a multi-disciplinary composition is quite rare, since the organization of government laboratories and university departments tends to force a compartmentalization of talent. Either users are

well-represented, or hardware systems developers, but rarely both. Data handling specialists are seldom involved in the project formulation stages, although the feasibility and success of the project often resides in an adequate understanding of the data needs and data limitations.

(2)   The complete systems study approach, described here, for carrying the program from user requirements through to sensors, platforms, and hardware, with all necessary intermediate steps, should be adopted more widely. Such a process isolates bottlenecks, clarifies roles, points up weaknesses, and improves general understanding. Thus we recommend that such an approach should be used as a standard procedure, and that the corresponding multi-disciplinary group should become a standard association.

(3)   With rare exceptions, computer processing power and data storage systems now available are adequate to handle most applications discussed for the 1980 time frame. Users should in general concentrate attention on models, data collection definition (with increased use of statistical sampling methods), and refinement of data final uses (as processed information), since these are more likely to prove to be the limiting factors. Techniques for measurement of parameters using remote sensing, and the way in which these parameters may be used as surrogates for data directly-collected by ground-based methods, should be emphasized.

(4)   Users should be made aware of the rapid development in low-cost microprocessors for data analysis. The use of special purpose small computers, compact but with substantial computational sophistication, should be borne in mind when designing in-situ measurement platforms. Many functions that a few years ago required human intervention in the measurement loop can now be performed entirely by low-cost, fully programmable microprocessors.

(5)   Cost projections rarely take adequate account of the economies of scale. Users should bear this in mind when assessing the economics of sensors, particularly in-situ sensors. In a fully operational phase, the costs of in-situ sensors and of information extraction from earth resources data will decrease dramatically.

(6)   Because the data needs and handling are so different in different application areas, there should be attention paid to the possibility of setting up *application-specific* data collection and analysis systems, with the meteorological satellite system as perhaps the natural model. LANDSAT and SEASAT, precisely because they are multi-purpose instruments, make the problem of estimating user markets and user benefits very complicated.

The remainder of this report is given over to the discussion of two case studies in systems methodology, from user requirements through to sensor and platform definition.

4.   Case Study of Crop Assessment and Forecasting of
     Crop Production Using Remotely Sensed Data Inputs (Summary)

This case study carried the systems methodology shown in Figure 1 through a complete analytical cycle. The study is of particular interest, since at least parts of this overall system are being implemented by a number of workers, and since the timely forecasting of crop production is an important world-wide problem.

4.1   System Goals

The goals of the final system are as follows:

(1)   Reduction of the standard deviation of the crop
      production estimate, before harvest date, on a
      national, and regional base (global statistics
      alone will not suffice).

(2)   Improvement of the present early warning system
      of impending crop failure.

(3)   Improvement of the quality of ancillary information
      relevant to the industries serving the food produc-
      tion process (fertilizer, pesticides, transportation
      of products, etc.).

(4)   Upgrading of the overall quality of the existing
      food information system, in timeliness, accuracy
      and precision.

4.2   Limiting Factors

There are practical constraints on the system that can be developed. These constraints are both hardware and processing functions. No earth resources satellite system is projected in the next decade that can offer LANDSAT resolution, spatially and spectrally, on a daily basis. In addition, the processing load implied by any attempt to run an exhaustive, world-wide system at the necessary resolution would be enormous. For these reasons, a multi-stage sampling approach must be used in dealing with the LANDSAT data. Meteorological data, on the other hand, can be treated exhaustively, and should be.

4.3   Summary of System Outputs

Determination of production is conveniently divided into two parts: determination of yields, and determination of crop areas.

Yields are determined, and forecast, using metsat data plus ground observation to compute radiation and precipitation budgets, and these are to be used to drive phenologic models. The consensus was that satellite-based estimates of precipitation, supported by ground-based information, can be satisfactorily performed using the present generation of weather satellites. LANDSAT supplements this information. Some panel members felt that improved thermal infra-red sensor resolution is still desirable. Suitable phenological models do not exist for most crops, and tuning of most models to geography is still needed.

Crop areas under cultivation may be found using LANDSAT data, as in the U.S. LACIE program (Large Area Crop Inventory Experiment), or from historical and collateral data. The LACIE approach, based on multispectral analysis of reflectances derived from LANDSAT's MSS sensor, is viewed as satisfactory only in areas where the fields are large and homogeneous. In areas where fields are small, crops are mixed, or historical data are lacking, the multispectral approach is considered unreliable. In such cases, one approach that has been used successfully is the use of LANDSAT data only to separate agricultural from non-agricultural land, and the use of any available ground truth or collateral data to apportion agricultural land among different crops.

4.4   Conclusions

The combination of meteorological and earth resources satellite data, supplemented by ground truth and weather station data, can be used in a multi-stage sampling framework to provide highly valuable information on crop production, well before harvest date. Although only selected crops and sites have so far been considered, this whole area is of such great importance that delay is undesirable. The working group was unanimous in urging that practical systems for regular estimates of regional production should be instituted as soon as possible, preferably under the guidance of an international organizing body.

5.   Case Study of Use of Satellites for Position-Fixing in Ocean Applications

This case study established the technical feasibility of using presently available technology in position-fixing, for a variety of applications. The key satellites for this work are polar vehicles of the TIROS-N and NIMBUS-6 types. Single passes over a ground location permit position-fixing to better than one mile. The most important applications were estimated to be:

(1)   Search-and-Rescue operations, on land or sea

(2)   Tracking of free-floating buoys in the southern
       hemisphere, as part of the FGGE program (beginning
       in 1978)

(3)   Tracking of free balloons with on-board beacons

(4)   Mammal tracking, as required, for example, by the
       U.S. Marine Mammal Preservation Act (relevant
       species include whales, porpoises, polar bears).

The system requires a net of satellite receiving stations if
the Search-and-Rescue operations, in particular, are to be effective.
It is estimated that a net of 100 stations, each containing a suitable receiver, a mini-computer for position calculation, and a data
storage unit, would suffice for world-wide coverage. With present
technology, each station could be purchased for about $30,000 U.S.

Seven-day coverage would be an essential, and this is not presently planned for example, in the case of TIROS-N. However, no
technical bottlenecks were encountered in the analysis. A position-
fixing system as outlined here could be implemented in one or two
years, using off-the-shelf hardware for the ground stations, and
piggy-back data collection systems on existing polar satellites.
The primary problems would be purely organizational.

SUMMARY REPORT C

MANAGEMENT INFORMATION SYSTEMS

Paul G. Teleki, Group Chairman
Eric C. Barrett
Gerhardt Hildebrandt
Eric S. Posmentier

R.A.G. Savigear
Sigfrid Schneider
Stanley E. Wasserman
Paul M. Wolff

1. Introduction

During the next two decades, increasing numbers of spacecraft will be launched for the study of Earth processes. There will probably be an accelerated growth in the amount of data processed onboard and returned from spacecraft sensors. Data will become more sophisticated, because sensor technology will improve. The result of these developments is predictable: the acquisition, processing and dissemination of information will have to become routine and the costs of data products will have to be reduced. Yet, there is little coordination on the issues of what kinds of data are desirable, how data from different systems complement one another, or how users can access varied types of data from divergent sources. While raw data are useful to some, more commonly raw data have to be transformed by users into "information products", and it is these second generation products which are directly useable for decision-making purposes.

As new remote sensing technology is invented, involvement of new users is expected, but in practice the visible community of users of remote sensing data does not seem to enlarge at the same rate as technology advances and as data acquisition rates increase. The questions arise therefore: Who are the potential users of remote sensing data? What is the best way to involve potential users in designing remote sensing experiments? Where are the markets for the information products of remote sensing systems? Are there differences among geographic and national entities in how raw and interpreted data are applied?

Users' needs for remote sensing data are essentially two-fold:
data for purposes of *research* and data for management decisions in
an operational mode. Nonetheless, in many instances there is considerable crossbreeding of these purposes. For instance, weather
data is considered to be perishable in operational weather forecasting, but it also has archival value for climate studies. Who
is to say what to keep and what to discard? Are data being assimilated from different sources to serve either the operational requirements or research needs? If not, how should data be integrated from
several remote sensing systems and from "surface truth" to be truly
useful to an investigator or to a manager?

Many depositories of remote sensing data are identified with
organized input and output procedures. These depositories, or
"data-banks", vary from an individual researcher's files to sophisticated institutional systems which process, store, archive and disseminate data. The latter are usually operated to serve regional,
national and in some cases global needs of users. For data streams
from *individual* sensors, some form of standardization for the input
and output procedures is commonly adopted. Often, however, such
procedures are not applied to data from sensor *groups* or to data
acquired earlier from older remote sensing systems.

The projected rapid growth in data acquisition from space makes
it necessary to plan for the management of remote sensing data, and
the information derived from them. The questions appropriate to
this are: What is the proper management scheme for these data and
information? Should data be managed for the sake of orderliness
and systematization alone or for identified purposes? What sort of
decisions will be made by users of the information? Should data en-
tering an information system be standardized? Should information
leaving such system be standardized? And, what methods (algorithms)
are useful to data and information management?

Experimental systems serving individuals, particularly research
needs, are quite different in scope than operational systems designed
to manage information for institutional programs. Should separate
information systems be designed for each community, or will a multi-
purpose system suffice? Where does the design of any management information system start, with the feasible array of sensors or with
the data base requirements?

2.   Concepts of Management Information Systems

2.1   Information Management and Time Frame

In identifying a preferred course of action, given several possible alternatives, the usual procedure is to conduct analytical investigations. These investigations seek information and knowledge
relevant to an identified problem. Information consists of data and

facts, which if assembled, weighed, integrated and analyzed, enable
people to decide a course of action. Information thus is "managed"
to address problems which may range in scope from short-term to
long-term. Short-term decision problems are usually associated with
continual decision processes, and require frequent "real-time" information updates (e.g., storm warning decisions based on half-
hourly cloud images from GOES). Medium-range problems are often
periodic in occurrence and require less frequent "near-real-time"
information updates (e.g., crop production forecasts and stock carry-
over decisions based on periodic sampling, including 18-day repeated
images from LANDSAT). Long-range problems tend to be unique, and
require "non-real-time" information updates (e.g., offshore engineering design decisions based on wave climatological data obtained over
many years from oceanographic satellites).

Any data, to have utility in the decision-making process, must
have a physical form and practical value, must be timely (i.e. appropriate (and not outdated) and must be available to management in
analyzed form (i.e. assembled, correlated and presented in an easily
comprehensible format). The means to integrate data from various
sources and transform those to information is ususally the computer,
which can be assigned two functions: (1) operate a data bank and
(2) process data into tailored information products for a management information system.

2.2 Importance of the Data Base

The management information system (MIS) is a manager's tool to
utilize information to achieve his goals and control his actions. An
MIS contains prescribed procedures for transforming and transferring
information. Structurally, it is formulated to have access to data
bases, to coalesce data from varied sources, to reinterpret and re-
package data, and to extract the desired information content. Its
efficacy depends on knowing where it starts and where it ends, on
documentation, on knowing who its users are and what information
they require, on timely information extraction and on proper procedures for formulating and transferring the needed information.

Management information systems rely heavily upon data bases. In
establishing such bases, one may ask: What data should be in the
data base? How should data be collected? How should data be entered? Should it be a centralized or decentralized data base? What
is the best way to communicate with the data base? It is generally
agreed that, rapid and painless input and output are desirable. However, data often have dynamic properties; therefore, adding, modi-
fying or deleting data has to be routinely accomplishable. Because
raw data or certain combined sets of data are often destroyed, back-
up data should be available. And, to ensure utility to a wide variety of users and to avoid redundancy of data elements, data coding
should be carefully designed and implemented.

3.   Design Considerations for a Remote Sensing
      Information System

   3.1 Data and Information Products

      The complexity of a remote sensing information system depends
on the level of sophistication of both the user and the remote sensing data accessible to him. Until now, photographic products have
dominated remote sensing because airborne cameras have been the primary sensors. Delays between the exposure of film and the effective utilization of photographic products have restricted their application to mapping static phenomena and monitoring slowly changing phenomena. Photographic products will probably continue to
dominate in the future, even though cameras are replaced by imaging
multi-spectral scanners and radars. Now the images can be generated
by computers utilizing analog and digital data from spaceborne sensors. Digital data from LANDSAT multi-spectral scanners can be used
to produce images that are geometrically corrected, enhanced and the-
matically highlighted. The applications now include monitoring phenomena that may change quickly. The GOES satellite, for example,
now produces time-lapse "movie" photographs of clouds in visible
and infrared spectral bands, with half-hour intervals between frames.
Now the data stored on tapes can be used to generate a variety of
soft information products such as transient video displays, and hard
products such as false-color photographs, statistical tables, and
thematic charts.

   3.2 Guidelines for Design of an MIS

      In designing a remote sensing information system, the following
guidelines are important:

   (1)   For the data base

      (a)   access to all relevant data sources

      (b)   a mechanism to integrate data from various sources

      (c)   versatility in data space relationships

      (d)   capability to assess and implement improvements
             in data types, in formatting and in data flow

      (e)   capability to purge valueless or nonessential data

      (f)   on line processing.

(2)   For the interaction between the information system and
       the user,

    (a)   responsiveness to specific (multipurpose) needs
           rather than universal needs

    (b)   responsiveness to queries of widely divergent
           nature and degrees of complexity

    (c)   accessibility to users by means of economical
           interactive systems containing algorithms de-
           signed to derive greatest utility and fastest
           decision

    (d)   data formatting and information standardization
           which enable users to assimilate information
           quickly

    (e)   flexibility in the system to allow its expansion
           or modification without suspending operations as
           new requirements are identified

    (f)   instructive value to the uninitiated user, or
           when changes in information content occur

    (g)   information on system reliability (evaluative
           models or testing procedures.

(3)   For the interaction between the information system and
       remote sensing technology,

    (a)   have an impact on the design of future sensors
           and data types

    (b)   influence designs for data collection methods
           and the flow of information.

3.3   Centralization versus Decentralization

Among many MIS in use, both centralized and decentralized systems are common. For the management of remote sensing data, central
processors are acceptable, if they are modular in design and as such
permit accommodation of new types of data or new output requirements.
Modular systems also allow users to access those portions of the MIS
relevant to their interests. Many successful MIS include interactive subsystems, natural language and simple vocabulary, all of
which assist users in establishing direct dialogue with the MIS on
their own terms of communication.

3.4   Standardization

Because various types of data in different formats are common
in the field of remote sensing, standardization, such as standard
georeferencing of earth resources data, is of fundamental importance.
If geometric and radiometric standards are universally adopted, algorithms can be developed to merge different types of data from different sources (e.g., LANDSAT and weather satellite images could be
correctly overlaid). Such algorithms would be especially powerful
in assisting operational decision-makers if they are designed to
serve multiple purposes and hence many different users. Properly
implemented, they should weigh information to account for differences in the value of data, which can range from descriptive to numerical and from simple to sophisticated.

3.5   User Instruction

Any information system should have the capability to instruct
users about methods of access to it, about the type, quality and
quantity of information contained therein, and about means of application. These attributes are also important to information systems
containing remote sensing data. Regular channels for disseminating
information to users are essential, but even more critical is that
information used for management be evaluated periodically, and comparisons among data interpretation algorithms and models be provided to the users.

3.6   Reliability

For information to be useful to decision-making it has to be
rated for reliability. An MIS is but a blackbox transforming sets
of data into a communicable product (output) which has to be accompanied by ratings of its reliability. Evaluative models or on-line
testing procedures can be used to test performance of the system.

3.7   Aid to Future System Design

To ensure its survival, meaning long-term utility, a remote
sensing management information system should influence what information becomes available in the future, including type of data, methods of acquisition and character of data flow. Information contained in the MIS can be used to design future experimental or operational spaceborne data collection systems. By representing users'
goals, an MIS should influence the times, modes and kinds of remote
sensing data acquired in specific regions or for specific applications, as well as supporting "ground truth" data.

3.8   Data Lifetime

Much data are valuable only for short periods of time. Conceptually, an efficient MIS should purge itself of dated information, at the same time preserving those elements which are necessary to long-term observations and activities. Selecting the right kinds of data to keep is difficult even with the best perspectives on future information requirements. The decision has to be based on two considerations. One is how the data are ultimately used; this is governed by disciplines, environments, and geographic regions. The other is the organizational complexity involved in the use of data which can range from local/agency levels through national to multi-national levels.

3.9   Limitations of Current Knowledge

Operation of a remote sensing MIS is constrained by several factors. One of these is our current knowledge and understanding of technical capabilities. Systems are efficient only when all relevant parts function equally well at the designated times and for designated purposes. Environmental factors could hinder an efficient operation. But, so can economic and social constraints, or those attributable to organizational problems. At present, our knowledge of such constraints and our ability to design management information systems is very limited. Consequently, there is a learning gap to be filled. It is an evolutionary process, in which technology produces new information, which has to be integrated into the system, to be followed by learning and formal training in how to apply it. This identifies the need for flexibility in accepting new technology, such as we have witnessed in the development of sensors and data products, and the need for flexibility in making known what new information exists and how to use it. Feasibility studies should be conducted in selected environments to test the concept of an MIS and evaluate its flexibility, that is, its varied levels of utility.

3.10   A Note of Caution

Finally, a note of caution on the concept of a remote sensing MIS. It should, perhaps, be restricted to serve well-specified purposes alone. An unbounded, unchecked system could easily delve into the privacy of individuals and nations to such an extent that its use would result in more harm than derivable benefits.

SUMMARY REPORT D
SOCIO-ECONOMIC SYSTEMS ASPECTS

S. Galli de Paratesi, Group Chairman
Peter W. Anderson   Hans Heyman
Wolfgang Baier   J.A. Howard
Charles Buffalano   James J. Gehrig
Robert N. Colwell   Lawrence W. Morley
John M. DeNoyer   Richard Mühlfeld

1. Introduction

Improved resource development planning and appropriate monitoring of the status of resources can be realized only if adequate information systems are developed to provide accurate, comprehensive and timely information on resource availability, use and potential. The present lack of such systems throughout the world is an impediment to the formulation of important resource management policies and is an obstacle to important developments in the less developed countries. (See, e.g., the background papers by Colwell and Howard.)

2. Essential Contribution of Remote Sensing

Current conventional methods for the repetitive monitoring of atmosphere, land and ocean surfaces are costly. At lower cost, remote sensing can significantly contribute to effective global resource management and enviromental control. Remote sensing can provide information for monitoring and predicting effects of human activities and natural phenomena.

The main conclusion of the Working Group on Socio-Economic Aspects (WGSEA) was that satellite and airborne remote sensing presents an unprecedented technological opportunity to improve mankind's information about the environment. It also presents a great challenge in an economic and political sense. The costs of operational remote sensing satellite systems will be high, and therefore international cooperation on systems development and operation seems imperative. Political leadership and diplomacy will be required to ensure that a broad international community can obtain maximum benefits from the new technologies.

3.   Need for Continuity in Monitoring Systems

Repetitive airborne and spaceborne remote sensing coverage can contribute cost-effectively to the recognition of environmental variables such as snow cover, soil moisture, sea state, vegetation cover, weather, etc.   Continuity of information is now essential to vital functions such as weather forecasting and disaster warning, and will become essential to functions such as land use planning, crop assessment, etc.

Half-hourly repetitive information produced by the GOES-1 satellite, for example, has already proven to be essential for accurate prediction of hazardous weather events in the U.S., on an operational basis.   (See, e.g., the background paper by Wasserman.) Eighteen-day repetitive information produced by LANDSAT-1 and -2 satellites, for example, have demonstrated the importance of information for crop assessments and other applications, on an experimental basis.   (See, e.g., background papers by Hildebrandt and Park.)

The detection of subtle environmental and cultural (land use) changes can be performed by repetitive airborne and spaceborne remote sensing under a wide variety of conditions.   It is anticipated that information about *temporal changes* will become increasingly important as operational remote sensing systems evolve.   Even certain disciplines that have not traditionally required information about short-term dynamic changes, such as geology and cartography, may be served in the future by the availability of repetitive remote sensing information such as produced by LANDSAT.

4.   Technical Feasibility and Benefits

The technical feasibility of both space and ground components of satellite earth observation systems has been amply proven by such satellites as GOES and LANDSAT.   Future systems can be designed to have high reliability and long expected lifetimes.

The wholly new capabilities represented by these satellites are presently being assessed by many nations and communities of users.   Scientific applications are being assessed.   Industrial and governmental applications are being planned.   Economic costs and benefits are being evaluated.   Organizational aspects are being examined.   There is no doubt that the new satellite systems have captured international attention at many levels, and that interest and activities in remote sensing are intensive and growing.   The consensus of the working group is that the ultimate technical, scientific and economic impacts of such systems are probably underestimated by the conservative assessment procedures now in use.

A dramatic increase in the use of LANDSAT data seems to be occurring at the present time, some five years after the launch of

LANDSAT-1 (mid-1972). Investigators around the world have tested the validity of LANDSAT data and the relevance of the data to resource management and environmental control.

The potential benefits to society appear to be significantly greater than the costs of deploying and operating an international satellite remote sensing system.

On all of the above-mentioned grounds, the working group has confirmed that current remote sensing technology can be expanded into an international operational system.

However, economic and political problems will have to be addressed, and it will take time.

5. <u>Transition from Experimental to Operational Systems</u>

The technical success of the experimental LANDSAT system, discussed above, has created new demands for information. But the new demands cannot be satisfied because of the obvious limitations of the experimental program. There now exists a substantial international demand for early transition from an experimental to a fully operational program.

At the same time, there is some reluctance by responsible agencies to embark on a costly international program until it has been confirmed that users are "willing to pay" for services provided. At this point in the evolution of a program, "willingness to pay" has not yet been demonstrated in a market sense or in the sense of cost-sharing commitments by national and international agencies. (See, e.g., the background paper by Gehrig, which implicitly poses questions about commercial-industrial involvement and the viability of a market for information products of both national and international programs.)

Earth resources satellite systems are different from communication satellite and weather satellite systems, in terms of the problems of transition from experimental to operational use. Communication satellites have made the transition successfully because of the homogeneity of the service, available market, obvious cost-effectiveness, and commercial viability. Weather satellites are presently in a state of successful transition (e.g. GOES), because of the prior existence of extensive national and international organizations and infrastructures to provide homogeneous weather information services (e.g., National Weather Service of NOAA in the U.S.A., and the World Meteorological Organization of the U.N.). Earth resources satellites are different, and consequently will face different problems in the transition from experimental to operational use. They do not produce homogeneous data products or services,

the "market" is fragmented and not easy to define, commercial opportunities are uncertain, and there do not exist extensive national
and international organizations and infrastructures, as in the case
of the weather satellites.

It was the consensus of the working group that an organized
international effort is required by the community of remote sensing
scientists and users to ensure progress toward an operational system.
The aim of this effort would be to bridge the gap between experimental and operational applications.

6.   Improvement of LANDSAT Experimental Capabilities

The working group concluded that the technology embodied in
the LANDSAT satellites could be extended and improved as a basis
for an operational system. The general features of such a system
might be the following:

(a)   Incorporation of the planned improvements of LANDSAT-C,
with the addition of a *Thematic Mapper* to meet a broader segment of
user requirements. The Thematic Mapper has a strong generic relationship to sensors now being flown, but has greater spatial resolution and further improvements in the multispectral characteristics.

(b)   High reliability in both space and ground segments of the
system through redundant subsystems (e.g., multiple satellites in
orbit, back-up ground subsystems).

(c)   Adequate frequency of repetitive coverage to satisfy a
variety of applications. A nine-day cycle appears to be desired by
many users (e.g., European agricultural applications).

(d)   Rapid availability of useful "data products" through

- a more extensive network of ground receiving
  stations and data relay satellites

- faster digital processing and duplicating of
  more specialized products

- rapid distribution of specialized products by
  transmission of digital data (rather than transportation of tapes, mailing of photo products)

- standardizing data product formats.

7.   The "Public Domain" Policies of the United States

The United States is a major player in the development of remote sensing satellite systems, and would probably be looked upon

as a leader in the development of an international operational system. The policies of the U.S. in this regard are therefore of paramount importance. The background paper by J.J. Gehrig, Professional Staff Member of the U.S. Senate Committee on Aeronautical and Space Sciences, provides some useful perspectives on U.S. policies and their implications for development of an operational system. Proposed legislation is discussed in the Gehrig paper.

8. Organization of Users Nationally and Internationally

The financing of experimental systems to date has been undertaken by national governments (e.g., the LANDSAT satellite by the U.S., ground receiving and processing facilities by Canada, Brazil and other nations). These national governments now seem to be asking the crucial question: "If there is a real demand for data on an operational basis, why cannot an operational system pay its way, without government subsidies, through charges levied against users?"

The question of subsidies and user charges is not easily answered, for two main reasons: (1) Data and information derived from the operational system may be considered as a "public good", in the conventional sense of economic theory. In this case pricing becomes a matter of political judgement. (2) Users want to see a national or international commitment to continuity of an operational system before they will invest in expensive facilities and personnel. Thus user demands may lag behind system developments; it is a "chicken-or-egg" problem.

The working group recognized that formal mechanisms are needed, at both national and international levels, to provide for the identification of operational users and to estimate the rate of growth of demand. The timing of the introduction and growth of an operational system, and financing based on subsidies and user charges, must be carefully geared to the growth of user demands.

9. Need for an International Organization

At the present stage of evolution from experimental to operational earth resources remote sensing systems, there is a clear need for an international body to plan and coordinate the acquisition, distribution, analysis and use of global resource and environmental information. The working group concluded that the time has come to promote the devevelopment of an international organization as vigorously as possible.

International discussions are required to reach agreement on the distribution of remotely sensed data and information pertaining to national territories. International coordination is required concerning standardization of data products, compatibility of subsystems, arrangements for exchanges of data and information, com-

missioning of international surveys, commitments of resources, and
financial arrangements.

Existing international organizations must be considered. For
example, the U.N. World Meteorological Organization (W.M.O.) has
already established communication networks, analysis centres and
information exchanges to serve participating nations on a global
basis. The W.M.O. could serve as a model for a comparable new organization to coordinate an international earth resources remote
sensing system.

Alternatively, existing organizations such as the W.M.O. and
the U.N. Food and Agriculture Organization (F.A.O.) could expand
their capabilities and areas of concern to include an earth resources remote sensing system.

10. Summary Conclusions and Recommendations

As a result of the working group's deliberations, a number of
conclusions and recommendations were put forward and discussed in
plenary session. There was general support for the recommendations,
but there were also a few cautionary comments, as noted in brackets
at the end. The term *system*, which appears in several recommendations, is meant to be inclusive of technological components such as
space and ground facilities, communications networks, organization
strucutres, and various elements of supporting infrastructure. The
term does not imply joint ownership or any particular division of
ownership of parts of the system. Nor does it imply joint international control or any particular division of control over parts of
the system. However, it does imply some form of international coordination of national efforts to provide integrated operations on
a voluntary cooperative basis.

Conclusions

(1)   Mankind is placing increasing demands on the earth's
resources and environment. Global monitoring is required to provide information for the better management of the earth's resources
and environment. Remote sensing can provide essential information
as part of a global monitoring system.

(2)   Reliability and continuity of any monitoring system are
of paramount importance. Reliability and continuity of remotely
sensed data are especially important to functions such as water
management, land use planning, agriculture and forest management,
disaster warning and damage assessment.

(3)   Experimental satellites cannot provide remotely sensed
data with required reliability and continuity. Operational satellite systems are required.

(4)   Operational satellite remote sensing systems are technically feasible. There is substantial evidence that significant
economic and social benefits will be derived from the use of such
operational systems.

(5)   To use remotely sensed data effectively in the interests
of participating States, there must be an understanding and commitment by each State. Though many States have demonstrated an understanding, none have yet indicated a national commitment to participation in an international operational system.

(6)   International mechanisms are now needed to facilitate
these understandings and commitments.

Recommendations

It is recommended that: -

(1)   an international operational Earth remote sensing sys-
tem be established in which participating States obtain certain
rights and accept certain responsibilities,

(2)   each State or region be encouraged to establish an ad-
ministrative focal point for coordinating earth observation acti-
vities,

(3)   each State or region be encouraged to establish one or
more centres for the collection, evaluation and dissemination of re-
motely sensed data of interest to it, to aid all States and regions
in making effective use of the data,

(4)   States or regions, through their academic, scientific
and technical organizations, develop the individual skills required
for the use and application of remotely sensed data,

(5)   international mechanisms be established to:

   (a)   assist in the world-wide collection of
            remotely sensed data

   (b)   disseminate remotely sensed data throughout the
            world

   (c)   analyse world-wide user needs

   (d)   provide assistance and training in the use of
            remotely sensed data,

(6)   existing international organizations should be considered with reference to recommendation number (5), and

(7)   an international conference be convened to consider the establishment of a global operational earth remote sensing system.

(It should be noted that there was unanimity in the plenary session about recommendation (1), at least as a goal to be pursued. But there were differences of opinion about a timetable for achieving the goal. A majority favoured immediate action to promote the development of the system. A minority took the cautionary view that a "market" for the data products of such a system must be assured before nations would make the necessary commitments. Realistically, the space segment of such a system would probably be based on LAND-SAT technology and development would have to be based on a national judgement relative to national goals. Possibly, an *international* operational system may have to follow development of a *national* operational system by the U.S.)

(It should also be noted that there was unanimity concerning all the other recommendations. Recommendation (7) was given a great deal of emphasis in plenary session discussion. It was envisioned that the proposed international conference would be at a *political* rather than simply a *technical* level, perhaps under the auspices of the United Nations or of some smaller grouping of interested States. Planning for such a conference should start immediately, so that it could be held about 1980.)

SUMMARY REPORT 1
ATMOSPHERIC APPLICATIONS

Vincent J. Cardone, Group Chairman
Wolfgang Baier
Eric C. Barrett
Charles Sheffield
Stanley E. Wasserman
Paul M. Wolff

1. Introduction

Whereas most working groups at the ARI deliberated on possible *future* operational earth observation systems, based on existing experimental systems, the working group on atmospheric applications (WGAA) began by considering an *existing* operational system, which for over a decade has included a remote sensing component. The existing global meteorological observation network also includes an associated global communications system, many regional information processing systems and an international coordinating body, the World Meteorological Organization (WMO) of the United Nations.

The WGAA examined the remote sensing component of the system, the operational meteorological satellite system, in particular and assessed its demonstrated strengths and potential, as well as its inadequacies in several applications areas. The applications areas considered in depth included atmospheric monitoring and observation, weather forecasting on time scales ranging from hours to days, climatology and certain aspects of agricultural meteorology.

The basic components of the operational meteorological observation system are reviewed in the background papers prepared by members of the WGAA. Those papers identify the successes of remote sensing, as well as its failures, and suggest ways in which the present systems should evolve, if future operational systems are to satisfy optimally the data and communications needs in the applications areas considered. The deliberations of the WGAA were eventually distilled to distinct issues from which were derived the conclusions and recommendations presented to the ARI. The issues can be expressed in terms of needs in the areas of sensor and platform technology, communications, education and training, and integration of satellite systems and in-situ systems.

2.   Sensor and Platform Technology

One of the most significant developments in remote sensing is the Geostationary Operational Environmental Satellite (GOES) system, which provides frequent, on-demand, computer-generated, visible and infrared images to local weather forecast offices. There is increasing evidence that where GOES data is available (presently in the Western Hemisphere only) it soon becomes the dominant data component of operational man-machine mix forecast systems for short range forecasts. It has improved forecast accuracy in the short range (1-12 hours) for most hazardous weather phenomena. A more complete discussion of this development is given in the background paper by Wasserman.

The success of GOES has identified further desired system improvements such as 1 kilometer resolution visible and infrared pictures, improved satellite positioning to provide grid accuracy of 5 kilometers or better,and a ground data processing system that provides the forecaster greater flexibility in choice of picture coverage, resolution, grid and enhancement.

The benefits of the experience of GOES can be extended to higher latitudes by the development of a polar orbiting satellite system that provides higher temporal resolution ($1\frac{1}{2}$-3 hours).

The impact of remote sensing data on environmental forecasts beyond the short range appears to be slight. Satellite imagery is ill suited to the specification of initial conditions for numerical weather forecasting models. Atmosphere sounding data from operational vertical temperature profiling radiometers (VTPR) on polar orbiters, when combined with conventional data in the initial value specification, have failed to increase skill in numerical predictions. This failure is due to deficiencies in coverage and accuracy of remote sounding data and, to some extent,to limitations in model physics and resolution which cause rapid growth of forecast error, irrespective of errors in the initial state.

Improved sounders with the ability to measure temperature and humidity with accuracies of 1° C and 10% respectively, through clouds, and with higher vertical resolution than present systems, are required to significantly improve the initial value specification for numerical weather prediction. Such data should be supplemented by systems which measure wind, pressure, air temperature and sea temperature in oceanic regions. Active and passive microwave remote sensors such as those to be flown on SEASAT, may be able to provide the surface wind and sea surface temperature data, but it appears that ocean surface platform systems will be required for pressure, air temperature and subsurface data.

Within the near future, remote sensing data will be available with much higher spatial resolution than that of present operational numerical forecast models. Presently available computers and data transfer systems allow the implementation of global forecast models of higher resolution, and such models should be implemented at the primary forecast and analysis centers.

In the applications area of agricultural meteorology, remote sensing data is most useful as a supplement to on-site measurements that are needed to apply weather based crop-yield models. The meteorological variables of most interest include precipitation, surface maximum and minimum temperature, relative humidity, solar radiation and wind. Clearly, ground based platforms will remain the primary data source, unless accurate remote sensing of rainfall and soil moisture becomes possible. Thus, the development of an automated agrometeorological station should be given high priority.

3. Communication of Earth Observation Data

The real-time data base of the operational meteorological earth observation system has not grown substantially since the introduction of satellite systems, because most satellite data cannot be reduced to quantitative geophysical variables. However, the contribution of satellites to the data base will increase substantially, as global cloud tracking winds become available from GOES and as sounder coverage increases.

The present system used for global communication of such data, the Global Telecommunication System (GTS), has not been totally effective in transmitting data in a complete and timely manner. As a result, some primary centers have established special bilateral communications links. The proliferation of such links could lead to a further deterioration of the GTS.

The WGAA feels strongly that the GTS should be retained and augmented by a system which centralizes all real time earth observation data into a communications satellite. Such a satellite system would be useful in two ways. First, it could act as a primary collection center for observations from isolated surface locations from land and sea, and serve to relay the entire global weather data base to all national and regional data processing centers at computer data rates. Second, such a system could combine meteorological and oceanographic data with whatever LANDSAT-type data are required, in real time, for certain applications in earth resource management.

4. Education and Training in Remote Sensing

Despite the fact that the meteorological satellite system has been operational for over a decade, there remains in the inter-

national meteorological community considerable ignorance of the
capabilities and limitations of remote sensing. It is often such
ignorance, rather than economic factors, that causes under-utiliza-
tion of available systems for earth observation.

Within the next few years, the GOES system will be extended
to global coverage. The techniques that have been developed in
the Western Hemisphere from the experience of the first GOES system should be transferred to personnel elsewhere, who will be exposed to GOES type data. Such a technology transfer will require
an intensive and directed program. The WMO Voluntary Assistance
Program should encourage such a program. More generally, governmental and academic institutions should increase educational and
training programs in remote sensing.

5.   Systems Integration

The present earth observation system could be better integrated in at least two ways. First, satellite systems need to be integrated so that sensor packages can provide data for a wider range
of applications and users. For example, it is well known that the
utility of remote sounding over oceanic regions is limited by uncertainty in the pressure reference level. Marine surface wind measurements from microwave systems to be flown on SEASAT could improve
the specification of the surface pressure over the oceans. If so,
future satellite sensor packages should include or combine both
types of sensors on the same satellite. Ideally, one satellite,
with a central sensor package that includes a simple multispectral
scanner, could serve a wide range of users in environmental sciences in general. More esoteric needs could be met by limited purpose sensors that augment the basic sensor package, or by special
data handling and processing systems.

The above recommendation does not preclude the use of operational satellites in various orbital modes and the continued deployment of research and development satellites. However, more
careful planning and further systems integration in remote sensing
should lead to greater stability in the operational system. That
is, operational satellites can be planned to operate for a specific number of years, with changes made only after technology has
been improved and proven with separate research and development
satellites. Such planning will require a strong international
coordinating body.

A second aspect of systems integration concerns the combined
use of remote sensing and *in-situ* observing systems. While remote
sensing systems have most often been developed as alternatives to
ground based systems, it is becoming increasingly evident that the
utility of remote sensing systems is enhanced when they are combined with in-situ systems. The relationships between the two

classes of data are often complex, as shown in the WGAA member
background papers in the applications areas of daily rainfall mapping (Barrett), numerical weather prediction (Cardone, Wolff),
hurricane forecasting (Cardone), short range prediction of hazardous weather (Wasserman) and crop yield prediction (Baier). The
utility of remote sensing data could therefore be decreased by the
recent trend toward diminished conventional data. Of particular
concern to the WGAA is the decrease in the number of conventional
meteorological reports from ships at sea and from continental land
areas of the tropics. It is urged that those elements of the conventional earth observation system which most effectively complement remote sensing systems be retained or expanded.

6.   Conclusions and Recommendations

      (1)   Improvements in monitoring the atmosphere are required
in both *in-situ* and remote sensing modes. The most important remote sensing improvements required in the near future include more
accurate vertical temperature and humidity profiling of the atmosphere, higher resolution (1 km) infrared data from geostationary
satellites, improved temporal resolution at high latitudes of data
from polar orbiting satellites with automatic data transmission,
and improved passive and active microwave sensors for over ocean
rainfall mapping and surface wind measurement. The most important
in-situ system improvements include the development of simple ocean
surface platform systems for air and sea temperature and sea level
pressure measurements and the development of an automated agrometeorological observing station.

      (2)   There exists an urgent need for better global communi-
cation of earth observation data, even meteorological data for
which a communication system has been in effect. There could be
a considerable improvement, if in addition to the present Global
Telecommunications System there were a centralization of real-time
data, including meteorological satellite data, into a communication
satellite system such that no real time data received would be more
than two hours old. Such a system would provide needed protection
against possible failures of the GTS and could provide a basis for
broader participation by inclusion of non-atmospheric earth observation data.

      (3)   There remains in many parts of the world a considerable
lack of knowledge and understanding of the utility of present meteorological satellite observing systems, particularly of the application of GOES-type data to short range environmental forecasting.
Countries should be made aware of existing educational and training
opportunities, as might be supported through the WMO Voluntary Assistance Program and NATO Science Committee programs, and their
commitment to these programs should be encouraged. Academic institutions are urged to strengthen their programs in remote sensing.

(4)   The earth observation satellite systems need to be integrated. Ideally, a central sensor system should be able to serve a wide range of users, through the production of specialized output at the data processing level. Macro and meso-scale products could be prepared at central facilities, and micro-scale products at direct read-out stations that receive full resolution data for restricted areas and specific local application. The system should include satellites with different orbital parameters and could be subject to improvements on the basis of a research and development program in remote sensing, carried out in parallel with the operational system.

(5)   Remote sensing systems are most effectively utilized when they are used in combination with *in-situ* observation systems. It is, therefore, of concern that there has been a steady diminution of meteorological data from conventional sources in recent years, especially from the oceans and in continental land areas in the tropics. This trend should be stopped, and the utility of the earth observation to operational weather, climate and agrometeorological programs enhanced by the judicious integration of in-situ data and remote sensing data.

(6)   An operational remote sensing system should be planned to function for a predetermined number of years. Subsequent changes should necessitate only that additional expenditure which is unavoidable if improvements are to be achieved.

SUMMARY REPORT 2
LAND APPLICATIONS

Gerd Hildebrandt, Group Chairman   James J. Gehrig
Robert N. Colwell   J.A. Howard
John M. DeNoyer   Richard Mühlfeld
S. Galli de Paratesi   R.A.G. Savigear
Hervé Guichard   Sigfrid Schneider

1. Introduction

    The Working Group on Land Applications (WGLA) focussed on the
significance and possible benefits of an international *operational*
earth observation system based mainly on remote sensing satellites
such as LANDSAT. Data from such a system would be available in all
parts of the world for global, regional or national purposes. Such
a system should be at least as efficient as existing conventional
systems.

    The WGLA emphasized that economic studies are necessary to
estimate the costs and benefits of such systems relative to

-   improvements in land use policy

-   better planning for resource management (e.g.,
    agriculture and forestry)

-   improvements in the exploration of minerals,
    oil, ground water and other natural resources

-   more effective protection and control of the
    ecological balance, productivity of the soil,
    and the endangered environment.

However, one has to recognize that economic benefits, which result
from better information, have multiple long-term aspects. Often
there are very important benefits not clearly definable in monetary
terms.

With regard to organization and management, an international earth observation system would have to be "centralized" to some extent, particularly in the operation of the space segment. Ground data reception, processing and data distribution could be centralized at national or regional levels. Data would be distributed to local users by national or regional centres, in the form of single band images, color composites, enhanced images, tapes, etc.

Further processing to meet specific needs would have to be "decentralized" and accomplished by the user, or by institutions close to the user. The same is true for the development of a suitable methodology to transform remotely sensed data, combined with other data, into project-related information. In this way local conditions, experience, knowledge and capacities, as well as the nature and properties of the objects to be observed, can be appropriately considered.

2.   Requirement: Flexibility of the System

The large variety and diversity of the users' data and information requirements must be considered. This diversity is related to the physical properties of things being observed, the scientific disciplines involved, the natural and socio-economic environments of regions and nations, and the specific problems and decisions being faced.

Therefore a global observation system, especially the satellite sensor packages, should be flexible enough to allow adjustments in data acquisition to fit various specific needs for data and derived information. Flexibility is also necessary in view of still unknown future needs. These needs can be extensively defined in terms of content, time, accuracy, frequency of observation, etc., for the most significant current applications in the fields of land resource management and environmental control. However, the future needs are more difficult to define and require an organized examination by international working groups.

3.   Requirement: Vertical Integration of Methods

Not all existing informational needs for land resources management and environmental control can be obtained, developed, or interpreted from data acquired by satellite sensors, because

- some vital data cannot be collected by remote sensing at all (e.g., subsurface parameters)

- the satellite system will not have the technological capability to accomplish the specific data requirements (e.g., in terms of resolution, repetition frequency)

- the accuracy or reliability of the data in
  respect to the resulting information is not
  sufficient.

Purpose-related vertically integrated methods, using data from a
variety of platforms, have to be developed. The working group
identified a number of gaps in the development of purpose-related
methodology, and recommends the stimulation of further research in
this field.

4. Requirement: Maximization of Multi-Purpose Benefits

Any global operational earth observation satellite system,
whether it will consist of one satellite type or whether it will
be composed of several types, should provide maximum benefits
based on multiple purposes.

In view of the already mentioned variety and diversity of the
users' requirements, and in order to develop a global earth observation system with the maximum multipurpose benefits, it seems to
be necessary to identify purposes with world-wide significance in
the various fields of resource management, environmental control
and land use planning. Once the purposes are identified, data
needs and technological requirements can be specified.

Some local environmental monitoring requirements can be met
most effectively by high, medium and low-altitude aircraft. But
national, regional and international requirements can be met most
effectively only through the use of remote sensing satellites.

The working group concluded that the development of a global
operational earth observation satellite system ought to be based
initially on the needs of national, regional or international organizations for *specific resource surveys*. Early applications of
an operational system should include well-planned, detailed, and
integrated resource surveys.

Early candidate applications for a global operational system
include the following:

(a)   Those segments of an agricultural crop forecasting
system which can derive advantages from remotely sensed data,
namely

- acreage estimation and

- estimation of losses caused by disease,
  insects, pests, meteorological effects, etc.

(b)   Integrated resource and environment surveys and monitoring, namely

- land use inventories

- land capability surveys, including estimation
  of soil productivity and vulnerability

- surface water, snow, and glacier inventories

- ground water exploration

- pre-investment surveys of forest lands

- range land inventories and estimation of
  animal carrying capacity

- mineral and fuel exploration

- monitoring of coastal and river water
  pollution

- monitoring of air pollution in densely populated
  and industrialized areas.

(c)   Topographic mapping and thematic mapping including map
revisions up to scales of 1/50,000.

A global operational satellite system would contribute to economic
and social benefits relative to improved productivity in agriculture and forestry, improved environmental control and improved resource management.

5.   Data Requirements for an Operational System

The data requirements for various applications are not identical but are similar enough to fit into a common specification.

With regard to *spectral bands*, some compromises are necessary
to accommodate a wide variety of applications. As a result of experiences with LANDSAT and airborne multi-spectral scanners (MSS),
improved interpretation and classification is possible with bands
as narrow as 0.04 μm. The background paper by Hildebrandt cites
work of Anderson (1976) and Heller (1976), for example, concerning
the usefulness of particular spectral bands in agriculture and
forestry, within the region from 0.53 μm to 2.60 μm, and with bandwidths from 0.04 μm to 0.60 μm.

With respect to *resolution*, while the working group was impressed by the current LANDSAT-MSS resolution, the group agreed

that it was not adequate for many important applications, including important agricultural applications in Europe for example. The group agreed to call for the best possible resolution that is technologically possible with respect to data acquisition, processing and transmission.

Dr. Archibald Park was seconded from another working group to prepare a brief note on resolution limits, which is appended to this working group report. It was concluded that there are limitations concerning the trade-offs between spatial resolution, swath width, frequency of coverage and data transmission and processing rates for a LANDSAT-type MSS. It was also concluded that no consideration be given to global satellite MSS coverage with resolution smaller than 30 metres, given present LANDSAT technology.

Considering the field sizes, land use patterns and management problems of land resources in Europe, East and Southeast Asia and some other parts of the world, it seems necessary to provide a ground resolution of 30 or 40 metres in many cases. However, a resolution of 10 metres is necessary in some cases and aircraft systems are required to complement satellite systems.

Regarding *frequency of coverage*, the group concluded that a 9-day cycle would be acceptable for the majority of applications under consideration. This would require two satellites of the same type as LANDSAT-2. In retrospect, the 18-day cycle of the original LANDSAT was an excellent compromise to satisfy a large variety of applications. However, a 9-day cycle would be much better in view of the frequent cloud cover in some areas of the world.

There are of course, some applications which require specific high frequencies of observation. For example, the control of the desert locust population in endangered regions requires a twice-daily observation in order to monitor the precipitation which affects the desert locust population. J.A. Howard (1976) pointed to the successful use of NOAA imagery for that purpose.

Regarding *time of availability* of remotely sensed data, very fast delivery of data products is needed for a few monitoring purposes in the field of pollution control, flood control, etc. Data availability within a few days is highly desirable for surveys and monitoring where phenological aspects must be considered, as for instance in the crop forecast application. Availability within a few weeks is sufficient for surveying and monitoring applications dealing with "slow-changing" features. Archival availability of some data products is required for long-term studies of changes, e.g., land use patterns.

6.   Additional Considerations and Recommendations

Besides the conclusions presented in the foregoing chapters, which relate chiefly to LANDSAT-type systems augmented by weather satellite systems, the group proposed some further technological and organizational recommendations for future remote sensing in the fields of land resources management and environmental control, as follows:

(a)   *All-weather, day-or-night* imaging systems, with global coverage capability, should be developed for several critical applications such as disaster monitoring (e.g., floods and landslides), snow cover measurements during the melting season, etc. The same imaging systems may also provide the capability to identify soil and rock types and patterns through short-term vegetation changes, and to provide information on agricultural crops and soil moisture.

A satisfactory system will require passive microwave and I.R. imagers, and synthetic aperture radar (SAR). Radars with improved radiometric capabilities need to be developed. Signal processing methods that preserve the radiometric quality, that can handle high data rates efficiently, and that can provide data products within a few hours of acquisition, need to be developed.

Use of the L-band SAR on SEASAT-A will provide the first opportunity to evaluate such a system from a satellite. SEASAT-A should provide data from land as well as ocean regions to test its usefulness and to develop interpretation methods and application models in related fields. Attention should be drawn to the fact that the interpretation of microwave images and data is still in the early stages of development. This means that a strong research effort is needed to identify how the data can be transferred into useful information what may be complementary to information derived from LANDSAT and other systems.

(b)   In cloudy regions the provision of high-resolution, large-area imagery during the very short cloud free periods or at specific times (e.g., for crop surveys, pollution monitoring or disaster surveys), cannot always be guaranteed by remote sensing from satellites or high-altitude aircraft. In some special cases, a solution could be the provision of cheap *rocket photographic platforms* (e.g., Black Brant, Skylark, Petrel, Skuarockets) capable of providing resolutions of 1-2 m, depending on camera arrays and focal lengths used.

(c)   Full use should be made of all possible methods of obtaining *stereoscopic images,* using photography, scanner imagery or radar images, from satellite as well as aircraft platforms. Stereoscopic coverage combined with multispectral remote sensing can provide a very important aid to interpretation. Specific examples are in geology: to describe land forms, to observe the relation between landforms and linear features, and to estimate the direction and magnitude of slopes. Other examples are in forest surveys: to classify forest types and other vegetative cover. High resolution stereoscopic images are also essential for the preparation of topographic maps.

Plans and programs need to be developed to use the capabilities of the *space shuttle* to obtain stereoscopic images for specific areas. Consideration should also be given to developing a system to provide global stereographic coverage with a resolution of about 20 metres and a base to height ratio of 0.6 to 1.0.

(d)   The time between data acquisition and *data availability* ranges from weeks to months for most users in the current LANDSAT program. Development of "quick look" technology has been demonstrated in Canada. Research efforts have shown that management decisions could be based on the data if the data had been available earlier. Examples include rangeland management, measurement of the extent of major forest fires, and flood mapping; in such applications rapid information is important for damage assessment and for planning recovery.

Strong efforts need to be made to reduce the time between data acquisition and availability for the LANDSAT program and for future experimental and operational imaging satellite systems. Capabilities need to be provided for data availability within five days for all data, with a special provision to provide data within hours for some critical applications.

(e)   *Multitemporal* approaches, as well as the matching of data or imagery from *different remote sensing sources,* require the best possible geometric accuracy in a unique geometric system of reference. The group emphasized the importance of developing operational methods in this field. In spite of encouraging progress, further research should be stimulated.

(f)   The group recommends that attention be given to the problem of developing and supplying *low-cost imagery* and data analysis *equipment* for use in developing countries and in small laboratories.

(g)   The group recognized that development of a global remote sensing system is limited by a lack of public understanding of the potentials of remote sensing to provide data and information of value to many national and international organizations. The group therefore recommends that as a matter of high priority, steps be taken -

(i) to inform governments and organizations of the value of remote sensing methodologies and techniques

(ii) to provide the press and other public media in diverse countries with "popular" illustrations of the nature and value of remote sensing

(iii) to encourage the presentation of lectures to appropriate learned societies throughout the world and to stimulate discussions of the usefulness of remote sensing in environmental research and development

(iv) to undertake pilot studies which demonstrate the usefulness of remote sensing for land resources management and environmental control

(v) to offer courses and seminars in remote sensing in "strategically" placed and regionally important universities, technical colleges and other organizations concerned with education and training, to demonstrate the value and stimulate interest in the subject

(vi) to provide scholarships and bursaries for remote sensing studies in universities and colleges that offer post-graduate diploma and higher degree courses in remote sensing, or in subjects in which remote sensing constitutes an essential component of the course (e.g., integrated survey or soil survey, forest inventory).

(h)   The group considered that research into the application of remote sensing in natural resources management and environmental sciences should be stimulated by the provision of research funds. Such funds ought to be adequate to provide laboratories and departments with both the equipment and the research and technical staff required to support remote sensing programs. This is considered particularly important in the earth sciences, where budgets are low in relation to those required to establish the laboratory and field facilities required for remote sensing research.

APPENDIX TO SUMMARY REPORT 2

A Note on Resolution Limits

Archibald B. Park

It is instructive to note that in LANDSAT-1 and 2 the MSS has a bandwidth requirement of 16 megabits whereas the STADAN telemetry network has a capacity of only 20 megabits. It should come as no surprise therefore that it is technically impossible to improve the resolution of LANDSAT without giving up some of the capability of the system in other ways. The following table shows the relationships that exist and the effect of changing various parameters.

MSS Resolution (IFOV)		Swathwidths	Frequency of Coverage
MSS 1-2	80 m	185 km	18 days
Option	160 m	370 km	9 days
Option	40 m	92.5 km	36 days
Option	8 m	18.5 km	180 days

Note that all these relationships are linear and, as stated before, are all limited by telemetry bandwidth.

If one assumes that the daily acquisition of LANDSAT-1 (200 scenes) is completely cloud free and therefore available for analysis, and that the analysis is fairly simple (e.g., agriculture compared with non-agriculture), and that a single scene can be formatted and analyzed in two hours on a fairly large computer (IBM 360-50), then on a 3-shift basis it would take 17 computers to do the job. A very large computer (Illiac IV), which can perform these analyses approximately 20 times faster, could just barely keep up.

Obviously this sort of demand on computing power does not exist. Note however that the IBM 360-50 is completely occupied doing 12 LANDSAT scenes a day. If one assumes that the resolution is changed to 10 metres and all else remains the same (4 bands), then we could do 1/64 of 12 scenes or less than 20% of *one* scene.

A recent study performed for NASA indicates that if resolution is reduced to 30 metres (LANDSAT Follow-on), in order to perform an agricultural analysis of even a 5% sample of the wheat growing areas of the world, each 9 days, a minimum of 10 interactive analysis stations of the Bendix, G.E., or I^2S type would be required. For this operation the existing design computer must be upgraded, i.e., from DEC 11-35 (45) to 11-70 machines, and a large machine must be used as a facility manager and for some bulk processing tasks.

On a more optimistic note it has been reported that with the use of carefully resampled, enhanced and registered multidate imagery, a factor of two (app.) improvement in LANDSAT-1 resolution has been achieved. It is expected that similar results can be achieved on LANDSAT Follow-on.

It is the opinion of the author that no consideration be given to requests for complete coverage of the earth from space with an IFOV of smaller than 30 metres. In the first place the burdens placed on communication, computer power, storage facilities and manpower are inordinately large and expensive. Second, available processing techniques may improve the resolvable features down to 15 metres where necessary in special applications. Third, any requirement below these numbers can only be afforded, and in fact used, for areas that are sufficiently small as to be covered by aircraft.

SUMMARY REPORT 3
OCEAN APPLICATIONS

Eric S. Posmentier, Group Chairman   A.D. Kirwan
Peter W. Anderson   Robert E. Stevenson
Donald J. Clough   Paul G. Teleki

1.   Introduction

The individual background papers written by the members of
the Working Group on Ocean Applications (WGOA) deal with the state-
of-the-art of selected aspects of oceanic remote sensing and information systems, and of operational applications of these systems.
Based on these papers, it is possible to determine a set of objectives which deserve high priority in the research and development
necessary for the operational applications of oceanic remote sensing and information systems. In this report, after a brief re-
capitulation of selected oceanographic uses of satellite technology,
recommendations for further research objectives, development specifications and applications are presented.

Oceanic activities which require the direct or indirect support
of oceanographic data include fishing and aquaculture, transportation, mineral, gas, and oil exploration and production, recreation,
and waste disposal. Some of these depend indirectly on oceanographic data through their use of weather, sea state, or sea ice forecasting, navigational route planning, fisheries location and surveillance, marine communication, pollution monitoring, search and
rescue operations and ocean engineering or architecture. However,
before any information systems can be successfully applied to the
problems of ocean operations, it will be necessary for the value of
oceanographic information to be convincingly demonstrated to managers of the ocean operations. In particular, it is important to
document and evaluate the incremental economic value of improvements
in data timeliness, frequency of availability, synopticity, spatial
density, spatial resolution, and accuracy achievable by systems
including remote sensor and data collection platforms.

Obviously, there do exist needs for a wide variety of oceanographic data, including data with the improvements mentioned above.
These needs are a challenge to sensor technology, data transmission
capabilities, data base methodology, management information systems,
international cooperation, and oceanographic research. This challenge can be met with the greatest economy and reliability by remote sensing and information systems used in conjunction with more
conventional techniques. To some extent, the development and implementation of remote sensing and information systems have satisfied the operational information needs of some oceanic (and other)
activities. However, further research and development will be required for certain other applications to become operational.

2.   Information Needs

Wave information is invaluable to several oceanic operations.
Routing of commercial transport vessels and fishing vessels, protection of coastal and ocean construction activities, determination
of design criteria for coastal and ocean engineering and architecture, and search and rescue operations, are but a few. The need
for wave information can be divided into two categories: *near real-
time, and climatological.* As more ocean operations develop the ability to utilize directional wave spectra, such data will become recognized as more valuable than significant wave heights and periods.

Many oceanic applications of remotely sensed data require
boundary information. As an example, a commercial fishing operation might be helped by knowledge of the location of a linear thermal boundary, more than by sea surface temperature information as
such. Similarly, there is a navigational need for near real-time,
precisely located ice-water boundary information. Other applications, including oceanographic research, require delineation of
boundaries between regions of different colours, roughnesses, wave
directions, etc., as indications of the boundaries between different water masses, of an upwelling region, or of a current. In addition to the intrinsic operational value of boundary information,
the technological feasibility of detecting boundaries and of transmitting, storing, and processing boundary information is frequently
much greater than the feasibility of obtaining the corresponding
broad area sea surface data.

Prognostic ocean and atmosphere models, including the forecasting of sea state, weather, climate, marine fisheries, and pollutant dispersion, require the *fluxes* of water, heat, and momentum
through the air-sea interface. The sea-surface-temperature and
current speed are less valuable than fluxes as boundary conditions
for such models, although they are often of greater ultimate interest and easier to measure directly.

For the purposes of prognostic models, the desirable spatial and temporal resolution of an observation's location and time, the measurement precision, and other system specifications are shown in Table I. Additional data which have potential for measurement from satellite sensors are water mass identification, surface currents, chlorophyll (type A,B, or C), colour and sediment load, organic slicks, shallow bathymetry, marine mammal tracking and tides.

3.   System Design Recommendations

While remote sensing from satellites is in many cases the best or only way of obtaining the information summarized in the preceding section, other platforms and sensors, including conventional research vessels, *in-situ* sensors, and remote position-fixing and data collection platforms, all have their respective advantages. In general, the optimum system for ocean observation will be a complementary *combination* of different platform, sensor, data transmission, and data handling technologies. Information systems combining remote sensing data with *surface truth* data will almost invariably be less limited than remote sensing systems alone. Remote sensing systems also have obvious limitations with regard to deep subsurface information. On the other hand, satellite sensors can provide uniquely synoptic data and satellite telemetry and position-fixing make possible the real-time acquisition of Lagrangian current information and data from *in-situ* sensors at the surface or at depth. Further research and development of oceanographic sensors should be justified on the basis of their potential role in an information system combining different technologies.

Information and management decision systems for many oceanic applications extend to shipboard operations. Such systems can be made more efficient by the development of *shipboard receivers* for data transmitted from remote sensor platforms or retransmitted via satellite from surface data collection platforms, and by the development of *computing hardware* and *software* suitable for use by shipboard personnel.

The need of prognostic models for air-sea water, heat, and momentum fluxes, discussed in the preceding section, dictates the need for an ocean information system to support an urgently required *research and development* effort for techniques of finding fluxes from combinations of instrumented buoy and remote sensor data. This research and development effort will certainly use concurrent data buoy, remote sensors, and other data. It may also include the use of   spacelab and shuttle technologies.

A number of technological capabilities can be used by future ocean information systems with little or no further development, and should be given adequate consideration in the design of such systems. These include the use of several types of aircraft as sensor and/or data collection platforms: lighter-than-air craft,

TABLE 1 - REMOTE SENSING OCEAN PARAMETER MEASUREMENT REQUIREMENTS

PARAMETER	ACCURACY	PRECISION	AREA	REPETITION
• Temperature, Surface				
Weather	1.0°C	0.25°C	200 km	< 24 hr
Ocean Condition	1.0	0.25	200/20	1 day
Climate	0.2	0.1	200	1 day
• Ocean Profile				
Temperature to 300 m	1.0	0.25	10	12 hr
• Surface Air Temp	1.0	0.1	200	12 hr
• Vertical Atm. Profile				
Temperature	1.0			
• Wind Speed	1 m/sec	0.5 m/sec	200/10 km	3 to 6 hr
• Wind Direction	10 deg	5 deg	200/10	3 to 6 hr

Waves				
Height Spectra	0.5 m/10%	0.5 m/10%	100/10	12 hr
$\overline{H}_{1/3}$	0.3	0.3	25/10	3 to 12 hr
Length Spectra	10%	10%	100	3 hr
Direction	10 deg	10 deg	100	3 hr
Ice Cover	0 to 100%	12%	1 km res	1 day
Leads	100 m	100 m	100 m	<3 hr
Thickness	0.5 m	0.25 m	20 km	3 days
Salinity	0.01 ppt	0.005 ppt	100/5 km	1 to 30 days
Chlorophyll	1 to 100 µg/ℓ	0.3	400 m	3 days
Surface				
Currents	2 to 5 cm/sec	10%	20/1 km	6 hr
Direction	10 deg	5 deg	20/1 km	6 hr
Ocean Topography, Research	2 cm	2 cm	500 m	6 hr

manned aircraft, and remotely piloted air-vehicles (RPV's). In
addition, geostationary platforms such as GOES and the proposed
Stationary Earth Observation Systems (SEOS) can potentially be used
for on-demand remote observations and on-demand relay of data from
in-situ sensors. The use of multi-spectral scanners in the visible
and near-visible range with greater spectral resolution is feasible
and potentially valuable. (This is presently under consideration
for future systems.) This improvement in spectral resolution can
be facilitated by using less spatial resolution, which would not
interfere with many applications, but which may not be necessary.

4.   Organizational Recommendations

     Several design concepts incorporated into SEASAT-A, particular-
ly the synthetic aperture radar (SAR), involve innovative experimental sensors. The further development of these sensors, of data
interpretation techniques, and of data applications would benefit
from the availability of SEASAT-A data for experimental use by research groups outside the United States. This would require the
organization of international cooperative efforts. (See the background paper by Miller.)

     To avoid wasteful duplication of effort, the development of
future spaceborne microwave sensor systems, including radar, should
be a multi-national effort.

Charles Buffalano, Group Chairman
J. D'Hoore   B.P. Miller
Hans Heyman   Lawrence W. Morley
Philip A. Lapp   Archibald B. Park

1.   Introduction

While the scientific and remote sensing communities are convinced of the present and potential utility of the information produced by satellites, national policy and decision making communities
are not similarly persuaded. Since governments must be convinced
that any new technology is in the national interest before they can
incur the costs of making it operational, it is of critical importance for the scientific and remote sensing communities to understand what motivates national decision makers and to make those
arguments and provide those analyses which will be found convincing.
Unfortunately, there is no single line of reasoning which *a priori*
is convincing. There are many motivating themes in any governmental
decision, some highly rational and quantitative, others subliminal
and non-quantitative but all extremely important.

The purpose of this report is to help extend and enlarge the
role that objective anlaysis plays in government decisions about
remote sensing programs. The Group drew on its personal experiences
to describe several decision rationales that had been used successfully in the past. Next the Group recognized the central place that
economic rationales have had in the evaluation process and discussed the analytic elements which contributed most to success and the
forces which made analysis difficult. The Group recognized that
there is no lack of professional literature on cost benefit studies
available to practitioners and did not intend to recapitulate that
work. Rather the Group felt that practitioners needed a description of the environment in which their analyses would be used to
understand better how to make the analyses convincing to others.
Finally, based on these observations, the Group agreed to several
recommendations.

2.   Rationales for Technology Implementation

Every technical program needs to answer the question, "Why", asked by government decision makers who must consider the program in competition with other programs for public support. These decision makers are motivated not only by quantitative considerations but also by unquantifiable and sometimes seemingly irrelevant considerations which can overwhelm the best quantitative analysis.

There are at least four answers to the question, "Why", some of which are quantitative, some not.

(1)   An important argument for the implementation of a new technology is that it provides *benefits to society well in excess of its costs*. Theoretically, a government decision maker would choose those new projects which showed the highest return on the citizen's investment. In actual practice, however, economic analyses are required only to show that the benefits exceed the costs and other, non-quantitative, considerations are used to make decisions between projects. This makes the economic analysis a necessary but not sufficient condition for new technological initiatives.

There are two major types of economic analyses: (a) benefit/cost studies where the new technology produces new benefits and (b) cost-effectiveness studies where the new technology replaces an older technology and does its job at a lower cost. Both types are highly quantitative and use decision theory, optimization methods, econometrics, and social process modelling.

Spacecraft feasibility demonstrations should be seen not only as tests of the *engineering* but also as tests of the *economics* of the system. Unfortunately economic analyses are often seen as a device for selling a new technology. Once it is sold there seems to be little motivation to determine whether or not the benefits actually materialized. This is unfortunate for two reasons. First, demonstrations of economic viability and the testing of the economic hypotheses give greater validity to subsequent economic analyses for the operational system. And second, the economic feedback can help direct new technological developments into more lucrative areas.

(2)   There is a *progress imperative* which demands that if it is possible to do anything new it should be done because any technical progress is a positive societal good. This is based on the view that serendipity often produces unexpected benefits which are neither predictable nor demonstrable *a priori,* and doing nothing assures no serendipity. Furthermore this view is reinforced by the fact that all present societal benefits are products of some previous technological venture.

(3)   *Countering external threats* demands doing something
new before the enemy does it first. This is a very important and
effective rationale in the military and intelligence communities.
A similar rationale exists concerning sovereignty protection.

(4)   There is a *knowledge imperative* which posits that
knowing everything about everything is itself a positive social
good.

3.   Economic Studies

Benefit/cost, cost effectiveness, and threat studies have
been quantified with varying degrees of success. Economic studies
are widely used because they can be done in terms acceptable to a
large audience. However, the fact that they are necessary but not
sufficient means they must be extremely well done, otherwise their
incompleteness can be used to justify negative decisions. Unfortunately since no economic analysis can be entirely beyond criticism because there is a very large number of assumptions and con-
jectures about the social processes under consideration they are
seldom fully compelling. However, criticism should not be based
on the absence of the following three important considerations which
should be part of every economic analysis.

First, the analysis should contain a clear statement of the
end objectives of the technology program. For example, remote
sensing in agriculture might have an objective to predict crop production earlier with greater accuracy. The goal is in terms of the
improvements in crop production forecasts rather than engineering
parameters which characterize the technology such as resolution,
swath width, signal to noise ratio, or the wavelengths of the spectral bands. While these technical parameters clearly influence the
extent to which the objectives can be achieved, they are not themselves the objectives.

Second, every analysis should have a clear and complete statement of the decision environment in which the information will be
used. For example, how do present ground observation systems function in crop production forecasting, what is the state-of-the-art
in agrometeorological modelling, and what are the specific time
constraints on information production? What private or public decisions will the information affect and how are these decisions
currently made? Are they made by the private sector in the market-
place or by the public sector through executive directive, provision of public information, or legislation?

Third, an analysis must contain the criteria by which the
improvements introduced by the new technology are measured. In a
purely economic analysis, the measure is simple: the ratio of the
benefits to the costs using a reasonable discount factor for the

future worth of money. If one wishes to consider any values other than economic, it can be very difficult because suitable measures are almost entirely missing.

4.   Limitations to Good Economic Analyses

It is easier to say what is required for good analysis than it is to do it. There are many limitations to the validity and usefulness of economic studies.

(1)   No real remote sensing system has a single purpose. Space systems are used in combination to provide information on a single physical parameter and a single system provides information on more than one physical parameter. For example, a satellite configured for agricultural purposes can also be used for forestry and land use management. Furthermore, agricultural production estimates require information from many satellite systems to estimate planted acreage and yield. This is a complicating factor which is difficult to fit into a rigorous analysis unless one takes the approach that only those benefits should be considered which a system produces uniquely and alone and that the benefits so estimated are a lower bound.

(2)   The real decision environment in which remote sensing information is used is diffuse and extremely complex in both democracies and centrally planned economies. Only the simplest of decision models for important economic and social questions exist. The value of information is an area in which economists are still doing basic research.

(3)   It is often not in the interest of program managers to make the objectives of a remote sensing system precise. Even when the system has only civilian uses, clear objectives for the new technology permit opposition to be crystallized and make it difficult for supporting coalitions to form at a high policy level. Coalition members who could agree on goals loosely stated may withdraw over the specifics.

(4)   Any technological system will produce unintended and secondary effects which were not well understood initially. To the extent that analysts attempt to include these externalities or second order effects for completeness, the analysis becomes more and more intractable. To the extent they are excluded, the analysis can be criticized for incompleteness.

(5)   Government decision makers are understandably reluctant to commit to operational space systems because they recognize operational systems imply a commitment for a long period of time and establish an institution whose own inertia makes it difficult to turn off the technology if it does not work. Remote sensing systems

could better be viewed as system feasibility demonstrations having limited duration. This has two advantages. First, governments which may be unwilling to begin operational activities may find it more acceptable to begin with limited duration commitments. The second advantage is that it gives the remote sensing community an opportunity to test the economic hypotheses which were used in justification. However, many users of remote sensing data are also understandably reluctant to replace less effective information gathering technologies with remote sensing when the new technology is not guaranteed to be available beyond the test period.

5. Recommendations

Based on these observations and discussions, we recommend:

(1)   Future economic studies contain precise statements of the end objectives of the technology, a description of the decision environment in which the new data will be used, and proposed criteria by which the improvements of the new technology are to be measured when compared with the conventional methods.

(2)   Applications scientists in remote sensing should be encouraged, as part of their technical studies, to report the way the results of their research fit into the decision making process and to suggest how to quantify the way in which their research improves the information available to resource managers.

(3)   Economic field tests should be designed into remote sensing projects. These economic field tests should not be appended to a project, but should be considered an integral part of project evaluation. These tests should provide validation for the sociological and economic assumptions in the supporting analyses.

(4)   Case studies of successful and unsuccessful technology transfer should be made. There is a need to review the process by which technology is transferred from the public to the private sector. If an experiment proves successful and seems to promise economic benefits, there is the problem of transferring not only the technical but also the managerial capability into the private sector quickly and without a loss of service. This process is not at all well understood, but understanding is necessary if benefits are to be realized for a nation.

(5)   More effort should go into modelling decision processes in the user community. There are cases today in which resource managers already employ numerical models to assist in decision making. For example, the forest industry has a few vertically integrated companies which maintain massive business data banks. In some cases, numerical models recommended harvest amounts for sustained yield. There are similar decision models in water resource management and electric power distribution.

EARTH OBSERVATION AND INFORMATION SYSTEMS

Donald J. Clough
Professor of Engineering, University of Waterloo, Canada
President, Systems Engineering Associates Limited

Lawrence W. Morley
Director-General, Canada Centre for Remote Sensing
Department of Energy, Mines and Resources, Ottawa, Canada

Introduction

Enormous strides have been made in recent years toward the
development of operational satellite and airborne systems for
the remote sensing of atmosphere, land and ocean surfaces. Environmental data buoys have been developed for deployment in
continental shelf and deep ocean areas, to provide continuous
monitoring of the atmosphere, ocean surface and subsurface. Land
stations, radars and atmospheric probe devices have been developed
to monitor a variety of environmental factors. All of these developments have been based on staggering advances in platform technology, electromagnetic sensor technology, computing and communications systems, and theoretical models. Mankind is developing the
potential to take the physical pulse of the earth at any time and
place and to whatever accuracy may be desired. The main limitations are those of economics and politics.

As the world's population increases, international observation
and information systems will become more important for the management of food, fuel and other scarce material resources. Such
systems will be costly. But they will provide great economic
benefits through the monitoring of environmental factors such as
weather, climate change, crop conditions, sea state and sea ice.
The LANDSAT satellites, for example, can provide information about
food crop and forest inventories, crop disease, and general soil
and biomass conditions. The SEASAT satellites will be able to
provide sea state and sea ice information to improve operations
in ocean transportation, offshore oil and gas production, ocean
mining, fishing and aquaculture. The weather satellites and ocean

data buoys have proven to be of great value in monitoring and predicting weather patterns, hurricanes and damaging sea conditions.
The potential economic value of better environmental information
seems obvious. However, in most cases the potential has yet to be
demonstrated through controlled economic verification experiments.

The development of international observation and information
systems poses some political problems. Who will build the systems?
Who will operate them? How will the information be distributed?
Some of the less developed countries have expressed concern that
satellite systems will provide information that will help the
more technologically advanced countries to exploit them further.
They are afraid of a new "technological imperialism" or "information
imperialism". Some of the industrial nations have expressed concern
about military "spy in the sky" aspects of satellites, and about
commercial advantages of one country over another in the buying
and selling of commodities such as wheat. The potential political
problems are obvious. The involvement of the United Nations is
important. However, some individual nations or groups of nations
may be prepared to develop costly international systems with or
without full UN participation.

The NATO Science Committee and its subsidiary Special Programme
Panel on Systems Science have evidently developed a particular
interest in the technical, organizational, economic and political
problems involved in developing operational earth observation systems
at both national and international levels. As one small step toward
elucidating the problems and specifying some objectives, these NATO
bodies have sponsored the NATO Advanced Research Institute (ARI)
on Earth Observation and Information Systems. The idea is to bring
together a small group of scientists, engineers and administrators
who are involved in various ways in the development and use of
such systems for environmental control and resource management.
The participants are expected to advance their views and opinions
as individuals, rather than as representatives of organizations.
The emphasis is to be on ultimate operational systems, rather than
experimental systems. It is hoped that the participants will be
able to reach concensus on some research and development directions
relative to operational systems, and to put forward some conclusions
and make recommendations for the benefit of a wider community. All
aspects are fair game for discussion, including scientific, technological, organizational, economic and political aspects.

This introductory paper is designed to help provide a starting
framework for workshop organization and discussion at the ARI. It
is meant to be suggestive, and touches on only a few of the many
possible aspects that may arise in discussion. The paper is organized as follows:

- Statement of purpose.
- Structure of working groups.
- Some background notes on systems.
- Some background notes on applications.

The authors are aware of the arbitrary nature of this approach to structuring the discussions. It is offered simply to facilitate the orderly reporting of proceedings, not to constrain the discussions. It is expected that some radical departures will emerge from the discussions.

ARI Purpose

Earth observation is a general term adopted for purposes of this NATO Advanced Research Institute. It embraces both monitoring and mapping the natural environment, and surveillance of human activities.

The general purpose of the ARI is to address the critical problems of integrating the relatively new science and technology of remote sensing into operational earth observation and management information systems. The main problems of concern are those related to systems design, organization, development of infrastructure, and use of information in decision processes.

In an attempt to bridge some of the gaps between the scientific/technological community and the resource management community, it is proposed that the ARI undertake the following:

- Identification of main land, sea and atmosphere information needs of resource managers, deficiencies of present decision models, and data/information gaps.

- Identification of existing earth resources observation, communication and information processing technologies, and their present limitations as a basis for operational systems.

- Identification of decision modelling requirements to take advantage of the new technologies.

- Identification of education and training requirements to take advantage of the new technologies.

- Identification of main management problems of organizing, planning and implementing national and global earth observation and information systems.

- Identification of critical systems research
  objectives to bridge the gaps between technological
  capability and operational utilization.

- If possible, preliminary specification of objectives
  and design criteria for some selected pilot operational
  systems to bridge the gap between experimental and
  fully operational systems, taking account of learning
  curves, time-frame for implementation, costs, and
  potential benefits.

This is obviously a large undertaking for a relatively small
group meeting for a short time. However, it seems worthwhile to
try to put it all together in one broad sweep at this time.

Working Group Structure

Because of the constraints on time available for workshop
group meetings and on the number of participants, it is convenient
to set up eight groups - four systems working groups and four
applications working groups. Each participant is assigned to one
systems working group and one applications working group. This
provides for some cross-linking of points of discussion. The
working groups are designated as follows, with a few discussion
topics listed as examples:

List of Systems Working Groups

A. Sensor/Platform Systems Group

    - Experimental and operational platforms
      (satellites, aircraft, buoys, etc.)
    - Experimental and operational sensors
      (optical, I.R., microwave, radar).

B. Data Handling Systems Group

    - Data telemetry and retransmission
    - Ground receiving stations
    - Signal processing, onboard and ground
    - Data distribution networks.

C. Management Information Systems Group

    - Data products, standardized geo-referencing
    - Physical systems models
    - Automated interpretation and forecasting systems
    - Management and economic decision models
    - Archiving and data base management.

D. Social Impact Systems Group

   - Organization and institutional processes
   - Politics
   - Economic benefit/cost models
   - Social change models.

List of Applications Working Groups

1. Atmospheric Applications Group

   - Weather forecasting
   - Climatology
   - Agro-meteorology
   - Pollution monitoring.

2. Land Applications Group

   - Natural resource inventories
   - Agricultural crop and forest monitoring
   - Soils, vegetation, hydrogeology assessment
   - Snow cover, ice, glaciology, water assessment.

3. Ocean Applications Group

   - Offshore oil and gas, mining, engineering
   - Ocean and coastal transportation
   - Fishing operations and fisheries management
   - Pollution monitoring.

4. Applications Methodology

   - Air-sea interaction
   - Radar signatures for land and sea applications
   - Integration and correlation of various platforms
   - Critical methodological issues.

The assignment of participants to working groups and the scheduling of workshop sessions and plenary reporting sessions are designed to provide for cross-linking of ideas and working group discussions.

Some Notes on Platforms

There is an immense variety of things to be observed. They can be grouped conveniently into ten main categories, as depicted in the following table:

	Natural Environment	Human Activities
water surface	x	x
water subsurface	x	x
atmosphere	x	x
land surface	x	x
land subsurface	x	x

Examples of things to be observed are given in the following sections of this paper.

There is also an immense variety of observation platforms that can be used, a veritable technological smorgasbord. The most important platforms - aircraft, satellites, ships, ocean stations and land stations - have varying capabilities relative to the monitoring of natural environmental phenomena, as depicted in the following table:

Monitoring Environment	Air-Craft	Land Stations	Satellites	Ocean Stations	Ships
land subsurface	x	x			
land surface	x	x	x		
atmosphere	x	x	x	x	x
sea surface	x	x	x	x	x
sea subsurface	x			x	x

It is important to note that various platform systems may be mutually complementary or competing. Integration of various platforms into mixed systems is a major challenge.

The same platforms have varying capabilities relative to the surveillance of human activities. For purposes of national sovereignty control, such surveillance generally involves three main tasks:

- detection (D)
- identification (I)
- collection of physical evidence (E).

The various platforms have varying capabilities to carry out these tasks effectively. A radar satellite or a system of underwater sonar devices, for example, may be useful for detecting the presence of fishing vessels and merchant vessels within a nation's coastal economic zone, under all weather conditions, day and night, and for determining the location of the vessels. However, a fixed wing aircraft or helicopter may have to be dispatched to identify vessel types, nationalities, ownership and activities. A coast guard or fisheries patrol vessel may then have to be dispatched to collect evidence of violations of fishing and pollution laws - to carry out boarding, cargo inspection, pollution sampling, and to provide eye-witness accounts. In matching surveillance platforms with missions, it is important to consider these three key tasks. The following table depicts the surveillance capabilities of various platform types, in terms of detection (D), identification (I) and collection of physical evidence (E).

Surveillance of Human Activities

	Air-Craft	Land Stations	Satellites	Ocean Stations	Ships
land subsurface		D,I,E			
land surface	D,I,E	D,I,E	D		
atmosphere	D	D			
sea surface	D,I,E	D	D	D	D,I,E
sea subsurface	D			D	D,I,E

Aircraft are the most flexible of all platforms for many observation tasks. For purposes of discussion of present and future
capabilities it is useful to consider the following aircraft types:

- manned airplanes, fixed and rotary wing
- remotely piloted vehicles (RPVs)
- balloons.

The general capabilities of conventional fixed and rotary wing
manned aircraft as remote sensing platforms are well known. They
are flexible in terms of payload arrangements and flight scheduling.
They are well suited to surveillance of human activities. However,
they are costly as platforms for broad-area repetitive environmental
monitoring tasks. Satellites will probably be much more cost-effective than aircraft for most broad-area environmental monitoring tasks
that involve high data rates and near-real-time processing. Satellites can generally provide broader sensor swath widths for frequent
repetitive coverage, have superior line-of-sight for data telemetry,
and are highly reliable once past the "infant mortality" stage in
orbit. However, aircraft platforms will be required for at least
another decade before operational satellite systems are available
to reduce the aircraft burden relative to such tasks as ice reconnaissance (e.g., International Ice Patrol). Even when operational
satellite systems come on stream, a high-low mixed system of satellites and aircraft may be most cost-effective for many tasks.

The capabilities of Remotely Piloted Vehicles (RPVs) are less
well known. A variety of RPV developments started in the 1950s on
several major drone aircraft programs in the U.S., of which few
survived to the 1970s. The failures of early drone programs occurred because of the lack of miniaturization and the inadequacies of
data transmission and data processing technologies. The drones
were too big, too heavy, too costly, and not suitable for remote
control at long ranges. The failures have had a profound impact on
recent developments. The present thrust is in the direction of
miniaturization of system components, small aircraft, and low cost.
To achieve these results, it is evidently the current philosophy
that one should not base an RPV design on the aircraft technology
of the day; instead, one should start with the existing model air-
plane technology and build upon it. This approach is leading to

novel designs of craft having very small cross-sections, small
piston engines, using new materials (e.g. fiberglass bodies) and
having unprecedented payload-to-gross-weight ratios and endurance
characteristics. There is a pressure to transfer spacecraft technology to aircraft (e.g., T.V. cameras weighing under one kilogram,
and forward looking infrared units weighing under 3 kg).

As a consequence of the new approach, a new mini-RPV technology
is rapidly emerging. Based on recent studies and prototype tests
it appears that RPVs weighing up to 200 lbs can be designed to cover
up to 1200 nautical miles in about 12 hours at altitudes up to
20,000 feet. They may be launched and recovered from devices that
do not require major investments or large operational manpower.
They may carry sensors to perform surveillance tasks of detection
and identification under variable weather conditions, day and night.
They include a man-in-the-system loop, at a ground or ship-board
control console. The discrimination and judgement of a human con-
troller is essential for many surveillance missions, and humans
will not be replaced by computers in the foreseeable future. (See,
e.g., "RPVs - Exploring Civilian Applications", T.J. Gregory et al,
Astronautics and Aeronautics, September, 1974.)

Though the concept is exciting, mini-RPVs have a tough development period to go through before one can confidently predict
that they will become operational. However, they may be available
to replace manned aircraft for some tasks by the mid 1980's.

Manned balloons (airships) may also be able to replace manned
aircraft for a variety of surveillance missions by the mid 1980's.
The main advantage of manned balloons is that they have longer
endurance than manned airplanes, yet they can hover in a stationary
position. Though extensively used for maritime patrol during the
second world war, they lacked a peace-time mission and fell into
disuse. Now, however, the problems of managing the 200-mile coastal
economic zones have created a renewed interest in using manned
balloons for new surveillance missions. A balloon may be particularly useful for gathering evidence of violations, since it can
descend to sea level to deploy a small boat, to take water samples,
or to facilitate eye-witness observation and boarding of vessels.
They are being used in Venezuela for land survey applications.

Stationary tethered balloons are presently available for
communications systems (e.g. the TCOM System), with balloons carrying stabilized equipment packages at altitudes of 10,000 to 15,000
feet. Such balloons are promising for special surveillance applications. For example, a balloon carrying a stabilized radar of
conventional design could be tethered to a buoy near an off-shore
oil drilling rig. It could provide surveillance of icebergs out
of more than 60 miles from the rig. It could also carry atmospheric
sensors for weather prediction purposes.

Atmospheric probe rockets (sounding rockets) equipped with cameras and other sensors are presently available (e.g. the Black Brant rockets of Bristol Aerospace Limited, Canada). Such rockets are promising for certain specialized environmental monitoring applications. They offer an instantaneous view of a large area with minimum delay, provided that the sensor payload is easily recoverable at the jettison point. They also offer direct national control with respect to time and security of data (unlike some satellite systems).

Land station platforms. The variety of possible land stations seems unlimited. It includes seismic stations, pollution monitoring stations, weather monitoring stations, coastal radar stations, and agriculture and forestry sampling stations,to name only a few. Extensive networks of such stations exist, and the newer satellite remote sensing technologies may be seen as extensions to the existing systems. It is important to relate the contributions of satellite data acquisition systems to existing land based data acquisition systems, and to keep the relative contributions in perspective. In particular, the investment in existing organizations and the technical infrastructures has to be kept in mind. The time required for the adoption of new technologies and possible displacements of old technologies has to be taken into account.

Satellite platforms are of special interest to participants of this ARI, and may be described in some detail in other papers. The following are particularly important, but by no means exhaust the list.

Earth Exploration: Experimental and Follow-on Satellites

- Nimbus, meteorological satellites
- LANDSAT and ERSOS System
- SEOS, synchronous earth observatory satellites
- SEASAT, oceanographic satellites
- GEOS and Geopause, earth geometry satellites
- HCMM, Heat Capacity Mapping Mission satellite.

Meteorological Satellites

- NOAA polar orbiting satellites
- SMS/GOES geostationary satellites
- TIROS-N, replacement for NOAA series.

General Purpose Spacecraft

- Skylab
- Spacelab.

Several satellites under study or in construction by the European Space Agency (ESA) are also of interest, including:

Earth Exploration Satellites

- SARSAT, synthetic aperture radar satellite
- PAMIRASAT, passive microwave radiometer satellite.

Meteorological Satellites

- METEOSTAT, meteorological satellite.

ESA is also involved in the U.S. Spacelab experiments. In addition, several countries are studying or developing satellites for various earth observation purposes, including the following:

Canada

- Cansat, synthetic aperture radar satellite
- Search and Rescue Satellite (transponder)
- Fishing Vessel Surveillance Satellite (transponder).

France

- Earth Resources Survey Satellite, similar to LANDSAT.

Federal Republic of Germany

- Earth Resources Survey Satellite, similar to LANDSAT.

Japan

- Earth Resources Survey Satellite, similar to LANDSAT
- GMS, geostationary meteorological satellite, similar to SMS/GOES.

Netherlands

- ARTISS, operational LANDSAT type.

U.S.S.R.

- Earth Resources Survey Satellite, similar to LANDSAT.
- Meteorological Satellite, similar to SMS/GOES.

Through these various satellite programs, a basis may be developing for international cooperative earth observation systems under reasonable partnership terms.

The Nimbus satellites are essentially research and development satellites to serve the needs of meteorology and weather forecasting. The seventh in the Nimbus series, scheduled for launch in 1978, will obtain information about large scale atmospheric dynamics (e.g., temperature and water vapour profiles) and about the ocean surface (e.g., chlorophyll concentrations, temperature, currents).

LANDSAT-1 and 2 (formerly ERTS) operate in near-polar orbits at an altitude of about 900 km. A four-channel multispectral scanner in the visible and near-infrared bands provides imagery with high resolution (80 metres), along a 185 km swath. LANDSAT-C will have a fifth channel of lower resolution (240 m) in the thermal infrared band, which will give it a night-time clear-weather imaging capability. An Earth Resources Survey Operational System (ERSOS) has been proposed for the late 1970's or early 1980's, having two satellites based on the LANDSAT series, but its status is uncertain. LANDSAT-D, scheduled for 1980, and the LANDSAT Follow-on Program will initiate new programs in the study of environmental quality and oceanographic phenomena, as well as land surface phenomena. Sensors for LANDSAT-D may include a seven-channel thematic mapper scanning a 185 km swath with 30 metre resolution (compared with 80 metres for LANDSAT-1 and 2). A four-channel high resolution pointable imager was planned to cover a 48 km swath with 10 metre resolution; but its status is uncertain. It would have the capability of detecting small surface vehicles.

SEOS, the Synchronous Earth Observatory Satellite, has been proposed for launch in the mid 1980's to investigate remote sensing techniques from a geostationary satellite. The objective is to provide ultimate continuous coverage, as opposed to the intermittent coverage by satellites in other orbits. SEOS would be followed by a series of Operational Synchronous Earth Observatory Satellites (OSEOS) in the late 1980's. The status of SEOS is uncertain, but applications should be investigated thoroughly in view of the success of GOES (below).

SEASAT-A, to be launched in 1978, will be dedicated to ocean dynamics applications. It represents an important step because of its microwave sensors, particularly the synthetic aperture radar (SAR), which will have all-weather day and night surveillance capabilities. The SAR will have a resolution of 25 metres, and will be able to detect trawlers and small merchant vessels, for example, as well as mapping sea ice and providing sea state information. Though SEASAT-A could provide repeat SAR coverage only once every 14 days at any location, a proposed Operational SEASAT System of the late 1980's could provide twice per day coverage with six satellites, each carrying two SAR sensors. Such an operational system would have an unprecedented ability to provide surveillance of coastal economic zones and the arctic regions, as well as open ocean regions.

GEOS-3, the third in the Geodetic Satellite System series, was
launched in 1975. It will map the topography with an absolute
accuracy of 5 metres and a relative accuracy of 1 or 2 metres,
Geopause, scheduled for launch in 1984, would satisfy a requirement
for a satellite tracking accuracy of 10 cm, needed for several earth
and ocean dynamic studies; it would also be used in investigations
of the earth's gravity field.

The NOAA satellite series followed the ESSA series in 1972. A
single polar orbiting satellite can provide complete global coverage
twice daily. Products include visible and infrared images of the
surface, and are useful for sea ice mapping with a resolution of 0.9
0.9 km at the sub-point and about 8 km out near the horizon. Vertical temperature profile radiometer measurements are used to estimate
temperature profiles from the earth's surface to about 30,000 metres,
and to estimate total water vapour content.

GOES (Geostationary Operational Environmental Satellites) is
the first operational system of weather satellites in geosynchronous
orbit.   It was preceded by SMS (Synchronous Meteorological Satellites). SMS-C launched in Oct. 1975 became GOES-1 in operation.
GOES provides observations of clouds and earth/cloud temperatures
with an 8 km resolution at the nadir. Its images are produced at
half-hour intervals day and night. GOES-B will be launched May 1977.

The first of the TIROS-N series is scheduled for launch in 1978.
It is a replacement for the NOAA series and will provide improved
accuracy of sea surface temperature measurements, and improved ability to differentiate clounds, water, solid snow and ice and melting
snow and ice.

The satellites described above are unmanned, automatic data
acquisition platforms, the prototypes for operational systems of
the future. The role of manned earth observation satellites seems
uncertain, but it is important to remember that a great deal was
learned about the earth's environment from manned spacecraft such
as Apollo and Skylab. The value of man in the spacecraft is obvious: he is able to detect and comprehend the unexpected. However,
the costs of manned satellites for routine earth observation are
high.

Ocean data buoy systems are complementary to satellite systems.
The development of ocean data buoy systems in the United States
has passed through a major experimental phase since 1967, and has
now entered into a pilot operational phase: The feasibility of
an advanced technology has now been demonstrated by the NOAA Data
Buoy Office (NDBO), including

  - buoy hulls that can withstand extreme environmental conditions

  - continental shelf and deep ocean moorings that are effective

- sensors of proven reliability for atmospheric and
  oceanic measurements

- reliable battery power supply

- position location of drifting buoys by satellites
  (e.g., Nimbus-6)

- data telemetry via satellites (e.g. Nimbus, SMS/GOES),
  with high reliability of data acquisition

- onboard computer processing and control of data
  acquisition and telemetry

- shore-based command and control of remote buoy data
  acquisition and telemetry

- routine operational deployment, replacement and
  maintenance of buoys by conventional coast guard vessels

- real-time delivery of atmospheric and oceanic data
  to user agencies.

The evolving U.S. ocean data buoy system provides a basis of
experience and a model for other nations and the international
community.

Although the U.S. data buoy program has been cited because
of its advanced state of development, other countries have data
buoy systems under study or under development. For example, Canada
is developing a Canadian Ocean Data System and has recently deployed
several experimental data buoys of various shapes and materials
off the east coast. Several countries, including the U.S. and
Canada, are contributing small drifting buoys to the First GARP
Global Experiment (GARP = Global Atmospheric Research Program).

The NDBO has specified detailed requirements for an operational
Prototype Environmental Buoy (PEB). These buoys are designed to
support weather monitoring and prediction data needs in severe
environments. Some 15 medium and large prototype operational data
buoys have been deployed or are planned to be deployed by the end
of 1976 in various locations off Alaska, the west and east coasts
of the U.S., and in the Gulf of Mexico. They are equipped with
sensors to measure atmospheric parameters and such as wind velocity,
air pressure and temperature, and oceanic parameters such as wave
spectra, sea surface temperature and subsurface temperature.

The NDBO is also developing a Water Quality Indicator System
for buoy installations. It will measure six parameters judged
to be of greatest interest to users: chlorophyll, conductivity,

dissolved oxygen, pH, temperature and turbidity.

Sensors

Human observers on manned aircraft are still necessary for some earth observation tasks, mainly because of the requirements of gathering "eye-witness" legal evidence in the case of fisheries management and pollution control, for example. However, the human visual detection range is limited and has to be augmented by electromagnetic sensors.

Radar is one of the most important sensors because it can penetrate fog, cloud and nominal rain and is therefore the only true "all-weather" sensor. Airborne radars include conventional scanning radars, side looking radars and synthetic-aperture radars. As the name implies, side-looking radar uses a single antennae, one on each side. An image of the surface alongside the aircraft track is built up by the motion of the aircraft. Synthetic aperture radar (SAR) is an extension of side-looking radar, in which the motion of the aircraft is used to generate artificially a very much longer aperture by storing coherent signals for an appropriate length of time while the aircraft travels a distance equal to the desired aperture.

The resolution of a synthetic aperture radar, as opposed to a real aperture side looking radar, is independent of range, and it therefore can have high resolution right out to the edge of the swath.

Since a synthetic aperture imaging radar has a resolution that is independent of range, its resolution from a satellite platform is as good as from an aircraft. However, this does not necessarily mean that its performance in detecting certain targets is just as good. More random noise (e.g., sea clutter) has to be dealt with in signal processing satellite data. The radar beam-width on the ground is greater from a satellite than from an aircraft. Because of the way the synthetic aperture radar collects energy reflected from the entire beam width, more noise is produced to complicate the signal processing and detection of small targets. However, the added noise is a small price to trade off against other satellite advantages. A satellite platform is more stable than an aircraft, has superior altitude to produce a broader swath width, has superior line-of-sight range for real-time telemetering of broad band data.

Spaceborne synthetic aperture radar may be designed to have swath widths of over 400 km (200 km each side of the satellite). However, there is a power limitation, so that present designs are limited to L-band frequencies. Resolutions of 10 metres are

feasible, but data handling requirements are severe. The "proof
of concept" SEASAT-A satellite SAR has a designed swath width of
100 km and a nominal resolution of 25 metres, using L-band frequency.

Over-the-horizon radars (OTHR), based on land, make use of
reflections from the ionosphere in order to increase the virtual
height of the antenna. The choice of frequency is, therefore,
limited to the HF-band that has successfully been used in oceanography for remote measurements of wave amplitude and direction.
Recent applications include air surveillance. Owing to the temporal
and spatial variations in the ionospheric conditions, the broad
antenna diagram, multiple signal paths and backscatter from the
sea surface, the received signal is very complex and requires extensive processing before extraction of desired information. Reflectors
at known positions, or geographical landmarks, are often used for
continuous calibration and monitoring of the state of the ionosphere.
OTHR is not expected to be useful in northern latitudes because of
ionospheric disturbances. Ranges of several thousands of kilometers
are feasible for detection of targets of large aircraft size, or
for location of surface ships carrying transponders.

Radio-frequency radiometers are sensitive receivers used to
take advantage of the natural radiation of electromagnetic energy.
The intensity depends upon the frequency, the absolute temperature
and the electromagnetic properties of the material and the surface
structure of the radiator. Both emitted energy at a given temperature and achievable resolution increase with frequency, but the
atmosphere and the weather set an upper limit to the increase in
operating frequency that can be used effectively. SEASAT-A, for
example, has a microwave radiometer designed to operate at five
frequencies: 6.6, 10.69, 18, 22.235, and 37 GHz.

Microwave radiometers with scanning and imaging capability
provide a valuable "all-weather" supplement to infrared line scanners (IRLS), since they can see through fog and cloud while IRLS
cannot. However, it should be noted that the available energy and
detector bandwidth in the microwave range is much less than in the
IR range, and therefore the same resolution and detection capability
cannot be expected.

IR sensors take advantage of the self-emitted radiation of
natural and man-made objects. This thermal radiation, described
by Planck's law, depends on the temperature, the emissivity and
the surface structure of the object. The atmosphere is reasonably
transparent only through certain absorption windows. The windows
of major interest are: 1.8 to 2.7, 3 to 5 and 8 to 13 micrometres.

To achieve a high thermal resolution the IR sensor should
operate in the 8 to 13 μm wavelength region, since a given small
temperature difference between a target and the background will

yield a higher radiation gradient than in the 3 to 5 μm region.
To achieve long range, the 8 to 13 μm region should be used since
the amount of energy emitted in this region is higher for target
temperatures below a few 100° C and also the scattering losses in
the atmosphere are smaller for most of the meteorological situations
than in the 3 to 5 μm wavelength band. ERTS-C, for example, is
designed to have a thermal IR channel in the 10.4 - 12.6 μm band,
with a spatial resolution of 240 metres. Several other satellites,
including HCMM, Nimbus-G, SEASAT-A, SMS, and METEOSTAT are designed
to have IR channels in the 10.5 - 12.5 μm band.

Optical sensors working in the visible and near-infrared bands
of the electromagnetic spectrum are commonplace, but technological
advances are still occurring. In particular, further improvements
are expected in the resolution and swath width of spaceborne sensors
in these bands. Improvements in airborne low-light television
(LLTV), combining an image intensifier with a TV camera tube, are
also expected. Airborne laser illumination, taking advantage of
the fluorescence properties of targets (e.g., oilslicks on water),
is expected to become operational.

Though satellite and airborne sensors seem more esoteric, the
possible arrays of sensors that can be mounted on an ocean data
buoy platform seem limitless. They include wind, temperature,
barometric pressure, humidity, solar radiation and other atmospheric
sensors; and chemical nutrient, chlorophyll, colour, current, oxygen,
pH, pollution, pressure, salinity, sound, temperature, transparency,
upwelling, and other subsurface sensors. Like satellite and airborne sensors, they are representative of advanced technologies.

Some Notes On Data Handling Systems

Data handling systems include data telemetry and retransmission,
ground receiving stations, signal processing computers, and data
distribution networks.

The trend is toward real-time or nearly real-time transmission
and processing of environmental data, and retransmission of data
products to user agencies. The communication network and computing
facility requirements are growing, and the technology is developing
rapidly.

Major ground receiving station and signal processing facilities
have been used for LANDSAT, NIMBUS, and high resolution meteorological satellite data (such as NOAA Very High Resolution Radiometer
data). These facilities are generally centralized under the control
of a government agency because of the costs involved. (For example,
Canada's LANDSAT ground receiving and data processing facilities are
centralized under the Canada Centre for Remote Sensing.)

In future, small low-cost decentralized facilities, providing
real-time data directly to various users, will probably become
economically feasible. LANDSAT-D, for example, may contain two
communications subsystems capable of simultaneous operation, with
one transmitting complete data to major facilities and one transmitting reduced data to local users.

Data from future earth exploration satellites may be relayed
through a geostationary tracking and data relay satellite system
(TDRSS), rather than directly to ground stations on line-of-sight.
This will facilitate the optimal location of ground stations and
reduce onboard storage requirements. NASA's TDRSS, planned for
1979 launch, will consist of two geosynchronous relay stations,
130 degrees apart in longitude, and a ground terminal in the U.S.
For reliability, it will also include spare satellites in orbit
and on the ground ready for launch.

Complex and costly data handling facilities are generally
required for real-time processing of remotely sensed data of any
form. For example, each LANDSAT image covering 185 km by 185 km
contains about 2×10^8 bits of data. Data transmission rates are
about 15×10^6 bits per second (15 Mb/s), and conversion to film
or computer readable tape must be performed at a comparable rate.
Geometric and radiometric correction of the data require complex
computing operations. However, there is a trend toward lower-cost
facilities. For example, a Portable Earth Resources Ground Station
(PERGS) has recently been acquired by the Canada Centre for Remote
Sensing at a cost of $1.25 million, for use on the east coast. It
is capable of receiving and processing data from LANDSAT and NOAA
spacecraft, and providing black and white 70 mm quicklook imagery
and computer compatible tapes. A digital FAX interface to land
lines is included, for transmission of FAX images to remote terminals. The PERGS station represents an order of magnitude reduction
of capital costs from NASA's first-generation facilities. It can
be readily modified to accept data from other satellites such as
SEASAT-A. It has a "quick-look" capability.

SEASAT-A data rates for the low-rate sensors are expected to
be in the range of 25×10^3 to 30×10^3 bits per second (25 to
30 Kb/s), but the imaging synthetic aperture radar (SAR) rates
will be in the 15 to 24 Mb/s range. For a full capability Operational SEASAT system (circa 1985) the rates are expected to reach
240 Mb/s.

Difficult technical problems have yet to be overcome in
developing a near-real-time processor of satellite radar imagery.
At present, most synthetic aperture radars use optical correlators.
The U.S. is evidently developing such a processor for SEASAT-A,
capable of processing 10 minutes of real-time SAR data in 4 hours
(i.e., the processing rate is 1/24 of the real-time data acquisition

rate). Preliminary Canadian estimates for <u>digital</u> processing indicate a processing rate that is 1/64 of the data acquisition rate.
A Canadian design goal is to get the processing rate down to 1/8
of the data acquisition rate by 1980. New approaches look promising.
Several countries are working on the problems and some optimism
seems to be in order.

Computer communications networks will be desireable, if not
essential, to support the information systems envisaged for the
future. For example, a nationwide communications network is being
built in Canada, featuring (a) digital transmission employing
packet switching techniques, and (b) very "intelligent" terminal
equipment. Such a network will facilitate the real-time use of
environmental data.

Some Notes on Management Information Systems

Environmental control and resource management require data and
information of various kinds. The data and information may be considered as <u>products</u>. A <u>production</u> <u>process</u> is required to generate
the products, and conventional production considerations may include:

- technology selection
- methods and process design
- maintenance
- replacement and upgrading
- production planning and control
- product specification
- quality control of product
- reliability control
- cost control.

A <u>market</u> is required to accept the data and information products, and conventional marketing considerations may apply, including:

- product line design
- quality of product
- reliability and quality of delivery
- structure of distribution system
- market structure
- promotion
- pricing.

The market for data and information products is different in
the case of <u>private</u> and <u>government</u> producers. In the private sector,
the sellers of information may claim exclusive proprietary rights,
the buyers may claim exclusive rights of ownership and privacy, and

the normal forces of economic competion may prevail. In the government sector, some of the data and information may be in the public
domain, the seller must provide equal access to all buyers, and the
normal forces of economic competition do not apply. However, some of
the data and information may be subject to security classification.

There is a tendency to overlook or underestimate information
management problems and the total costs of producing and marketing
data and information products. There is a tendency to over-emphasize
technology selection, particularly in terms of platform performance
and costs. This probably happens because the design of platforms and
planning of operations is relatively easy to do in advance, but the
development of management information systems is more evolutionary.
The users begin to line up _after_ the platform and sensor technologies are proven.

Most users obtain their environmental data from a variety of
sources, and the rates and volumes of data are evidently increasing.
Data from various sources such as airborne remote sensing systems,
satellites and ocean data buoys have to be correlated and reduced
to meaningful terms through the use of mathematical models and interpretation schemes. The information that is finally produced has to
be useful to decision-makers.

The _standardization_ of data products and information products
is of vital importance. Standardized geo-referencing is needed so
that images from various satellite systems, and from the same systems at different times, can be accurately superimposed. Standardized radiometric referencing is needed so that data from various
systems, and from the same systems at various times, can be reliably
compared.

The interpretation of environmental data is heavily model-
dependent. In some cases, the results of interpretation seem to
depend more on the model than on the data, and error rates are high.
In other cases, existing models do not readily digest new kinds of
data. For example, macro-scale numerical weather forecasting models
are often not useful for meso-scale forecasting and do not utilize
weather satellite data fully. Commercial operators who need real-
time meso-scale weather information are turning to "now-casting"
based on primitive data, rather than relying on "forecasting" based
on inadequate models. New sensors and measurement techniques will
probably force the development of new models of physical and biological processes.

Because of the high rates and volumes of data that have to
be handled, a great deal of research and development effort is being
spent on automated methods of classification and interpretation. At
present, machine methods are limited and generally cost more than
manual methods. It may take many years before automated methods of

interpretation catch up to the present data acquisition capabilities
of satellite systems such as LANDSAT and SEASAT.

The design of management information systems requires an
understanding of the decision processes of users, and how environmental data products and information products enter into such processes. In most cases formal decision models do not exist. However, it is important to formalize the description of decision processes to whatever extent may be possible. In particular, it seems
important to offer decision makers alternative samples of information products from which they may choose the best to suit their
needs. It seems worthwhile to simulate a variety of information
products for management consideration prior to the development
of costly information systems.

Archiving vast quantities of environmental data poses special
problems of data base management. The designation of appropriate
archiving agencies and the development of institutional arrangements for sharing data bases are among the more important considerations.

Some Notes on Social Impacts

Much can be considered under the rubric of "social impact".
The social impacts of earth observation systems may be measured
in terms of effects on organizational structures and modes of
decision-making, politics, economic benefit-cost analysis, and
rates of social change.

Military technologies are sometimes considered to be "bad"
and pollution control technologies "good". Satellite and data
buoy technologies seem "neutral". However, environmental information derived from satellites and data buoys may have different
value to different users, depending on the application.

The question of national versus international control of earth
observation systems is of great concern, particularly to nations
that do not have the technological capabilities. The question of
relative roles for private enterprise and public enterprise is of
concern to nations that have a tradition of fostering private
enterprise. The problem of organizing operational systems to
satisfy the missions of government departments has to be solved by
each nation involved.

Political questions arise. For example, should a country
that possesses advanced space technology consider it to be a
national resource or an international resource? Should such a
country offer unlimited free access to its observation and information systems, or should it offer access to the international

community on a shared-cost basis? What opportunities exist for
international cooperation in the development of shared systems,
including shared research, development and subsystems manufacturing
opportunities? Can the United Nations provide an umbrella organization for the development of international earth observation and
information systems, taking into account differing security requirements of its member nations. Can a treaty organization such as NATO
provide such an umbrella organization for the convenience of its
member nations? What kind of lead roles can be expected of NASA,
ESA and other space agencies? The political questions seem endless
and the answers are elusive. Some countries may become impatient
with the international machinery and may be inclined to "go it alone"
at a very high cost.

Pre-launch economic benefit-cost analyses and post-launch
economic verification experiments are becoming essential. They
are necessary as a basis for justifying programs to funding agencies.
More importantly, they help identify priority economic applications.
Verification experiments can be used to begin the process of technology transfer from the developing agency to the user agency. In
this regard it is interesting to note the difference in the approach
to SEASAT-A and to ERTS-A. In the case of ERTS-A, which was launched
in 1972, the NASA experiments were essentially scientific experiments
designed to test the technical feasibility of certain applications.
In the case of SEASAT-A, to be launched in 1978, NASA's economic
verification experiments will evidently be designed to test the
economic feasibility of certain applications, with a much earlier
focus on operational systems aspects.

With respect to models of social change, the first question is
simply this: Will operational earth observation and information
systems have any measurable effects on nations, in terms of such
factors as employment, pollution control, institutional changes,
economic growth, social justice and national sovereignty?

Some Notes on Atmospheric Applications

The main applications of atmospheric observation and information
systems include weather forecasting, climatology, agro-meteorology
and pollution monitoring.

Satellite imagery has been particularly useful in "now-casting"
and "forecasting" of hazardous weather events such as severe storms
and hurricanes. As pointed out by S.E. Wasserman, satellite pictures
fit in with other types of observations. Images from the Geostationary Operational Environmental Satellite (GOES) are used operationally by meteorologists at forecast offices because they are timely
and have computer-generated grids superimposed on the picture. Imagery from NOAA operational polar orbiting satellites are not used

extensively by these meteorologists because the intervals between
picture-taking are large (12 hours for IR, 24 hours for visible)
and because the pictures lack the superimposed reference grids to
facilitate interpretation.

Satellite data have not been utilized to their full potential
in numerical weather models. Satellites provide broad area coverage
but the numerical models require input statistics at specified grid
points. Sensors such as infrared and microwave radiometers provide
data on certain integrated properties of the vertical atmospheric
column (e.g., water vapour). But subsidiary data from other sources
and a process of mathematical inference are required to convert the
radiometric data into parameter values that can be accepted by the
numerical models. This incompatibility is evidently awkward and
there is some natural resistance to change. It is not clear whether
or not the convergence of data acquisition and weather modelling
efforts can be speeded up.

Some Notes on Land Applications

The main applications of land observation and information
systems have become well known since the launch of the ERTS-A
(LANDSAT-1) satellite in 1972, and it is not necessary to reiterate
them here. However, a few applications seem to be of great international importance; crop forecasting is perhaps the most important.

The U.N. Food and Agriculture Organization (FAO) is developing
a Global Information and Early Warning System on Food and Agriculture. As one input to this system, the FAO requires meteorological
data and forecasts from the World Meteorological Organization (WMO),
including satellite data, as described by Wolfgang Baier. At the
same time, the U.S. is conducting a Large Area Crop Inventory Experiment (LACIE), utilizing data from LANDSAT, SMS and other satellites
to measure crop areas and soil and weather conditions that affect
crop yields. It is expected that improved crop forecasting systems
will lead to better management of crop production and distribution,
both nationally and internationally.

Some Notes on Ocean Applications

National ocean management responsibilities will undoubtedly
expand as jurisdiction over resources is expanded to include the
200-mile economic zone or the edge of the continental margin. Both
governments and private enterprises will face growing opportunities
and concommitant responsibilities, and will be under increasing
pressure to work in concert, within the framework of the Law of the
Sea.

Ocean observation and information systems are required to provide information relative to various <u>primary</u> economic <u>activities</u> of both governments and private enterprises, including the following:

- fishing and aquaculture
- offshore oil and gas
- offshore mining
- ocean and coastal transportation
- communications
- coastal ports and harbours
- coastal agriculture and forestry
- coastal recreation
- industrial and municipal use of coastal waters
- ocean dumping.

The primary activities are supported by <u>secondary</u> <u>activities</u> (support services), including the following:

- weather, sea state and sea ice information and forecasting

- oceanographic information services

- vessel traffic management, route planning and navigation aids

- fisheries surveillance, location, quota control, hygiene and safety inspection

- offshore oil and gas and mining surveillance, location, quota control and conservation, safety inspection, logistic supply, maintenance

- emergency communications services, maintenance of stations, cables, etc.

- pollution surveillance, oilspill containment and cleanup, and related warning and logistic services

- ice breaking, convoy support and aid to distressed vessels and offshore installations

- search and rescue services

- ocean and coastal engineering and architectural services.

Operational meteorological satellites such as GOES, and oceanographic satellites such as SEASAT, are designed to provide weather, sea state and sea ice information that can be used to improve the effectiveness of the secondary activities mentioned above, thereby creating economic benefits relative to the primary economic activities. Ocean data buoys can also be used in conjunction with the satellites.

Ocean data buoys are in some respects competitive with

satellite observation systems, but in some respects they are complementary. They seem to be particularly suited to providing continuous micro-scale and meso-scale weather data and surface and subsurface data for several operational and research applications, as
follows:

Operational data for

- fishing operations and fisheries yield management
- offshore drilling and production operations
- pollution monitoring, subsurface
- design of vessels and offshore engineering, subsurface.

Research data for

- improved weather modelling, particularly meso-scale
- improved climatological modelling
- air-sea interaction models
- models of ocean dynamics
- models of ocean biological production and fish population
  dynamics
- models of fisheries operations.

Some Notes on Applications Methodology

Systems R and D seem to be driven largely by the scientists
and engineers involved in sensor and signal processing technologies.
Applications methodology by the natural scientists in user organizations tends to lag behind.

Remarkable advances have been made in LANDSAT-type multispectral scanners, and the technology has proven to be reliable.
Increases in resolution from 80 metres to 30 metres are on the
horizon, and pointable scanners having 10 metre resolution seem
feasible. Applications to crop classification have shown limited
but encouraging success rates since the 1972 launch of ERTS-A,
and intensive applications methodology work is presently underway
(e.g., the LACIE program of NASA, NOAA and USDA). It seems likely
that generally acceptable success rates will be achieved by the
time an Earth Resources Survey Operational System may be available
(late 1970s or early 1980s).

SEASAT-type synthetic aperture radar and signal processing
technologies are under intensive development, but will not be
tested until SEASAT-A is launched in 1978. However, theoretical
work, simulations and airborne tests have been promising. The
current design objective is 25 metres resolution, and 10 metre
resolution is within the state of the art. Applications methodology
and economic verification experiments are being planned by the U.S.

and Canada. One major application of interest to Canada, for
example, is sea ice and iceberg surveillance. The introduction
of spaceborne imaging radars may prove to be as momentous as was
the introduction of spaceborne multi-spectral scanners.

The list of desireable applications methodology projects seems
limitless. It includes, for example, air-sea interaction studies
utilizing active and passive microwave sensors, interpretation of
spaceborne radar signatures for land and sea applications, development of systems for integrating and correlating data and imagery
from a variety of sensor-platform types, and development of low-cost
facilities for local users. The drive to lower cost is essential.

It is important to recognize that continuing R and D can result
in continual changes which can be unsettling to potential users.
At some stage, operational system specifications have to be fixed
and held stable for a long enough period that users will be encouraged to make investments in their own facilities. It seems clear
that many potential users are wary about getting involved in earth
observation and information systems at an early date, because they
fear that their initial investments will be subject to rapid obsolescence. Some perceived stability in the adoption of innovations is
necessary. Some stability in the form and quality of data products
is essential.

EPA'S USE OF REMOTE SENSING IN

OCEAN MONITORING

Peter W. Anderson
Director, Marine Protection Program, U.S. Environmental
Protection Agency, Region II, Edison, New Jersey, U.S.A.

John P. Mugler
Acting Head, Environmental Quality Program Office
National Aeronautics and Space Administration
Langley Research Center, Hampton, Virginia, U.S.A.

Abstract

The U.S. Environmental Protection Agency (EPA) has legislative
responsibility to protect and enhance the quality of our air, water,
and land resources. In this regard, EPA has developed and conducted
a number of integrated monitoring programs, that effectively collect
and evaluate reliable environmental information needed to assess
ongoing programs, support enforcement actions, and evaluate the
overall status of the environment. The use of remote sensing
techniques from satellite, aircraft, and in-situ platforms has been
considered in a number of EPA's monitoring efforts.

This paper will discuss several cooperative programs being
conducted by EPA, the National Aeronautics and Space Administration
(NASA), and the National Oceanic and Atmospheric Administration
(NOAA), to evaluate and utilize in the decision-making process
remote-sensing techniques for monitoring ocean dumping of municipal
and industrial wastes in the New York Bight and its impact upon the
marine environment. Two areas will be specifically reviewed: 1)
evaluation of a NASA developed remote-automatic coliform bacteria
quantification device to monitor the marine waters of the New York
Bight, and 2) evaluation of the use of imagery from high flying
aircraft and satellites to augment other efforts to provide effective surveillance of ocean dumping practices.

Introduction

Public demand in the United States for increased protection
of their environment has resulted in a proliferation of federal,
state, and local legislation. For example, within the past fifteen
years, federal legislation has brought about the funding of publicly-
owned, wastewater-treatment plant construction, the provision of
enforcement procedures regarding the pollution of interstate waters,
the establishment of water-quality standards for interstate waters,
and the requirement for the preparation of impact statements whenever the environment is to be altered by federal actions. However,
in the history of environmental protection, concern for the ocean
is relatively new.

There were few direct legal controls on the ocean disposal of
wastes generated either inside or outside the United States, prior
to the passage of the Marine Protection, Research, and Sanctuaries
Act (P.L. 92-532) on October 23, 1972. Earlier federal legislation
was concerned only with pollution resulting from oil spills and from
the exploration and exploitation of natural resources. This legislation, more commonly called the "Ocean Dumping Act", assigned specific functions to the Environmental Protection Agency (EPA), as well
as to the Coast Guard, the Army Corps of Engineers (COE), and the
National Oceanic and Atmospheric Administration (NOAA). In general,
EPA administers and enforces the overall program; administration
involves the issuance or denial of dumping permits to municipal
and industrial applicants, the evaluation of alternative means
for handling wastes, and the selection and management of dump sites.
Administration of dredged material and issuance of permits for their
disposal in the marine environment is the province of COE. The
Coast Guard carries out police-type monitoring of sites, making sure
that vessels dump at the proper location and in accord with EPA or
COE permit conditions. Finally, NOAA is responsible for monitoring
the effects of ocean dumping on the marine environment, for conducting research on the effects of dumping and other pollutional sources,
and for identifying and establishing marine sanctuaries or areas
where no dumping should occur. Two other federal agencies have
limited involvement under the Act; the Department of Justice, in
enforcement cases which involve court trials, and the Department
of State, in situations where emergency dumping occurs in international waters. It should be noted, in addition, that criteria
developed by EPA under this Act for evaluating the characteristics
of the waste to be dumped are used also to evaluate point-source
waste discharges (outfalls) under provisions of the Federal Water
Pollution Control Act Amendments of 1972 (P.L. 92-500).

Since pollution tends to seek the level of least regulation,
industries and municipalities in coastal areas, hard pressed by
the increasingly more stringent requirements contained in new air
and water pollution statutes, and faced with a lack of suitable

facilities for land disposal, looked to the ocean as the ultimate
"sink" for their most persistent wastes. For example, in 1968,
the first year anyone really bothered to assess the volumes and
type of materials ocean dumped, it was conservatively estimated
1/ that 10.4 million tons were being dumped via barge/vessel annually, excluding dredged materials. Of this total, 45 percent was
industrial waste; 43 percent, sewage sludge; 5.5 percent, construction and demolition debris; and the remainder, garbage and explosives., Projections, on the basis of trends in coastal area population growth, and the potential diversion of wastes to the ocean by
more stringent regulation of air and water pollution, indicated
a doubling in volumes ocean dumped by 1980, unless strictly regul-
ated. EPA 2/ recently reported that by 1973 this volume had increased slightly to 10.9 million tons. Percentages, however,
changed somewhat: industrial wastes, 47 percent; sewage sludge,
45 percent; 8.9 percent, construction debris, and the remainder,
garbage. By 1975, after the effective date of the Act, the volume
decreased to 8.9 million tons, with 38 percent representing industrial wastes; 57 percent, sewage sludge, and the remainder, construction debris. The ocean dumping of garbage and explosives are
no longer practiced in the United States.

New York Bight - The Problem Area

In the United States, about 80 percent (1975) of all ocean
dumping, via vessel, of municipal sewage sludge, acid wastes, and
chemical wastes takes place in the New York Bight at six discrete
dump sites (figs. 1-2). The ocean dumping of sewage sludge in the
Bight began in 1924, when New York City rejected the disposal of
sludges generated by a northern New Jersey municipality in the
adjacent Hudson estuary. Dumping of industrial wastes, acids and
toxic chemicals, started in the late 1950's, when pollution-
abatement programs in New Jersey established ocean dumping as an
acceptable practice in lieu of discharge into streams or on land.

Unquestionably, implementation of the Act on April 23, 1973,
spurred industrial dumpers to develop and construct land-based
treatment facilities and/or implement other environmentally acceptable alternatives for handling their wastes. Of the roughly 150
industrial ocean dumpers in 1973, only 13 remain as of January 1977.
By mid-1979, this number will be reduced to four; DuPont, Allied
Chemical, NL Industries, and Merck. The ability of these four
industrial waste generators to effect technically feasible and
environmentally acceptable land-based alternatives prior to EPA's
goal for the complete phase-out of all municipal and industrial
ocean dumping by 1981 remains an unresolved question.

The record with regard to the ocean dumping of municipal wastes
is less favorable. It became clear, shortly after EPA received its

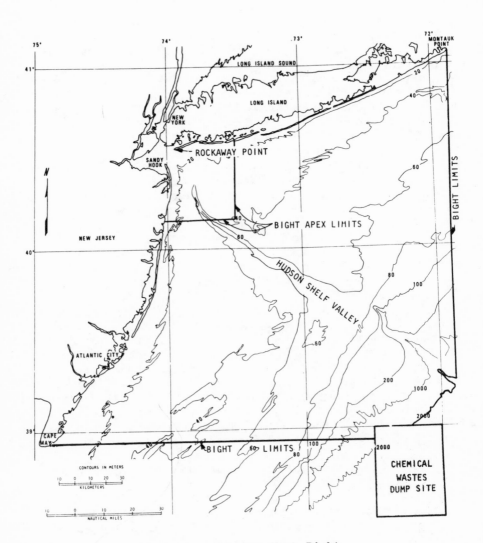

Figure 1.   The New York Bight

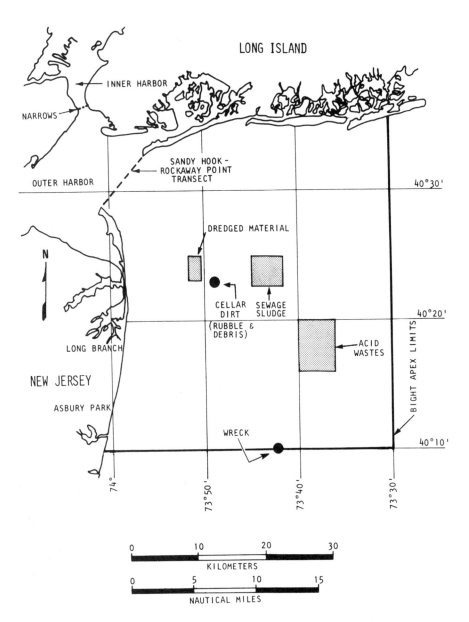

Figure 2.   Bight Apex and Existing Dumping Sites.

mandate under P.L. 92-532, that the construction of new and improved wastewater treatment facilities, authorized under P.L. 92-500
and scheduled for completion between 1977 and 1983, would increase
threefold the amount of sludge generated by municipalities that
practice ocean dumping. The problem of handling, in an environmentally acceptable manner, not only the present volumes, but also
the projected threefold increase is only now being resolved. Even
though EPA determined in August 1976 that acceptable land-based
alternatives were available, we recognize that the major municipalities in the New York-New Jersey metropolitan area cannot implement
these alternatives within the next five years. Thus, while the
number of municipalities has been reduced from approximately 250
in 1973 to 95 in November 1976, the large volume generators still
remain as ocean dumpers.

The fact that 80 percent of all dumping of municipal and industrial wastes occur in the New York Bight and scientific evidence
that this dumping, particularly that of sewage sludge, has adversely
impacted the marine environment has resulted in "high public visibility." Newspaper reports of sludge on the beaches of Long Island
and New Jersey are common, although unsubstantiated, every spring
and summer. Congressional hearings, normally conducted in the
nation's capital, are held in the New York City area to receive information just on ocean dumping activities in the Bight. At a recent
scientific workshop conducted by the National Science Foundation
3/ a statement was made that the New York Bight is the most widely
studied area in the nation and perhaps in the world. All of this
interest on the part of the public and scientific community reflects
the severity and magnitude of the pollution problems.

Recognizing the environmental problems in the Bight and its
responsibilities under the Act, EPA has sought to develop an effective, integrated ocean monitoring program involving several Federal,
State, and local agencies. This program, which is under continuing
review, is designed to collect and evaluate reliable environmental
information, to support enforcement actions, and to evaluate the
overall status of the marine environment. Because of unique problems associated with ocean monitoring, several innovative approaches
have been incorporated utilizing remote-sensing techniques. The
remaining paragraphs include descriptions of several cooperative
programs between the National Aeronautics and Space Administration
(NASA), NOAA, and EPA to evaluate and utilize such techniques to
input to the regulatory decision-making process.

Aircraft and Satellite Imagery

The initial program in the New York Bight utilizing remote-
sensing techniques for ocean monitoring was developed between NASA
and NOAA. This program, which commenced in April 1973, had as its

objectives the following three items: first, to investigate the
role of remote sensing in defining the circulation in the Bight
and the application of the information to monitoring and managing
ocean dumping; second, to investigate the capability of remote
sensing to obtain coastal zone information, such as baseline
information for environmental assessment; and lastly, to assist
NOAA in incorporating remote sensing and modeling capability
in the design of an integrated monitoring system.

To achieve these objectives, multispectral scanner and
photographic instrumentation have been flown aboard NASA aircraft
(C-54 and U-2) and satellite (LANDSAT 1) to remotely study major
circulation features and to detect waste materials dumped into
the ocean. NOAA made concurrent "sea truth" measurements which
were used to correlate remote measurements with in situ measured
water parameters. Since 1975, EPA has actively supported this
program by supplying logistical support with the waste generators
and in the collection of "sea truth" data. Figure 3 illustrates
data collected in April 1973. Clear weather conditions permitted
the detection of several strong surface features, an acid waste
dump and the sediment plume of the Hudson estuary. Data was
collected simultaneously to the LANDSAT imagery using a Multichannel
Ocean Color Sensor installed on a NASA aircraft (C-54). This
sensor provided additional and more detailed information on the
acid dump (Fig. 4). This program is continuing and results to date
demonstrate that remote sensors can monitor a variety of water
pollution parameters; e.g., chlorophyll, sediment, acid wastes,
and sewage sludge (Fig. 5). Additional studies, however, are
needed to obtain information on the extent to which remote sensing
techniques can provide quantified information on specific pollutants and assist in determining the proper mix of in situ and
remote sensors to meet ocean monitoring requirements.

As noted above, EPA became more active in this experimental
program in 1975. This interest in imagery from high flying aircraft and satellites stems from the fact that some water-surface
features are readily detectable with airborne multispectral scanner and photograph techniques. Examination of imagery (Fig. 6)
from the NASA Ocean Color Scanner (a forerunner of the Nimbus
G Coastal Zone Color Scanner) and multispectral cameras taken
from a NASA U-2 aircraft at 65,000 feet over the New York Bight
clearly shows the location of acid and sewage dumps. In addition,
imagery from this altitude contains shoreline features which can
be used as a reference in scaling distances and fixing locations
on the image. Using conventional techniques to interpret the
multispectral photographs from the U-2, it is estimated that the
location of dumped materials can be determined to 0.1 nautical
mile. Thus, by locating fresh acid and sewage sludge dumps on
the imagery relative to the designated dump sites, EPA can determine
if dumping operations are conforming to permit restrictions. To

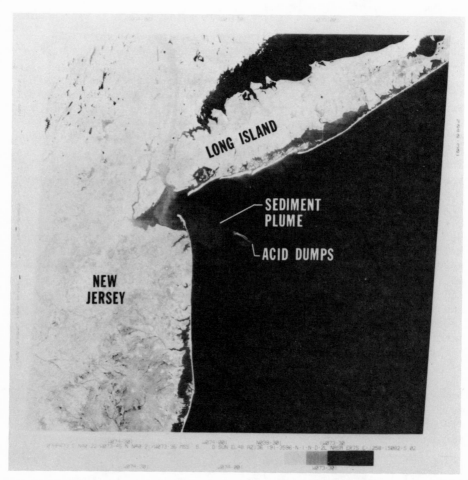

Figure 1.   LANDSAT 1 Image of New York Bight, April 7, 1973.

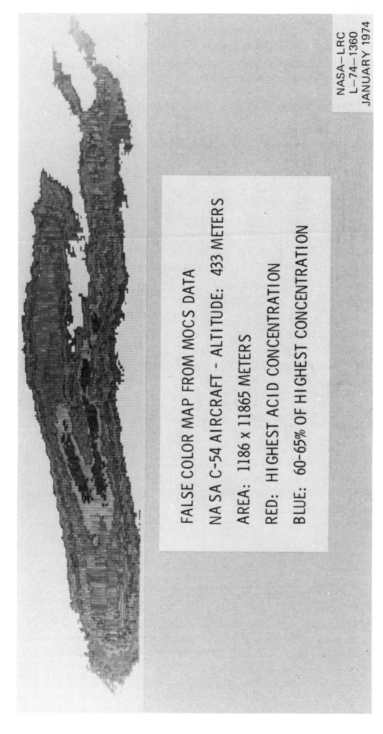

FALSE COLOR MAP FROM MOCS DATA

NASA C-54 AIRCRAFT - ALTITUDE: 433 METERS

AREA: 1186 x 11865 METERS

RED: HIGHEST ACID CONCENTRATION

BLUE: 60-65% OF HIGHEST CONCENTRATION

NASA-LRC
L-74-1360
JANUARY 1974

Figure 4. Acid Waste Dump - New York Bight, April 7, 1973. (The contrast of the original color photograph is lessened in this reproduction.)

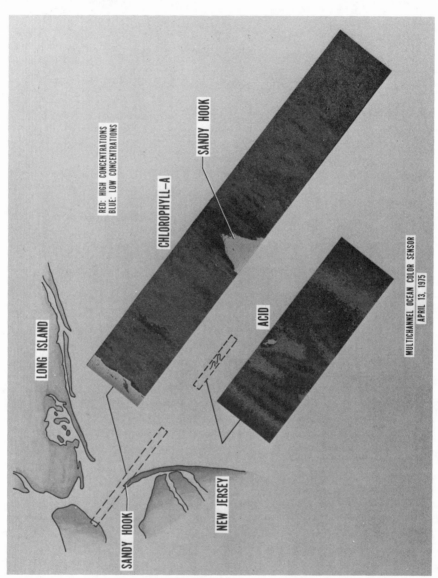

Figure 5. False Color Maps of Chlorophyll and Acid in the New York Bight. (The contrast of the original color photograph is lessened in this reproduction.)

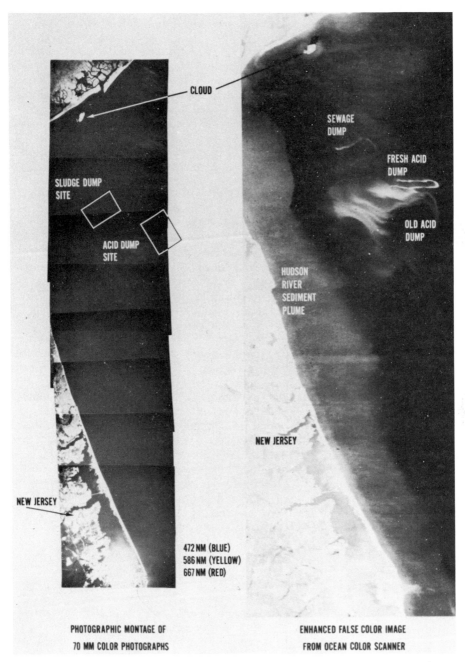

Figure 6. U-2 Coverage of New York Bight, April 9, 1975. (The contrast of the original color photograph is lessened in this reproduction.)

further explore the application of this procedure, EPA obtained
imagery from recent U-2 flights over the New York Bight conducted
by NASA during their routine research and development activity.
Imagery collected on September 15, 1974 indicated that a "short
dump" took place near the acid wastes dump site: i.e., acid waste
was dumped before the barge reached the designated site. Based
on this imagery and other supporting data, EPA is proceeding with
an enforcement action against the permit holder for alleged violation of the terms of the permit. Should the alleged violation
be upheld, the permittee is liable for a fine of up to $50,000.
However, since the use of remotely collected scanner data in an
enforcement process is somewhat of a precedent, the movement towards
a final determination on this violation has been slow.

Currently, EPA and NASA are studying ways in which remotely
sensed data from aircraft and spacecraft can be incorporated into
an expanded system for monitoring ocean dumping in the Bight. Some
of the systems under investigation are illustrated on Figure 7.
Of course, in the long term the system also could be used to monitor
other pollution occurrences as well. For example, EPA requested
NASA assistance on two occasions during the summer of 1976. In June,
many of the recreational beaches along the south shore of Long Island
were closed due to the wash-up of large quantities of floating debris. NASA conducted an overflight along 46 miles of coastline
with a C-54 aircraft. Photographs were taken with two aerial cameras, one using natural color film and the other, color infrared
film. The flight altitude was nominally 1,500 feet which provided
a ground tract 2,200 feet wide perpendicular to the flight path.
The information was intended to assist in determining the origin
and extent of the debris; however, in this case, beach observations
and observations from helicopters proved more useful. In August,
as a result of Hurricane Belle, a high density sediment and debris
plume, originating in the Hudson estuary, caused the fouling of
nearby New Jersey coastal beaches. A similar overflight was conducted by NASA along a 50-mile flight path. These data proved useful
in determining the origin and extent of the plume. Similarly, remote
sensors could play an important future role in monitoring proposed
offshore oil drilling operations in the Bight and, of course, the
occurence of an oil or toxic chemical spill.

In summary, results from the cooperative experimental programs
have shown that multispectral remote measurement systems coupled
with digital data processing, enhancement, and display techniques
can discriminate and map the surface extent of ocean dumped pollutants in the New York Bight. Continuing research holds promise of
measuring surface concentrations using similar systems. In view
of these recent advances, it is evident that incorporating remote
measurement techniques and data products into an expanded system for
ocean dumping monitoring would add significantly to system effectiveness.

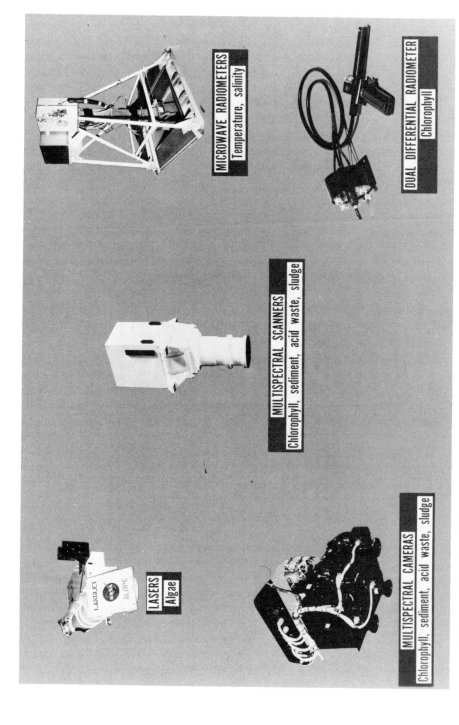

Figure 7. NASA/EPA Cooperative Program for Evaluation of Water Quality Sensors.

Coliform Monitor

As part of its mandate under P.L. 92-532, EPA is chartered to
consider the effect of ocean dumping on human health and welfare,
including economic, esthetic, and recreational values. Thus, within the Bight, EPA conducts a continuing monitoring program directed
towards defining water quality along the beaches of Long Island and
New Jersey and in the offshore ocean waters. One of the principal
water-quality parameters measured is coliform bacteria, a bacterial
group which is widely utilized in recreational and shellfish use
standards as a primary indicator of recent human waste discharges.

Interest on the part of EPA in a NASA-developed coliform sensor to monitor coliform levels in the Bight developed in early 1975.
The NASA coliform sensor is a new electrochemical method for rapidly detecting coliforms 4/, 5/. The method is based on the time of
hydrogen evolution from coliform cultures incubated in a nutrient
broth. The advantages of this method are that it can reduce the
time required to detect a given level of coliform bacteria in water
by a factor to three or more over standard techniques, and can be
fully automated and placed in remote locations for periodic unat-
tended operation. Under present methods for the determination of
coliform bacteria, results are not available for 24-72 hours. Utilizing the NASA-developed technique this time period can be reduced
to 3 hours, if bacterial levels are significant, or 7-8 hours, if
low. The fact that the technique can be automated and remotely
placed in the field is of particular interest to EPA. For example,
such a remote monitor could be placed at selected locations in the
Bight to provide an early warning of potential near-shore water
contamination by shore-ward transport of material dumped at offshore
dump sites.

EPA and NASA recently developed a joint project for the evaluation of this electrochemical method. The objectives of this investigation are: first, to develop a statistical data base for
comparison of the new coliform detection method with standard techniques in order to verify the validity of the method; second, to
determine the effects of elevated incubation temperature and salinity on sensor performance as related to monitoring fecal coliforms
in coastal waters; and last, to evaluate an in situ monitoring system in the New York Bight.

An early version of a laboratory unit was provided EPA in
January 1976 and tested aboard ship and in the laboratory to identify field engineering and interface problems. Considerable scatter
of the data was observed with the onboard tests due to power surging
and high humidity. Ninety-three percent of the laboratory endpoints
were within the error limits of the fecal coliform calibration
curve. In order to define and improve sensor performance in saline
waters, two research grants have been awarded to investigate physio-
chemical factors on the electrochemical/organism interface and the
effect of stressed coliforms on response times. The final stage

Figure 8. Automated Coliform Detector.

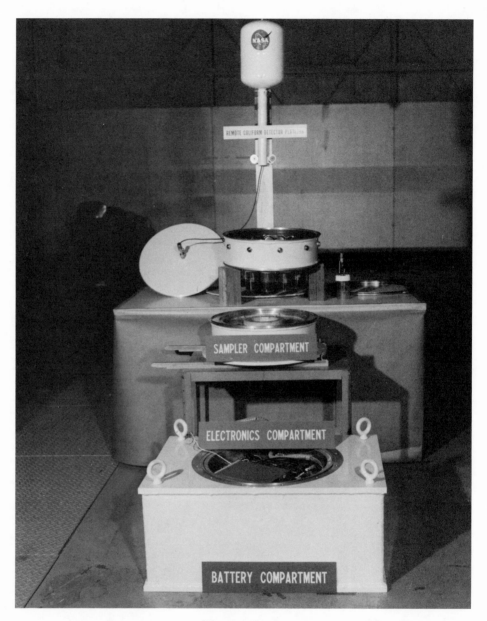

Figure 9

of the EPA/NASA project is to evaluate the remote sampling platform, first in water adjacent to the NASA Langley Research Center, and subsequently in a field demonstration of the system in the Bight area.

The design concept of the automated coliform detector is illustrated on Figure 8. It can be used at the surface, at the bottom, or at points in between. The unit is designed to receive a command from a central ground station to take a sample and record the results of the analytical determination. The results can then be queried and received by the command station. While the initial monitor is limited to a specified number of samples, because of growth-media restrictions, NASA scientists are confident that a continuous coliform sensor will be developed in the near future.

The prototype automated coliform monitor is illustrated on Figure 9. It weighs about 800 pounds. A floating transmitting beacon is located at the top. This beacon is connected to the instrument package by a hard wire. The instrument package is made up of three compartments, a battery compartment, an electronics compartment, and a sampling compartment with ports through which water samples are drawn into the test cells.

While the initial interest in this automatic coliform detector on the part of EPA has been in terms of its ocean dumping program, the detector obivously has applicability in the monitoring of coliform levels in lakes, rivers, streams, public water supplies, sewage treatment plant discharges, and many other similar situations. We look forward to its actual demonstration next year in a _real_ problem area - the New York Bight.

REFERENCES

1/   U.S. Environmental Protection Agency, 1973, Ocean Dumping in the New York Bight - Facts and Figures: USEPA, Region II, 14 p.

2/   _____, 1976, Ocean Dumping in the United States - 1976, Fourth Annual Report of the Environmental Protection Agency on Administration of Title I, Marine Protection, Research, and Sanctuaries Act of 1972, as amended: USEPA, Wash., D.C.

3/   Sharp, Jonathan H., 1976, Anoxia on the Middle Atlantic Shelf during the Summer of 1976: National Science Foundation, 122 p.

4/   Wilkins, J. R., Stoner, G.E., and Boykin, E.H., 1974, Microbial Detection Method Based on Sensing Molecular Hydrogen: Applied Microbiology, v. 27, no. 5, 949-952.

5/   Wilkins, J.R., and Boykin, E.H., 1976, Analytical Notes - Electrochemical Method for Early Detection and Monitoring of Coliforms: Amer. Water Works Jour., v. 68, no. 5.

INFORMATION REQUIREMENTS FOR REGIONAL AND GLOBAL

OPERATIONAL SYSTEMS IN AGRICULTURAL METEOROLOGY

Wolfgang Baier
Chief, Agrometeorology Research and Service Section
Chemistry and Biology Research Institute
Agriculture Canada, Ottawa, Canada

Introduction

Remote sensing and agricultural meteorology have progressed rather independently but have strong mutual interests in the development of operational systems especially in the area of crop surveillance. Research results warrant the operational use of remote sensing techniques to provide quantitative information on crop acreage, growth conditions and yields (Strome, 1975). Agricultural meteorology is concerned with the application of meteorological information to practical agriculture use. This discipline has also developed to the stage where operational systems for the assessment of vegetative conditions, crop yields and crop quality are being introduced (Baier, 1975). It is therefore appropriate at this time to review the data input requirements of such crop assessment techniques based on past and present meteorological data in relation to additional information provided by remote sensing techniques. Such an analysis is indeed essential for the proper coordination of research and operations of the various remote sensing and weather-based crop surveillance systems which are now mainly in the experimental stage.

Features of Crop-Weather Models

Extensive research into crop-weather relationships over the past two decades has provided the necessary background for the modelling of crop responses to environmental factors. The new world food situation, which developed since 1972 as a result of rather large variations in weather and crop yields on a regional scale, has led to renewed interest in the monitoring of weather patterns

123

Table 1.   Summary of selected statistical models suitable for operational crop yield predictions (Source: Baier, in press).

Region	Crop	Agrometeorological data requirements* Variable/Period	Reference
CANADA			
(Prairies)	S wheat	P_A; $P_{May, Jun, Jul}$	Williams & Robertson 1965
(Prairies)	S wheat	P_A; $(P, PE)_{May, Jun, Jul}$	Williams 1969
(3 soil zones)	S wheat, oats, barley	P_A; $(P, PE)_{May, Jun, Jul}$	Williams et al. 1975
(Prairies)	Barley	P_A; $(P, PE)_{May, Jun, Jul}$	Williams 1971
U.S.A.			
High Plains (Mid-west)	W wheat	P_A; $(P, T)_{Apr, May, Jun, Jul}$	Thompson 1969a
Midwest (Central)	Maize	P_A; P_{Jul}; $T_{Jun, Jul, Aug}$	Thompson 1969b
Midwest (Central)	Soybeans	P_A; $P_{Jul, Aug}$; $T_{Jun, Jul, Aug}$	Thompson 1970
U.S.S.R.			
(Chernozem)	W wheat	SM_S; $(No. stalks/m^2)_S$; P_{May}; P_{Jun}	Chirkov 1973
	S wheat	$(SM; P; D)_{Phases}$	Ulanova in press
	Maize	$(SM; T)_{LA}$	Chirkov 1969
(27 Soil-Climate Regions)	W wheat	Fertilizer use; $(T; P; SM)_{Oct-Jul}$	Zabijaka 1974
	S wheat	Fertilizer use; $(T; P; SM)_{Apr-Aug}$	Zabijaka 1974
AUSTRALIA			
	Wheat	$(P, PE)_{Weekly}$	Nix and Fitzpatrick 1969

Country	Crop	Variable	Reference
INDIA	Wheat	$P_{Weighted\ over\ season}$	Gangopadhyaya & Sarker 1965
IRAN	Wheat	P_{Annual} $P_{Growing\ season}$	Lomas 1973
ISRAEL	Wheat	P_{Annual}	Hashemi 1973
	Wheat	$P_{Weighted\ over\ season}$	Lomas 1972 Lomas & Shashona 1973
TURKEY	W wheat	$(P,T)_{Jan-Dec}$	Coffing 1973

* Agrometeorological data required for generating "independent" variables in prediction equations. These variables may be used as simple, derived, exponential or interaction terms or any combination and selection of them. Variables accounting for time trend, soil characteristics or geographical features are not considered here.

Variable

SM = Soil moisture reserves; No. stalks/m^2 = Number of stalks per $1m^2$;
P = Precipitation, T = Temperature; PE = Potential evaporation;
D = Deficit between actual and saturated water vapor.

Periods

W = Winter; S = Spring; Phases = Phenological phases;
A = Antecedent (weighted) P total for period proceeding planting;
LA = Period based on Leaf Area, observed or derived, e.g., from plant height.

in view of their potential impact on the world food supply/demand
situation. Even though reliable medium-range weather outlooks
(30 days) are not yet available, it has been demonstrated that large
weather-induced anomalies of grain yields can be identified about
mid-way through the growing season (Baier and Williams, 1974;
McQuigg, 1975). Weather-based crop production estimates derived
from suitable crop-weather models have been issued prior to harvest
in experimental systems (Baier, in press).

The success of weather-based crop forecasting is based on the
following premises (Robertson, 1974):

(1) Current crop conditions reflect the integrated effect of past
weather on the potential yields of the present growing season.

(2) Soil moisture can be estimated at any time from past and present
weather data and its distribution during the various crop development
periods determines to a large extent the expected level of crop yields.

(3) The probability distribution of future weather conditions which
may effect the current crop prospects can be used to indicate possible ranges of expected crop yields.

(4) A limited number of weather observing stations in the crop forecasting area provides adequate information to generate the timely
and spatial distribution of the weather elements which are most important for the growth, development and yielding of crops.

The use of crop-weather models in crop surveillance systems
has several advantages. Yield estimates can be made almost on a
day-to-day basis as the meteorological data required for some models
are available for every 24-hour observation period. Basic data
requirements of typical crop-weather models which can be used in
operational yield assessment systems are listed in Table 1. The
primary meteorological input data include precipitation, maximum
and minimum air temperatures, relative humidity, solar radiation,
sunshine or cloud cover, and wind. These data are sometimes used
to develop so-called derived or agrometeorological data including
potential and actual evapotranspiration, soil moisture, heat units
and their derivations, and indices of heat, cold or moisture stress.
Most of the primary meteorological data are available through the
SYNOP and CLIMAT codes carried over the Global Telecommunication
System (GTS) operated by the World Meteorological Organization.
Efforts are being made to improve the collection, transmission and
inter-regional exchange of data reported in the SYNOP and CLIMAT
codes as well as to develop a special AGMET code which would provide
those primary data that are of special agrometeorological interest
but presently not reported over the GTS. In addition to the meteorological input requirements, astronomical data such as daylength

or solar radiation at the top of the atmosphere (Q_o) are sometimes
needed; these data are available from tables or computer programs
(Robertson and Russelo, 1968).

The usefulness of crop-weather models is limited by uncertainties in the meteorological modelling process. The reliability of
the yield estimates is influenced amongst other factors by the design of the model the selection of the independent variables and the
number of variables to be used. For example, if too many weather
variables are used in multiple regression analysis, the coefficients
become unstable and the problem of multicollinearity arises. Even
though an acceptable set of coefficients has been obtained in the
analysis, the estimates can only reflect direct - and to some degree, indirect - influences of the selected weather variables on
crop yields. Other factors which also determine yields and production, but which usually are not predictable from weather data, include health and vigor of the stand, plant density, lodging, acreage
and other related crop/soil conditions.

The accuracy of weather-based yield estimates increases as the
growing season advances. In Canada, research into the development
of regression-type crop-weather models is well advanced and the usefulness of this approach for assessing Prairie crop production has
been demonstrated in feasibility studies (Baier and Williams, 1974).
The accuracy of weather-based yield estimates in relation to the
time of their issue prior to harvest was compared with the post-
harvest estimates released by Statistics Canada (Table 2). Very
little estimation skill was demonstrated for yield estimates of
wheat, barley and oats made at the end of April and May. Incorporating the June and July weather data in the equations reduced the
error in pre-harvest estimates of wheat, oats and barley yields
to 9%, 4% and 4% respectively. More important than the error is
the timeliness of these estimates which were made several weeks before harvest, whereas the yield estimates issued by Statistics Canada became available several weeks after harvest.

In Phase I of the Large Area Crop Inventory Experiment (LACIE)
yields are estimated using meteorological data from ground stations
and/or satellites. The models are of the classic multiple-variable
regression type as proposed by Thompson (1962, 1969) and widely reported in the literature (Baier, in press). The Center for Climatic
and Environmental Assessment, Columbia Missouri, introduced several
modifications such as improved modelling of the technology trend,
combining monthly temperature and precipitation into an aridity index thereby reducing the number of variables, and introducing the
variable "degree days above 90°F" (McQuigg, 1975). For use in forecasting yields throughout the crop season, truncated models are
available in which only climatic variables through a given cut-off
point are included (Table 3).

Table 2.   Means of <u>post-harvest</u> Prairie crop yield
           statistics (1961-1972) and average absolute
           errors of <u>pre-harvest</u> yield estimates based
           on weather data to end of April, May, June
           and July. (Source: Adapted from Baier and
           Williams, 1974).

Crop	Post-harvest mean	Absolute error of pre-harvest estimates made at the end of:			
	kg/ha (1)	April kg/ha	May kg/ha	June kg/ha	July kg/ha
Wheat	1527	241	248	134	134
Oats	1737	160	156	87	72
Barley	1906	167	162	103	81

(1)   Converted as follows: Wheat 1 bu/acre = 67 kg/ha
                              Oats 1 bu/acre = 38 kg/ha
                              Barley 1 bu/acre = 54 kg/ha

Table 3. Truncated models for Kansas winter wheat (1931–74).

(Source: McQuigg, 1975)

Variable	Trend	Time of Truncation February	March	May	June
Constant	10.383	10.471	11.407	13.263	13.347
Linear Trend 1931–55	0.250	0.268	0.213	0.208	0.225
1955–74	0.819	0.741	0.811	0.775	0.759
Aug–Feb. Prec. (in.) DFN	-----	0.521	0.343	0.293	0.284
March Prec. - P.E.T. DFN	-----	-----	1.875	1.487	1.591
(in.) SDFN	-----	-----	-0.170	-0.120	-0.139
May Prec. (in.) SDFN	-----	-----	-----	-0.369	-0.299
May Degree Days Above 90°F	-----	-----	-----	-2.424	-2.453
June Prec. (in.) DFN	-----	-----	-----	-----	-0.133
SDFN	-----	-----	-----	-----	-0.119
Standard Error (bu/acre)	3.68	3.48	2.90	2.53	2.48
R^2	0.77	0.80	0.86	0.90	0.91

Standard Deviation of Yields = 7.42 bu/acre

DFN = Departure from normal   SDFN = Squared departure from normal

The Phase I testing of the LACIE yield estimates, based on a number of tests in the U.S. Great Plains, indicated that the models can be expected to meet the 90/90 criterion in regions having characteristics similar (in geography and agriculture) to these of the states in the yardstick region. This criterion specifies that at harvest production estimates at a county level be 90 percent accurate 9 years out of 10 or 90 percent of the time. The best available yardstick value is used for comparison. In the U.S., these are Statistical Reporting Service (SRS) results. In tests of the yield models over the years 1965 to 1975, the Coefficient of Variation (C.V.) of the yield estimates was in the order of 2 percent at the national level, lower than the 4.25 percent required. When combined with SRS area estimates in these same years, the yield estimates would not satisfy the 90/90 criterion for production when errors of equal magnitude in the area estimates were assumed. A major source of the yield estimation error was the form of the model which resulted in unrealistically high or low yield estimates for extremely high or low values of temperature and precipitation. An improved form or so-called "flagged model" to overcome this problem has been developed and preliminary test results are encouraging.

It is recognized that the models may not perform as well in foreign areas where historical meteorological and yield data are lacking or nonexistent. Here it is appropriate to recall that LACIE is designed to meet USDA needs in areas where ground information is not readily available. LACIE is not designed to improve the accuracy of the U.S. crop reports, although the U.S. Great Plains has been chosen to test the design and accuracy of the yield, area and production estimation systems (NASA, NOAA, USDA, 1976).

Crop Development Models

In any weather-based simulation system of regional crop production it is necessary to drive the system by some kind of phenological "clock". Temperatures and photoperiod or daylength determine to a large extent the rate of crop development or progress towards maturity. Robertson (1973) reviewed the various models for relating crop development to meteorological temperature data such as degree-days and their derivations.

A typical example of such a model is the biometeorological time scale for a cereal crop proposed by Robertson (1968). It consists of three quadratic terms, one each for daily photoperiod, maximum temperature and minimum temperature. The model might be used along with historical climatic data for calculating crop calendars or with current weather data for assessing the day-by-day development of a cereal crop towards maturity. The coefficients for wheat development were calculated from data gathered at several stations across Canada over a 5-year period. The equation for which coefficients were published (Robertson, 1968) is of the general form:

$$M = \sum_{S_1}^{S_2} \left[\{ a_1(L-a_0) + a_2(L-a_0)^2 \} \{ b_1(T_1-b_0) + b_2(T_1-b_0)^2 \right.$$
$$\left. + d_1(T_2-b_0) + d_2(T_2-b_0)^2 \} \right]$$

where: M = degree of maturity or development,
       L = daily photoperiod,
       T_1 = daily maximum (daytime) temperature,
       T_2 = daily minimum (nighttime) temperature,
       a_0, a_1, a_2, b_0, etc. are characteristic
                         coefficients to be determined.

Reproducibility of results was demonstrated by using a second set of 5-year data and by using a third set of completely independent data from Argentina. Subsequently, the model was tested with growth stage observations collected from 7 of the USDA ground truth sites under the Earthsat Spring Wheat Yield System Test 1975 (Earth Satellite Corporation, 1976). The results together with preliminary tests over 66 location years of data in North Dakota seemed to suggest that the estimates of the date of jointing, heading and soft dough were +9, +5 and -1 days respectively from the observed. It was concluded that the model is working adequately and does give a reasonable estimate of plant growth stages.

On similar grounds, Williams (1974a) used Robertson's time scale model to develop the appropriate equations for barley. Test of the barley equations with independent data indicated that they could usefully be applied under the environmental conditions of most of Canada's farmland (Williams, 1974b).

Features of Satellite-Based Crop Surveillance Systems

The various types of earth observation satellite systems provide data which either supplement the information obtained from weather-based crop models or overcome some of the deficiencies encountered with such models. For example production estimates are a function of both area and yield estimates. Crop acreage-weather data have not been used so far for developing statistical relationships, although the suggestion has been made that a substantial portion of the acreage variability is weather related (USDA, 1974). In that report it was proposed to regress acreage against weather on a regional basis and thereby use climatological records and acreage statistics to estimate current acreage from trend, weather and possibly other variables.

However, results from research and from experimental satellite-based earth observation systems, such as LANDSAT, have already demonstrated the feasibility and accuracy of area estimates which can be expected from this approach. In LACIE Phase I, area is derived by classification and measuration of Landsat Multispectral Scanner (MSS) data acquired on a sampling of about 2 percent of the agricultural area in all regions where wheat is a major crop. The wheat area estimation at harvest over the U.S. Great Plains was deemed marginally satisfactory in view of the 90/90 at harvest criterion for wheat production estimates (NASA, NOAA, USDA, 1976). The area estimation system shows a tendency to underestimate (by about 10 percent) when compared to SRS results. The C.V. computed for the LACIE area estimator, when projected to the national level, is about 5.0 percent, slightly above the 4.25 percent required if production estimates are to meet the 90/90 criterion. LACIE results however did indicate a difficulty in differentiating wheat from other closely related small grains. The accuracy does appear to degrade some wheat in region of marginal agriculture where there are small fields with confusion crops.

Many studies have indicated that the separation of crop areas on the basis of their multispectral signature is dependent on the stage of the phenological development of each crop. Park (1975) in a report to FAO has compiled the state of art of crop species identification from aircraft imagery. Table 4 summarise this information. Park pointed out that some of these observations are already made from space and most, if not all, will be possible in the late 70's from earth observation satellite systems. In this connection accurate information on the crop development stage is very important. Historical phenological data have been used in the development of crop calendars which are useful for studying regional and global crop patterns. But for the assessment of current conditions in specific areas, weather-based crop development models can provide, on a day-by-day basis, quantitative information on the present stage of crop development. This knowledge is essential for the proper identification of remote sensing images of field crops.

On the other hand, information obtained from the various satellite systems can supplement meteorological and field observations taken on the ground in several ways. From systems such as LANDSAT an assessment of current crop conditions on a large area can be obtained for known soil/climatic regions. An early indication of much below or above average crop conditions for a particular date can be readily determined which then complements and verifies results from weather-based growth equations. LANDSAT imagery also provides spatial information on regions for extrapolation of site-oriented data or for sub-sampling purposes. For example, Park (1975) cites a general regression equation relating clouds and precipitation of the following form:

$$P_{sat} = k_0 + k_1 Cb + k_2 Cc + k_3 Ns + k_4 St + k_5 B1 + k_6 B2$$

where: P_{sat} is the satellite rainfall estimate,

Cb, Cc, Ns, St, are percentages of cumulonimbus, cumulus congestus, nimbostratus and stratus cloud types measured from the visible satellite images,

B1 and B2 are percentages of brightest and bright cloud cover occurrences in the infrared images,

$k_0 \ldots k_6$, are regression coefficients.

For those areas or so-called agromet cells (12.5 ml spacing between successive grid points) that do not have reports, the final estimate of precipitation is calculated by a combination of the satellite estimate P_{sat} and the ground estimate P_q. The ground estimate is obtained from the precipitation measured at the synoptic station assigned to that cell.

In the LACIE Program, visible and infrared imagery from the Synchronous Meteorological Satellites SMS-1 and SMS-2 were utilized in the estimation of cloud cover for potential evapotranspiration (ETP) and rainfall calculations. The ETP sub-element of the System produced estimates during the 1975 test that were within $\pm$ 2mm of Class A pan measurements 68% of the time. The precipitation estimation sub-element provided daily estimates accurate to within 3mm 72% of the time and 7mm 90% of the time (Earth Satellite Corporation, 1976).

Means[1] reported that progress is being made in the use of satellite data for estimating precipitation amounts by personnel of NOAA's National Environmental Satellite Service. Satellite precipitation did not occur over certain areas. Satellite imagery has been used to monitor the extent of snow cover over winter wheat areas and the advancement of the monsoon over India.

Operational Agrometeorological Systems

The objective of operational agrometeorological systems is to interpret current and immediate-past weather in terms that are meaningful in weather-sensitive agricultural situations. The information obtained forms the basis for management decisions pertaining to soil cultivation, planting and harvesting operations,

---

[1]Climate data as received and used in the United States. Unpublished report for WMO, April, 1976.

Table 4. Earliest time of identification matrix (Aircraft Imagery, Preliminary), (Source: Park, 1975)

Blank = Black & White Color
C = Color/IR Needed
I = By Inference/Context

Crop Group & Individual Crops	Early Pre-Flowering Inc. Planting	Flowering	Ripe	At Harvest	Post-Harvest	Comments
Root Crops						
Potatoes		Yes (in some areas)		Yes		Confused with sugar beets until harvest.
Cassava						Uncertain (I, principally).
Yams & Sweet Potatoes		Yes				
Sugar Beets	Yes (in some areas)			Yes		Confused with potatoes until harvest-descriptors may be developed.
Sugar Cane	Yes			Yes	Yes	Very stable in location easily monitored for acreage.
Fibres						
Cotton			Yes			Leaf-Fall, exposed Bolls, picking pattern all assist in identification.

Small Grains	Yes			
Wheat	Yes	Yes(C)		
Padi Rice	Yes (Only Winter)(I)			
Upland Rice	Yes	Yes(C)(I)		
Barley		Yes		
Rye		Yes(I)		
Oats			Yes	
Coarse Grains	Yes			
Corn	Yes	(Yes)	Yes	Possibility high of identification before Harvest as experience is gained in disc. between Corn and Grain Sorghum.
Grain Sorghum	Yes	(Yes)	Yes	
Both for Skage			Yes	
Millet		Yes(I)(C)		
Oil (Seed)				
Soybeans	Yes(C)	Yes(C)	Yes	
Ground Nuts			Yes	Leaf all, color change. Equipment important, may be possible pre-harvest.

Table 5. Relation between biological
and meteorological scales of
interpretation for environmental
processes. (Source: Haufe 1975).

Scale		Order of Spatial Dimensions		Realm of Interpretation
Meteorological	Biological	Surface Area	Height	
1. Macroscale	Extra-terrestrial	$(10\ km)^2$	10 km	Current continental and global weather synopses
2. Mesoscale	Near terrestrial[2]	$(3\ km)^2$	10 km	Forecasts for agrometeorological impact[4]
3. Microscale[1]	Niche[3]	$(1\ km)^2$	1 km	Ecological processes
4. Micro-microscale	Microhabitat	$(1\ m)^2$	1 m	Physiological processes

[1] According to Geiger's concept of micrometeorology.

[2] Approximate limits of the biosphere.

[3] According to ecological concepts.

[4] Also corresponding to the limits of environmental interaction in 'ecosystems'.

the control of weeds, pests and diseases, and the production performance of crops and livestock. These systems operate at the
microscale, mesoscale and macroscale levels (Table 5). The approach
used and the data input requirements differ according to scale and
realm of weather interpretation.

Those systems operating at the microscale and mesoscale levels
require rather detailed and accurate weather information. For example, hourly values of air temperature, precipitation and wetting
duration are needed for providing advisory services such as potato
blight, weather-hazards, hay drying conditions, integrated pest and
disease management especially in fruit production (e.g. European
red mite, apple maggot, coddling moth). Most of the necessary
meteorological data are available from the international code SYNOP
and from the hourly aviation weather reports METAR according to regional and national coding practices. There are however many problems in the practical use of these data in operational agrometeorological systems because of reporting, coding and transmitting
practices. Other shortcomings result from the inadequate density
of observing stations and the fact that the observations are taken
at airports. The answer here is probably on-line recordings of
the critical weather elements directly in agricultural fields,
orchards, vineyards or wherever the information is required.

The data acquisition problem is quite different in macroscale
operational systems concerned with regional crop yield assessments.
Although it has been shown that the reports from the existing weather stations in the major grain producing areas represent the
prevailing weather pattern over these areas fairly well, spatial
interpolation of the meteorological data used in these operational
systems would be preferable.

A typical example is the Soil Moisture Evaluation Project
(SMEP) conducted in 1976 by the Agrometeorology Research and Service
Section of the Chemistry and Biology Research Institute of Agricul-
ture Canada. (Project Leader: S.N. Edey). The test area has been
Southern Saskatchewan, the major wheat producing area in Canada.
Daily weather observations (maximum temperature, minimum tempera-
ture, and precipitation) from selected climatological stations on
the Prairies are obtained daily by telephone from the Atmospheric
Environment Service, mostly through the Climatology Section, Western
Region, in Winnipeg. These data are used in the Versatile Soil
Moisture Budget (Baier and Robertson, 1968; Baier et al., 1972)
to calculate daily soil moisture in 6 soil depths, for 3 soil textures, and 2 field conditions namely wheat crop and fallow. A
series of computer derived maps (Fig. 1) depicts current soil moisture reserves and snow cover when appropriate, in addition to tabulated data for individual stations giving both current, past and
30-year average values. Observed soil moisture data and reports
by field personnel are used as much as possible for verification
purposes.

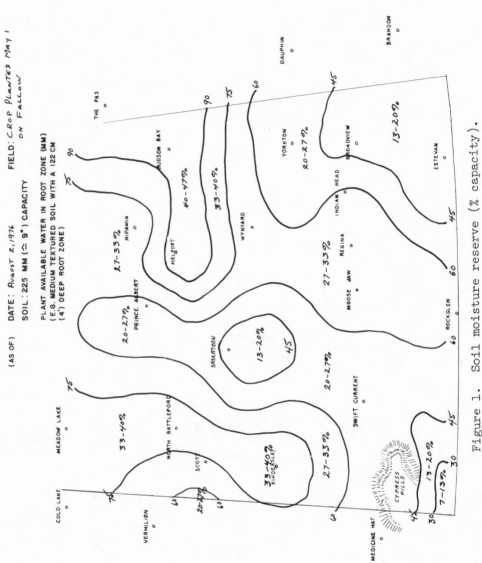

Figure 1. Soil moisture reserve (% capacity).

The reports use weather records up to Monday morning and are issued Monday or Tuesday on a weekly basis from mid-May to the end of the growing season and on a monthly basis during the fall-to-spring period. Copies are made available to the Canadian Wheat Board, Agriculture Canada research stations, extension services, and agencies involved in Prairie cereal production. A novel aspect of the project is the ability to determine at anytime during the year the soil moisture reserves for crop production. This information should be useful in assessing current prospects of seasonal wheat yields in Saskatchewan.

A system of weather/crop assessments by regions is operated by the U.S. Environmental Data Service's Center for Climatic and Environmental Assessment (CCEA). The weekly weather crop assessments use information from various sources including CLIMAT, SYNOP, upper air and satellites. Computer outputs are in the form of charts showing monthly temperatures, precipitation, departures from normal temperature and percentage of normal precipitation. This information for selected major grain producing areas in the world is used as input to sub-models providing estimates of soil moisture, crop development and yields. An interesting feature included in the weekly reports is a map of major crop-threatening weather anomalies. This feature could conceivably be improved by satellite-based surveillance systems.

The U.S. National Weather Service (NWS) also uses CLIMAT reports to produce charts for the major continents of the world. These charts include average monthly temperatures, total monthly precipitation, departure of average temperature from normal, and percentage of normal precipitation. They cover most of the Eurasian Continent, Africa, South America, and Australia.

These charts are published in the Weekly Weather and Crop Bulletin, a joint publication of NWS and the USDA Statistical Reporting Service. The international charts are published in an issue shortly after the middle of each month. Monthly U.S. temperature and precipitation charts are presented in an earlier issue each month. Weekly U.S. charts are presented in each issue.

A special write-up entitled "World Agricultural Watch" discusses weather and crops for major production areas. It is published in the international issue each month. Both domestic and international distributions are made to agricultural interests.

At the international level it should be noted that the Agrometeorological activities in aid of food production as decided by the 7th Congress of the World Meteorological Organization (April/May 1975) include a component: Agrometeorological Information(Baier,1975).

This area of activity comprises two parts: (i) agro-weather forecasts; and (ii) summarized past and present meteorological and climatological information. FAO has indicated to WMO that its immediate requirement for meteorological input to the Global Information and Early Warning System on Food and Agriculture is for CLIMAT data. Plans are being developed for the regular transmission of CLIMAT data via the GTS to FAO.

Conclusions

The significant features of operational systems in agricultural meteorology and remote sensing are such that data acquisition and interpretation mutually interact and supplement each other. The advantages of such a coordinated approach are not only to fill the observation voids of data-sparse areas but also to supplement ground observations in areas where data are more plentiful. Operational regression-type crop-yield models have been developed for a number of crops grown in the major agricultural regions of the world. The basic meteorological data needed as input to these models are obtained from on-site weather observations transmitted over the WMO Global Telecommunication System. This information can usefully be supplemented by satellite-based earth observation systems.

Research into the development and application of improved weather-based crop yield models needs to be strengthened. Some efforts into weather-related crop acreage models is justified but the major thrust of research and development should be towards determining crop acreage from satellite-based systems. The present operational agrometeorological systems suffer from problems related to inadequate data sources and irregular data transmission. Even though efforts are being made to improve this data situation, there still remains the tremendous task of coordinating regional weather/crop assessments towards a global crop surveillance system operated by international organizations such as WMO and FAO.

Acknowledgements

The author wishes to acknowledge the valuable comments on the first draft of this manuscript by two of his colleagues in Canada Agriculture: Dr. A.R. Mack, Soil Research Institute and Mr. W.K. Sly, Chemistry and Biology Research Institute.

REFERENCES

Baier, W., "Outline of the new WMO Agrometeorological activities
    in aid of food production" (with special reference to Canada's
    role). Misc. Bulletin 4. Agrometeorology Research and Serv-
    ice Section, Chemistry and Biology Research Institute, Agri-
    culture Canada, 1975, 11p.

Baier, W., "Crop-weather models and their use in yield assess-
    ments". WMO Technical Note (in press).

Baier, W. and Robertson, G.E., "A new versatile soil moisture
    budget". Canadian Journal of Plant Sciences, 1966, 46: 299-
    315.

Baier, W. et al., "Soil moisture estimator program system".
    Technical Bulletin No. 78. Agrometeorology Section, Plant
    Research Institute, Canadian Department of Agriculture,
    Ottawa, 1972, 55p.

Baier, W. and Williams, G.D.V., "Regional wheat yield predictions
    from weather data in Canada". Proceedings of the WMO-Sympo-
    sium, Agrometeorology of the Wheat Crop, Braunschweig, Federal
    Republic of Germany, 22-27 October, 1973, WMO No. 396, 1974,
    p. 265-283.

Chirkov, Y.I., "Agrometeorological conditions and maize yields".
    L. Gidrometeoizdat, 1969, 251p.

Chirkov, Y.I., "Agrometeorological forecasting methods and eval-
    uation of the productivity of agricultural crops. Paper
    presented at WMO Symposium, Agrometeorology of the Wheat Crop,
    Braunschweig, Federal Republic of Germany, 22-27 October, 1973.

Coffing, A., "Forecasting wheat production in Turkey". Foreign
    Agricultural Economic Report No. 85, U.S. Department of
    Agriculture, Economic Research Service, 1973, 38p.

Earth Satellite Corporation, "Earthsat spring wheat yield system
    test 1975". Final Report, Earth Satellite Corporation
    (Earthsat), 7222 47th St. (Chevy Chase), Washington, D.C.,
    U.S.A., 1976.

Gangopadhyaya, M. and Sarker, R.P., "Influence of rainfall distri-
    bution on the yield of wheat crop". Agricultural Meteorology,
    1965, 2: 331-350.

Hashemi, F., "Predicting the wheat yield of Iran with weather
    data". Iranian Meteorological Department Doc. No. 630.2515,
    1973, 38p.

Haufe, W.O., "Agrometeorological studies of dynamic ecosystems". 16th Annual Report of the Canada Committee on Agrometeorology to the Canadian Agricultural Services Coordinating Committee, Agriculture Canada, 1975, Appendix 10.

Lomas, J., "Economic significance of dry-land farming in the arid northern Negev of Israel". Agricultural Meteorology, 1972, 10: 383-392.

Lomas, J., "Forecasting wheat yields from rainfall data in Iran". State of Israel Meteorology Service, 2nd Progress Report, 1973.

Lomas, J. and Shashoua, Y., "The effect of rainfall on wheat yields in an arid region". Proceedings of the Uppsala Symposium, 1970, (Ecology and Conservation 5). Plant Response to Climatic Factors, UNESCO, 1973, p. 531-538.

McQuigg, J.D., "Economic impacts of weather variability". Department of Atmospheric Science, University of Missouri-Columbia, 1975, 256p.

NASA, NOAA, USDA, "Large area crop inventory experiment (LACIE)". Phase I Evaluation Report, LACIE-00418, 1976, 47p.

Nix, H.A. and Fitzpatrick, E.A., "An index of crop water stress related to wheat and grain sorghum yields". Agricultural Meteorology, 1969, 6: 321-337.

NOAA, "Application of meteorological satellite data in analysis and forecasting". Report prepared for NOAA by Anderson, Ralph K. et al., ESSA Technical Report, NESC 51, 1974.

Park, A.B., "Programme plan for developing the capability of forecasting crop production". Report prepared for FAO, 1975.

Robertson, G.W., "A biometeorological time scale for a cereal crop involving day and night temperatures and photoperiod". International Journal of Biometeorology, 1968, 12(3): 191-223.

Robertson, G.W., "Development of simplified agroclimatic procedures for assessing temperature effects on crop development". R.O. Slatyer, (Editor) Proceedings of the Uppsala Symposium, 1973 Plant Response to Climatic Factors, UNESCO, Paris, 1973, p. 327-343.

Robertson, G.W., "World weather watch and wheat". WMO Bulletin, 1974, 23(3): 149-154.

Robertson, G.W. and Russelo, D.A., "Astrometeorological estimator".
    Technical Bulletin 14. Agrometeorology Section, Plant Research
    Institute, Canadian Department of Agriculture, 1968, 22p.

Strome, W.M., "Remote sensing the future". Proceedings of Third
    Canadian Symposium on Remote Sensing, Edmonton, Alta.,1975,
    p.55-74.

Thompson, L.M., "Evaluation of weather factors in the production
    of wheat in the United States". Journal of Soil and Water
    Conservation, 1962, 17: 149-156.

Thompson, L.M., "Weather and Technology in the production of wheat
    in the United States". Journal of Soil and Weather Conserva-
    tion, 1969a, 24: 219-224.

Thompson, L.M., "Weather and technology in the production of corn
    in the U.S. corn belt". Agronomy Journal, 1969b, 61: 453-456.

Thompson, L.M., "Weather and technology in the production of soy-
    beans in the Central United States". Agronomy Journal, 1970,
    p.232-236.

Ulanova, E.S., "Agrometeorological forecasting". Guide to Agri-
    cultural Meteorological Practices, WMO Secretariat, Geneva,
    (in press).

USDA, "An investigation of the feasibility of developing a semi-
    automated system for monitoring spring wheat production".
    Report prepared for the U.S. Department of Agriculture
    (Contract No. 123341024) by Earth Satellite Corporation,
    Washington, D.C., 1974.

Williams, G.D.V. and Robertson, G.W., "Estimating most probable
    prairie wheat production from precipitation data". Canadian
    Journal of Plant Sciences, 1965, 45: 34-47.

Williams, G.D.V., "Weather and prairie wheat production".
    Canadian Agricultural Economy Journal, Vol.XVII, 1969,
    No. 1: 99-109.

Williams, G.D.V., "Physical resources for barley production on
    the Canadian Great Plains". M.A. Thesis, Department of
    Geography, Carleton University, Ottawa, 1971, 228p.

Williams, G.D.V., "Deriving a biophotothermal time scale for
    barley". International Journal of Biometeorology, 1974a,
    18: 57-69.

Williams, G.D.V., "A critical evaluation of a biophotothermal time scale for barley". International Journal of Biometeorology, 1974b, 18: 259-271.

Williams, G.D.V., et al., "Regression analysis of Canadian Prairie crop district cereal yields, 1961-1972, in relation to weather, soil and trend". Canadian Journal of Soil Sciences, 1975, 55: 43-53.

Zabijaka, V., "A preliminary weather-crop yield model for Soviet grain". Foreign Demand and Competitive Division Working Paper, Economic Research Service, U.S. Department of Agriculture, 1974, 12p.

IDENTIFYING THE OPTIMUM SYSTEM

FOR OBSERVING EARTH

Dr. Eric C. Barrett

University of Bristol

United Kingdom

The Author's Viewpoint

In reviewing the art of the reviewer, author Auberon Waugh
has recently written that "It is the duty of the reviewer to write
dispassionately. Since this is often not possible it is right that
from the outset, he should declare his special interest, including
any personal prejudice or antipathy". It seems that the reviewer
of practice and progress in a field of scientific enquiry would do
well to do likewise. Certainly, one's own experience in a particular field must inevitably effect his relation with it and colour
his judgement concerning the future developments he would like to
see.

Some personal notes seem, therefore, to be in order, especially
since my own experience with satellite systems probably parallels
a number of stages in the changing attitude of much of the scientific community towards remote sensing of Earth from Space.

Having undergone a standard form of undergraduate training in
a British university, majoring in Geography, I researched first in
historical climatology. This led me to an awareness of the mani-
fest deficiencies of conventional systems for observing the environment. I first encountered satellites and the potential of such
systems at a British Interplanetary Society Symposium in London
in 1963, on Meteorological Satellites. I attended this by chance,
but immediately appreciated that satellite imagery had a big potential role to play in educational situations, (Barrett, 1964).
Therefore, rather slowly, I began to investigate the usefulness
of satellites as contributors to our general pool of knowledge
of the Earth's atmosphere (Barrett, 1969, 1974) and then as sources

of information on selected atmospheric parameters (Barrett, 1971).
It seemed that certain types of environmental studies might be
based effectively, even advantageously, upon remote sensing data.
A tacit belief was that satellites were in competition with in situ
sensing systems to provide the data essential to various programmes
for monitoring the physical environment of man. More recently, the
important realisation has dawned that in many instances better results can be obtained by combining information from the two different
types of sensor sources (Barrett, 1975). Involvement in design
studies for European Space Agency (ESA) satellites including Pamirasat (the Passive Microwave Satellite), MEOS (the Multidisciplinary
Earth Observation Satellite) and Minimetsat has suggested - alas,
post-event in these three cases - that still better results might
be obtained were the design of both in situ and satellite data
collection systems to be rather fundamentally rethought. Some allu-
sion is made to this matter in Barrett and Grant (1976b). (See also
Barrett and Grant, 1976a.)

     This paper seeks to explore as objectively as possible the
scenario in which Earth observation systems are set. It does so
principally from an academic point of view, though reference is
made to some problems of an applied nature which I have been asked
by international agencies to try to solve. It is my feeling that
the Earth observation systems we have at present, like Topsy "just
growed". Political, financial, commercial and historical factors
have dominated their development, and ignorance and prejudice have
been further hindrances. The opportunity afforded by this Symposium
to consider in an essentially detached way the structure of an optimum system for observing Earth is too good to miss. The thoughts
that follow represent a personal view as to the appearance such a
structure might possess.

The Respective Merits of In Situ and Remote

Sensing Methods for Observing Earth

     The community of environmental scientists seems to have been
rather clearly divided into those who are enthusiastic about the
use of remote sensing techniques for environmental monitoring and
those who are not. This has not been conducive to a worthwhile
dialogue between the respective parties, nor to the objective recognition of a balanced system for global monitoring which pays heed
to the relative merits of the in situ and remote sensing approaches.
At the outset of any unprejudiced discussion it would seem necessary
to try to identify different categories of observations, classified
basically in terms of the best and the only means whereby such
observations could be obtained. Three categories spring readily to
mind: a) Those observations which could be obtained only by in situ
or remote sensing methods, not by both; b) Those observations which

could be obtained by either type of method, but more easily or
effectively by one rather than the other; c) Those observations
which would be obtained best by in situ and remote sensing methods
together. It is instructive to examine the constituents which
these categories presently include.

1. Observations Obtainable Only By One Method Or The Other

The, inescapable, though somewhat unexpected, conclusion which
emerges from the lists which follow is that remote sensing systems
seem to be more restricted in their applicability than "conventional", in situ, systems. The following classes of observations
may be recognised:

a) Observations from in situ systems

   i   Observations of phenomena which find no simple
       expression in the reflection or emission of
       radiant energy towards Space. Examples include
       sub-surface manifestations such as soil temperature
       and ocean water profiles and slight to moderate
       Earth tremors, and aspects of human behaviour
       enacted under cover.

   ii Observations of phenomena whose radiant energy
       expressions cannot (as yet) be isolated from
       the "noise" of radiation from surrounding and/
       or overlying sources. Such phenomena include
       most features below the resolution threshold
       of a remote sensing system, where resolution is
       considered to embrace spatial, temporal, and/or
       spectral considerations. Larger, more widespread
       features may be included if the signals they
       reflect or emit are weak, and/or are largely
       absorbed or back-scattered by media or objects
       between themselves and high altitude sensors.

   iii Observations of phenomena which, on account of
       their intrinsic nature, can be observed only
       within the Earth and its atmosphere and/or at
       the interface between the two. Examples include
       location-specific radiant fluxes reaching the
       surface of the Earth from its atmosphere and the
       Sun.

   iv Observations of some phenomena when continuous
       records are essential, for example rainfall and
       wind speed.

b)   Observations from remote sensing systems

   i   Observations of phenomena which, by their
very nature, could be monitored only from
outside the Earth/atmosphere system, for
example components of the Earth/atmosphere
heat and radiation budgets.

  ii   Observations of any phenomena whose patterns
of radiant energy reflections, emissions, and/
or transmissions through space and/or time
are subjects of enquiry for their own sakes,
for example the spectral reflectance characteristics of rocks and minerals.

Future development of sensors and sensor systems may be expected to redress the imbalance alluded to above. For example,
it should not be impossible to observe continuously in due course
some Earth phenomena. Since this would necessitate the use of
geostationary satellites, high latitude regions would be much
less amenable to this than the tropics.

2.   Observations Obtainable Better By One Method Or The Other

Here a much stronger case is evident for remote sensing of
the Earth from satellites, and it is here that value-judgement
decision-making becomes critical in deciding whether remote sensing
or in situ approaches would be preferable in given situations.

a)   Observations from in situ systems

   i   Observations of phenomena which are more
readily monitored in this way because of
the uncertainties surrounding the interpretation of the appropriate radiation data
obtained from satellite systems. All
satellite observations are affected by the
design and operation of the sensor systems
which obtained them, and by the means through
which the data are translated into conventional units. Many environmental phenomena
significant to operational activities are
monitored better by traditional means because
the physical indications monitored by remote
sensing systems are usually different and more
indirect. Precise physical interpretations
of the one in terms of the other are hard to
achieve on account of our inadequate knowledge
or understanding of the relationship between

them. Examples abound. One is air
temperature, which can be measured
directly by thermometers, but only
estimated indirectly through emitted
radiation.

   ii   Observations of phenomena whose radiation
signals are often, though not always,
obscured on account of factors of scale
and resolution, and influences of the
local environment. For example, the
recording of many Earth surface features
by remote sensing is hindered by cloud
cover, ice and snow, and vegetation cover.

   iii   Observations of manifestations of human
activity and belief which sometimes have
recognisable expressions in radiant energy,
but cannot be relied upon for these. Examples are as diverse as agricultural cropping
systems and results of personal allegiances to
political, religious and/or social doctrines.

b) Observations from remote sensing systems

   i   Observations of phenomena for which requirements of breadth and homogeneity of cover
are more important than the requirements
for more precise, but less complete, sets
of data from in situ sensor systems.

   ii   Observations of phenomena which would be
hard to monitor adequately using conventional
means, on account of factors such as cost,
accessibility, instrumental deficiencies, and
human behaviour. This last category is
depressingly broad, and quality control in
programmes of in situ observation is often
hard to enforce. Diverse problems are entailed,
for example, in assessing total cloud cover
(where personal (neuro-physical and psychological)
and geometrical (perspective) factors interplay);
in formulating agricultural land use and crop
returns (where educational, practical, financial
and political influences are significant);
and in rainfall mapping (where diligence, honesty
and care are required from the observer, and
tolerance from a potentially disruptive local
population).

iii Observations of phenomena which are
     destructive, or potentially so. In
     this case the maintenance of special
     observation systems to sound the alert
     can be very expensive indeed, and waste-
     ful in terms of men and resources.
     Satellite technology affords the opportunity
     to devise and operate a single system
     capable of being focussed rapidly upon a
     wide variety of environmental hazards
     for large-scale, short period studies
     (e.g. the U.S. hurricane watch).

Unfortunately the potential of satellite systems for Earth
observation purposes has been but little realised in areas like
these. Despite expectations, the principal reason would not seem
to have been economic, for cost-benefit studies often confirm the
promise of the environmental satellite, but the training, experience
and general outlook of the expert or administrator with responsibility for the key decision-making choice.

One example may be cited from personal knowledge. In the
Middle East weather analysis and forecasting is still heavily
dependent on traditional data and individual "skill" in interpreting
it, despite the fact that this is a region in which even the minimum
requirements of the World Meteorological Organisation (WMO) for
surface and upper air stations (W.M.O., 1967) are not completely
met. Satellite reception facilities are available at several stations, but most provide only standard resolution images (c. 4.5 km
at best), and interpretation of the data is often in the hands of
staff with no formal training in its interpretation or application.
At one station in 1975 the senior (American) staff member had recently concluded an agreement for the installation of a NOAA-SR
receiver (rather than a VHRR receiver which would have served his
purposes much better), on the advice of a WMO adviser, because no
one on the station had had training in satellite meteorology.
Unfortunately the expert adviser also lacked such training. Bad
advice was given and accepted through ignorance alone.

Clearly careful, critical, and informed thought should be
given to ways in which Earth observation data generally might be
put to good use in the service of the world community of nations.
In this, educational programmes will have a vital role to play.
It is sobering to realise that weather satellites have been with
us now for the better part of two decades, yet many meteorologists
from international bodies and local weather stations alike, still
have no proper awareness of their utility - or their limitations.

3.   Observational Programmes Effected Best By a Combination

Of In Situ And Remote Sensing Methods

As remarked in the introductory section, there has been amongst environmental scientists a tacit belief that in situ and remote sensing systems are competitors for the operational programmes of the future. Thus although it has often been necessary to invoke conventional data in the interpretation of satellite data, statements from these two different types of sources have been viewed largely as rival statements concerning individual phenomena, or, at best, supplementary statements of different phenomena. Perhaps the first movements towards a more fully realistic attitude were made spontaneously in meteorology, where satellite data began to be used in the early 1960's for cyclone tracking and frontal analysis in maritime areas, whence few conventional data were forthcoming. This has led to a conscious awareness that satellites may be more appropriate platforms for the monitoring of certain parameters in some areas than the full range of possible parameters everywhere, a fact now being capitalised upon through the withdrawal of weather ships - very expensive platforms for atmospheric sounding by radiosonde balloons - and the utilisation of satellite sounding data to fill the gaps. (Editor's note: the withdrawal of weather ships is controversial, and the effective substitution of satellite sounding data is not yet accomplished to the satisfaction of all meteorologists, as discussed in the Workshop in Atmospheric Applications and on Sensors/Platforms.)

Table 1 is an impressive list of statements now made available routinely to the meteorological community in the U.S.A. from weather satellite operations. It is worth remarking that, although in some cases conventional data have been employed in calibrating the satellite signals, these statements are almost exclusively satellite products. A few find their way into the data pools for weather analysis and forecasting (e.g. upper winds and vertical temperature profiles), but most do not. More research into the mutual compatibility of different data types, especially considering the needs for data intercomparison and interdigitation, would surely produce worthwhile results. Even in meteorology, after nearly twenty years of satellite operations and quite detailed appraisals of satellite pictures as sources of synoptic information (Anderson et. al. 1973), there is a fundamental need to rethink the relationships between in situ and satellite observing systems. Manifestly, the greatest possible benefits are being derived from neither in situ nor satellite observing systems individually, nor the two together. Such needs are even greater in other branches of environmental monitoring.

Table 1. Operational weather satellite products prepared by the
U.S. National Environmental Satellite Service (see Hoppe and Ruiz,
eds.) 1974.

Manual and Basic Products

1. Photographic imagery
    a) Geostationary full disc frames
    b) Geostationary mosaic loops
2. Facsimile
    a) Great Lakes ice charts
    b) U.S. cloud cover depictions
    c) Satellite Input to Numerical Analysis and Prediction (SINAP)
    d) Northern hemisphere snow and ice charts
3. Alphanumeric
    a) Satellite weather bulletins and tropical disturbance summary
    b) Two-layer moisture analyses
    c) Plume winds
    d) Astrogeophysical Teletype Network (ATN) messages

Unmapped Image Sectors

1. Facsimile
    a) Automatic Picture Transmissions (APT)
    b) Unmapped SMS/GOES WEFAX sectors

Man-Machine Combined Products

1. Alphanumeric
    a) Card deck, low and high level cloud motion vector field
        messages

Quantitative Computer-Derived Products

1. Alphanumeric
    a) Vertical Temperature Profile Radiometer (VTPR) soundings
    b) Experimental Global Operational Sea Surface Temperature
        operations (GOSSTCOMP)
2. Facsimile
    a) Unmapped SMS/GOES cloud motion vector facsimile formatted
        tape card deck for low level wind messages

Table 1. (continued)

Computer-Derived Image Products

1. Photographic imagery
    a) Gridded, unmapped, pass-by-pass SR images
    b) SR hemispheric polar mosaics
    c) SR mercator mosaics
    d) SR North America polar mosaic sectors
    e) Very High Resolution Radiometer basic images
    f) 5-day minimum brightness composites
    g) Augmented resolution map sectors
2. Facsimile products
    a) Visual and infrared Scanning Radiometer products for
        transmission on NWS facsimile networks
    b) Visual Scanning Radiometer products for transmission
        via WEFAX

Archival Products

1. Magnetic tapes
    a) Scanning Radiometer data tapes
    b) Sea surface temperature data tapes
    c) Vertical Temperature Profile Radiometer data tapes
2. Photographic images
    a) Scanning Radiometer data
    b) Very High Resolution Radiometer data
    c) SMS/GOES data

Maximising The Merits Of In Situ And Remote Sensing Observations

1. Generalities

Whether remote sensing and in situ data separately provide
the most appropriate statements concerning aspects of the world
in which we live, or whether such data are best woven together
for environmental analysis, much thought is required if the
advantages inherent in the different data types are to be maxi-
mised. Relatively few problems attend the design of programmes
in which satellites or in situ sensor systems are the sole sources
of certain types of information. More problems attend the deli-
berate selection of the one system or the other to undertake a
function largely or completely alone. Perhaps the most difficult,
yet potentially the most important, problems are those to be
solved in determining how in situ and remote sensing data might
best be integrated in programmes with stated aims. Here hard
work and imagination are required, plus a willingness on the part
of enthusiasts for either approach to collaborate for the common
good. One thing seems certain, namely that the conclusions regarding the most favourable combinations of in situ and remote
sensing data will often be complex. Two examples may be discussed
in some detail to illustrate the kinds of issues which are involved.
The first example is related to a long-standing programme of observation of a single parameter for general use of the results; the
second is drawn from the need to observe several environmental
features in order to achieve a single, clearly identified objective.

1 Example 1: Daily Rainfall Mapping

The global distribution of rainfall is a topic which has
long been of particular interest to the present writer, and one
to which he has given considerable thought. Daily rainfall data
are required for many purposes, by a large number and variety of
authorities ranging from some concerned with the global hydrological cycle to others whose concern is very local. Rainfall data
are basic requirements for many hydrological, meteorological,
climatological, agrometeorological, civil engineering, and water
resource studies and applications. Unfortunately for many users,
the conventional rainfall observation network suffers from numerous
deficiencies, despite the fact that rainfall is monitored in more
detail than almost any other environmental variable. The deficiencies include the following:

a) Rainfall is a notoriously "noisy" element as far as its
    spatial and temporal patterns are concerned: almost
    everywhere the present network of observing stations is
    inadequate to monitor even its principal variations.

b) The conventional raingauge network is highly irregular,
    with much higher gauge densities in some areas than others.

c) The types of gauges in common use are of many designs,
    and different regulations pertain to their exposure in
    different countries. The recognition of an International
    Reference Precipitation Gauge (IRPG) by the WMO has influ-
    enced the situation little: widespread changes of gauges
    would be expensive, and adversely affect the homogeneity
    of long-term rainfall records.

d) The standard times at which gauges are read differ with
    respect to GMT, local time, and local cycles of weather
    activity.

e) In many countries the standards of observations are
    manifestly below the threshold of acceptability. Gauges
    are read irregularly, and sometimes not at all.

Our research in the University of Bristol has revealed that in
many countries, especially in the old colonial areas, acceptable
rain data are less abundant today than they were even forty years
ago. Thus the need for assistance from remote sensing in mapping
rainfall is greater now than in an earlier decade. One suspects
that the same is true of other environmental variables also.

Some satellite-based possibilities for improving monitoring
of rainfall on a regional scale have been discussed in detail elsewhere (Barrett, in press). Here, that search may be extended in
order to identify a structure appropriate for a global rainfall
mapping programme. Such a scheme might involve:

a) The identification of those existing rainfall stations
    from which dependable data are, or could reasonably be
    expected.

b) The augmentation of the network in a) by as many addi-
    tional stations as circumstances might permit, with
    standardisation of observational method and quality of
    the resulting records as the critical criteria.

c) The mapping of areas which could not be represented
    adequately by the data from the augmented network; these
    are the areas for which remote sensing estimates might be
    usefully obtained.

d) The selection of appropriate remote sensing data for local
    rainfall estimation. On account of technical considera-
    tions the following types of satellite data seem the most

d) continued
    likely to yield the types of information needed in differ-
    ent situations:

    i   Passive microwave data, like those from
        the Electrically Scanning Microwave Radio-
        meter (ESMR) systems of Nimbus 5 and 6.
        Unfortunately, the resolution is coarse
        (at best 25 km), no such system has yet
        been accepted for operations beyond Nimbus
        G, and interpretational problems have restricted
        dependable mapping of rainfall rates from
        such data to oceanic areas, where the chief
        potential of such system would seem to be.

    ii Infrared data from the Visible-Infrared Spin
        Scan Radiometer (VISSR) of geostationary
        SMS and GOES satellites (resolution at best
        2 km). Until now rainfall estimation from
        satellite data has been based almost exclu-
        sively on data from polar-orbiting satellites,
        but results from arid and semi-arid areas
        have indicated the need for images to be
        provided more than once or twice daily if
        the results of the rapid growth of convective
        (tower) clouds are to be appreciated.

    iii Visible imagery from polar-orbiting satellites
        like NOAA, Tiros-N or the DMSP/Block VD series
        for use in areas beyond the reach of geosta-
        tionary satellites.

e) The development of a suitable paradigm for rainfall esti-
    mation from the remote sensing data for areas sparse in
    conventional observations.

    Thus a realistic scheme for global rainfall mapping utilising
both conventional and satellite data would be more complex than
might be anticipated at first. The practical problems could be
solved only in the context of an acceptable degree of international
cooperation. One suspects that without a centralisation of the
necessary facilities for data collection and processing the potential of such a scheme would be realised only in restricted areas
and for limited purposes.

2 Example 2: Desert Locust Control

    This example is different from the first in that here the
utilisation of remote sensing data is for a single specific

application. The desert locust (Schistocerca gregaria) still
poses a significant economic threat to countries within the tropical
desert belt of the Old World. The development of swarms has always
been difficult to monitor because of the sparseness of the population and the hostility of the terrain in which locust population
explosions may occur. Recently the Food and Agriculture Organisation of the United Nations have concluded the first stage of a
project involving the joint use of in situ and remote sensing
techniques for the planning of locust control operations. Fig. 1
illustrates the ideal structure that such a programme might possess.
Here remote sensing data are used as and when appropriate to compensate for deficiencies in conventional rainfall observations, and
observations of soil moisture and vegetation growth.

Once again the scheme is inherently complex, but there are
clear economies that might be expected in comparison with any
alternative based on in situ observations alone. Such economies
will be even more attractive when Landsat-type data become available
via local reception facilities costing substantially less than those
required at present.

General conclusions arising from these and other studies are
drawn together in the final section which follows.

Future Systems For Observing Earth

1. Key Desiderata

Bearing in mind the types of considerations discussed earlier
in this paper a number of key features for future Earth observation
systems may be identified. These include:

a) A vigorous education policy for environmental remote
   sensing so that more potential users of satellite data
   might have a balanced awareness of their chief merits
   and demerits.

b) Realistic and honest, yet imaginative appraisals of those
   aspects of man's activities and environment which might
   with benefit be examined routinely by remote sensing
   methods or methods in which remote sensing plays a part.

c) The promise of suitable satellite systems to meet the
   needs of as many potential users as possible, with special
   reference to resolution of the data in both space and time,
   the continuity of the system through a specified period,
   and data accessibility.

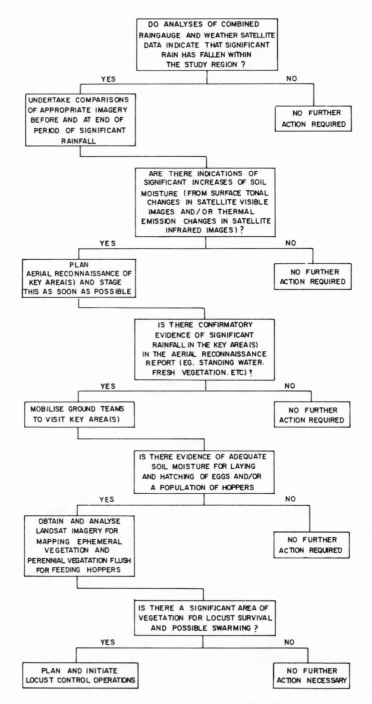

Figure 1.   Decision making flow diagram for desert locust control
            program (using both in situ and remote sensing data).

d) Suitable transmission/reception systems to permit the
acquisition of direct read-out data even by users operating
on tightly restricted budgets.

e) Adequate networks of in situ observatories (ground truth
control sites) for remote sensing data interpretation and
evaluation, and for suitable interdigitations of in situ
and remote sensing data in co-ordinated mapping programmes.

f) Research and development for data processing, interpretation and analysis, directed especially towards automatic
techniques yielding clearly-specified products.

g) Efficient schemes for the results of the monitoring programmes to be put to practical use in the service of the
world community. Particular care should be taken to avoid
substantial overlaps between the programmes of different
countries, agencies and authorities.

With such requirements in mind it is possible to proceed to a
sketch of an optimum system which can be envisaged for observing
Earth.

2. The Optimum System

This may be considered in terms of three of its basic attributes, namely its sensor systems, read-out systems, and the data
products it provides.

First, concerning satellite sensor systems, it may be said
that the practice of mission analysis which has been common hitherto
may have involved approaching the problem from the wrong angle. In
the design of a proposed family of multidisciplinary satellites, for
which I have served as mission consultant, those branches of environmental science which might be served thereby were identified first.
Then the scientific data requirements in those branches were identified. Finally, a compromise set of requirements was drawn up.
Perhaps a more successful approach might generally be to design the
sensor system first, and identify the full range of its potential
applications afterwards. Now the electromagnetic spectrum has been
rather thoroughly and systematically explored the realisation is
dawning that it affords far fewer opportunities for worthwhile observation of the Earth than once was hoped. Indeed, the ideal choice
of wavebands for many phenomena coincide. A further development
in recent months has been the tendency to feel that in multispectral
monitoring data from a small number of channels (two, three or four)
is often adequate - besides being much easier to use than data from
more elaborate multiband sensors. So a manageable set of sensors

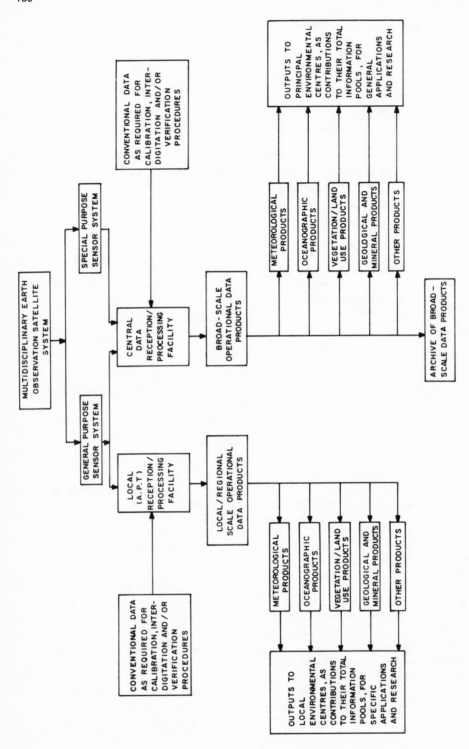

Figure 2

might be designed for very general environmental use - including
meteorological, oceanographic, and terrestrial - augmented only
by sensors for more restricted applications where important data
could not be provided by the general package (e.g. atmospheric
depth-sounding units for weather studies).

So the specific needs of different branches of environmental
science for satellite products would be met by the data processing
systems, not by design constraints in the satellites. Products of
the desired type and resolution would be prepared at major receiving
stations, subject only to the final constraints of the resolution
and frequency of the input from the satellites and elsewhere. For
this large, complex computing facilities would be required. However, supplementary data products for more specialised and/or more
local use could be prepared cheaply at local receiving stations
where cheap, hard-wired processors might be employed to exploit
those data received from the satellite by direct read-out links.

Consequently we may conclude with the following propositions
concerning the indicated satellite sensor systems, read-out methods,
and data products for an optimum Earth observation satellite system
operating in conjunction with an appropriate network of in situ
sensors (see Fig. 2):

a) A satellite sensor system comprised of a basic (general
   purpose) package, with ancillary (limited purpose) sensors.

b) A read-out system involving transmissions of recorded
   data to principal reception facilities for global or con-
   tinental-scale programmes and direct read-out data to local
   reception facilities for more local use.

c) A central data-processing system through which wide ranges
   of operational products for many different applications
   are prepared, a "total-systems" approach to the analysis
   of the satellite data. Simpler, and/or more specialised
   products could be prepared as required at local facilities.

Clearly such systems could not be implemented successfully
without a global plan to which many nations should subscribe. So
the over-riding need would seem to be for an international body -
say a World Remote Sensing Organisation - to plan, co-ordinate and
operate and maintain a suitable satellite service to meet the needs
of many other national and international bodies for that information
which is urgently required from remote sensing platforms to improve
our knowledge and care of planet Earth.

REFERENCES

Anderson, R.K. et al., "The use of satellite pictures in weather
     analysis and forecasting". W.M.O. Technical Note No. 124,
     W.M.O., Geneva, 1973, 275p.

Barrett, E.C., "Satellite meteorology and the geographer".
     Geography, 1964, 49: 377-386.

Barrett, E.C., "The contribution of meteorological satellites
     to dynamic climatology". Unpublished Ph.D. Thesis, University
     of Bristol, 1969, 280p.

Barrett, E.C., "The tropical Far East: high season climatic
     patterns derived from ESSA satellite images". Geographical
     Journal, 1971, 137: 535-555.

Barrett, E.C., "Climatology from satellites". Methuen (London)
     and Barnes & Noble (New York), 1974, 418p.

Barrett, E.C., "Rainfall in northern Sumatra: analyses of conven-
     tional and satellite data for the planning and implementation
     of the Krueng Jreue/Krueng Baro irrigation schemes". Final
     report for Binnie & Partners (Consulting Engineers), London,
     1975, 50p.

Barrett, E.C., "Applications of weather satellite data to mapping
     rainfall for the solution of associated problems in regions of
     sparse conventional data". Proceedings of the 26th Colston
     Symposium, Remote Sensing of the Terrestrial Environment,
     (eds. R.F. Peel, L.F. Curtis, and E.C. Barrett), Butterworths,
     London, (in press).

Barrett, E.C. and Grant, C.K., "Comparisons of cloud cover evalua-
     ted from Landsat imagery and meteorological stations across
     the British Isles". ERTS Follow-on Program Study, No. 2962A,
     Fourth Quarterly Report, NASA, Greenbelt, Md. U.S.A., 1976, 45p.

Barrett, E.C. and Grant, C.K., "An appraisal of Landsat 2 imagery
     and its implications for the design of meteorological observing
     systems". Journal of the British Interplanetary Society, (in
     press).

World Meteorological Organisation, "The role of meteorological
     satellites in 'the World Weather Watch'". WWW Planning
     Report, No. 17, W.M.O., Geneva, 1967, 27p.

BENEFITS OF LANDSAT FOLLOW-ON (OPERATIONAL SYSTEM)

Charles Buffalano
Chief, Applications Assessment Office
Goddard Space Flight Centre, NASA
Greenbelt, Maryland
U.S.A.

Introduction and Summary

This volume presents the results of a benefit and cost study
for the LANDSAT Follow-on system with both a Multispectral Scanner
and a Thematic Mapper. The analysis shows that the present value
of the benefits exceeds the present worth of the costs using a 10
percent discount rate and an infinite horizon for both. (Editor's
note: the cost data have been deleted from this paper since they
are under NASA review).

Although uses and applications of LANDSAT technology continue
to grow, as is characteristic of successful technology in its
introductory phases, this study focuses only on major applications,
conservatively evaluated. Benefits have been included where a
definite need for the information has been shown, a mechanism for
disseminating the information has been defined, a technical capability can be quantified, and a defendable method of evaluating the
economic worth has been developed. The study approach has been to
make the technical capability, the data reduction to management
information and the costs and benefits self consistant. This
means that certain applications with definite promise could not be
assigned benefits and do not appear here. For example, a satisfactory data system could not be found for range management so it was
dropped. Mention is made of these areas, however, either in the
appropriate subject chapter or in the final chapter on "Non-Quanti-
fied Benefits". The benefits shown here therefore are not exhaustive. The major benefit areas foreseen for the baseline global
LANDSAT Follow-on with a Thematic Mapper are summarized in Table 1.
Benefits total .5 to 1 billion dollars annually.

Table 1   Annual Benefits of LANDSAT Follow-on
(Fiscal Year 1976 Dollars)

	Estimated Annual Benefits ($million)
Agricultural Crop Information	336-581
Petroleum-Mineral Exploration	90-260
Hydrologic Land Use	22
Water Resources Management	13-41
Forestry	7
Land Use Planning and Monitoring	15-48
Soil Management	5-9
	488-968

Agriculture

Improved production and distribution of crops would be made possible by improved forecasts of world crop production. Estimated annual benefits, by 1985, are most likely to be $336 with an optimistic potential of $581 million/year. The present value of these benefits is 2.0 to 3.5 billion dollars.

The benefit range is from the most likely value to an upper bound based largely on the assumption that the Thematic Mapper will surpass its present technological goals. A more conservative benefit range will soon be available from two other studies. This range will be the spread of the most likely value. The most likely value of the benefit is obtained by choosing the most reasonable economic assumptions and then computing the benefits using the most likely values of the economic parameters. For example, it may be reasonable to assume that the demand functions are linear and then use most likely values of their elasticities. However, we recognize that there is a spread of assumptions and parameters. For example, the demand curves could be hyperbolic with elasticities that ranged from half to twice the "most likely" value. These spreads cause the benefits to spread. The first study will estimate the spread of the benefits when the economic assumptions vary and the second will estimate the spread when the parameters vary.

Marginal costs would be incurred to build and maintain a data processing facility for the Department of Agriculture. This facility would process digital tapes and produce information which the Department would use to produce foreign crop production estimates.

Oil and Mineral Exploration

The benefits result from exploration cost savings to United States producers. Annual benefits are estimated at $90-260 million/year. The present worth of these benefits is 300 to 800 million dollars. The marginal costs of processing the data from a product tape are incurred at the EROS Data Center but subsequently recovered from the users. This view is based on a major assumption that oil and mineral companies will incur only small additional recurring data processing costs beyond what they currently spend to reduce similar aircraft data. Furthermore, there will be only small non-recurring costs because they will have been sunk supporting the Multi-spectral Scanner.

The very wide benefit range reflects large uncertainties in many of the variables in the estimate. Unlike the agricultural benefits which are based on a sophisticated econometric model, the oil and mineral benefits are based completely on the opinions of a small number of experts. There are major uncertainties in the size of the market and the likely impact of the technology which cannot be resolved by analysis at this time.

Finally these exploration benefits disappear after roughly a ten year period of exploitation. The length of this period is highly conjectural but it is based on the observation that deposits susceptable to a new technology are found quickly so that productivity is initially high. After the simple finds are made the process becomes less productive.

Hydrologic Land Use

By providing cheaper surface cover maps, annual cost savings exceeding $22 million would be possible within the federal agencies and the regional, local, and special purpose water resource programs studied. In many cases, these benefits could be reinvested in extending or improving the work of the agencies involved, resulting in some further benefit which has not been included in the estimates. The present value of these benefits is 127 million dollars. Marginal costs would be incurred for a Federal data reduction facility.

Land Use Planning

Annual benefits of $15 to $48 million are estimated, based on the cost of obtaining equivalent planning data by the best alternative means, an aircraft-borne surveillance system. This estimate is based on estimates of the data requirements of federal, state, and regional agencies not included in the agriculture, oil-mineral exploration, water resources, forestry or soil management applications. The major data processing costs are incurred at the EROS Data Center and subsequently recovered from the users. No other marginal data processing costs are included because the user community already has the capability of processing aircraft data.

Water Resources Management

Annual benefits of $13 to $41 million from more irrigation water and hydropower are expected based on improved management of impounded water due to better information on river basin snow cover. The major marginal data processing costs are incurred at the EROS Data Center and subsequently recovered from the users. No additional marginal costs are expected because the user community already uses aircraft data.

Forestry and Soil Management

Annual benefits of at least $7 million are forecast by using

LANDSAT information to improve Forest Service timber inventories
and then using this improved inventory data to make better tree
planting and harvesting decisions.

Annual benefits of $5 to $9 million are also estimated from
cost savings in the preparation of soil base maps. Further benefits from detecting soil loss or nutrient deficiencies are identified but not quantified. No data processing costs were estimated
beyond those incurred at the EROS Data Center for these two applications because the user community already uses aircraft data.
Because these are the smallest benefit areas they could be uneco-
nomical if pricing policies at EROS or significant additional user
data processing costs are added.

System Costs

The cost of an operational LANDSAT Follow-on system is discussed in detail in the System Design Document (the results are not
reproduced here). The Space Segment cost includes the spacecraft,
telemetry, command and control, maintenance of the system in operational status, shuttle servicing, and refurbishment. The Basic
Processing System portion of the Data Management System includes
the cost of processing data to the point where a digital magnetic
tape is prepared. The unique user costs in Agriculture, Hydrologic
Land Use, and at the EROS Data Center include special processing
of the digital tape to obtain management information. The EROS
Data Center is expected to fill the needs of all but the Department
of Agriculture and the water management community.

Rate of Adoption

The rate of adoption of LANDSAT technology, and hence the
rate of achieving the potential annual benefits, is difficult to
estimate. We have assumed that 50 percent of the estimated potential benefits will be achieved within the first year that the
LANDSAT system becomes available; 80 percent of the potential benefits will be achieved three years after the system becomes available, and 95 percent of the potential benefits obtained three years
after that. Reductions in yearly benefits beyond 1991 reflect the
decline in Petroleum-Mineral Exploration benefits.

Benefit-Cost Comparison

In order to relate the benefits of the LANDSAT Follow-on
system to the costs, we compare the present value of the benefits
and costs. All benefits and costs are computed in FY 1976 dollars
and a discount rate of 10 percent is applied. This choice of

discount rate is based on the guidelines of OMB Circular A-94. An infinite time horizon is employed in this study. Clearly no one expects this technology to operate forever. However, the introduction of any new technology beyond this must also be justified as marginally cost beneficial with respect to the existing technology with which it competes. Since the premise in such a decision process is that an alternative is always to continue with the existing system an infinite horizon is theoretically sound.

Table 2 shows the present value of the benefits (costs are not shown). Based on these values the LANDSAT Follow-on system has a benefit/cost ratio greater than unity, even at the lower bound.

	Estimated P.V. Benefits ($million)
Space and Data Management Systems	--
Agricultural Crop Information	1950-3370
Hydrologic Land Use	127
Petroleum-Mineral Exploration	283-819
Water Resources Management	75-237
Forestry	41
Land Use Planning-Monitoring	87-278
Soil Management	29-52
TOTAL	2592-4924

Editor's Note: All costs have been deleted since they are under NASA review.

SATELLITE OBSERVATION SYSTEMS AND ENVIRONMENTAL

FORECASTS, PROGRESS AND PROSPECTS

Vincent J. Cardone
Assistant Director
CUNY Institute of Marine and Atmospheric Sciences
The City College, Wave Hill, Bronx, New York, U.S.A.

Introduction

One decade has elapsed since the inauguration of the world's first operational satellite system with the launch of ESSA 1 and ESSA 2 in February 1966. Since then research and development in sensor technology have been incorporated rapidly into improved operational systems. The current operational system consists of satellites in near earth polar orbits (the NOAA series) and in geostationary orbits (the GOES series). The former provide daily visible and twice daily infrared global coverage from medium resolution (4 km visible, 7 km infrared) and high resolution (1 km) radiometers. The NOAA-2 satellite, launched in October 1972, had a multichannel infrared radiometer that provided quantitative data on the vertical profile of atmospheric temperature. An improved version of this vertical temperature profiling radiometer (VTPR) on NOAA-4 currently returns about 1000 profiles per day. Data from the NOAA series in stored form reach central readout and processing facilities, while about 200 ground stations in about 100 countries receive medium resolution local data in real time from the Automatic Picture Transmission (APT) system.

There are two satellites in the GOES (Geostationary Observational Environmental Satellite) system which are designed to provide a nearly continuous view of the Western Hemisphere in the visible (effective 1 km resolution) and infrared (effective 8 km resolution). (GOES-1 was launched in October 1975, following on SMS-1 in May 1974 and SMS-2 in February 1975). Images made each 30 minutes reach most NOAA weather forecast offices within 20 minutes. Winds inferred from cloud motions in the upper and lower troposphere are transmitted through the global data processing communication system routinely in real time.

The NOAA and GOES satellites are supplemented by data from
the U.S. Air Force Defense Meteorological Satellite Program (DMSP)
and by NASA experimental satellites such as the NIMBUS series. The
current NIMBUS 5 and 6 systems carry scanning microwave radiometers
that are able to map sea ice through clouds, and measure rainfall
intensity and several advanced sounders including a microwave
spectrometer for vertical temperature profiling through clouds.
By the late 1970's the GOES system will become truly global and a
third generation polar orbiting system, TIROS N, will become operational.

The environmental satellite system represents a substantial
commitment of U.S. national resources. According to Jensen(1975)1/
nearly 1 billion U.S. dollars were expended on this activity between
1965 and 1975, or about 15% of the entire federal investment in
meteorological programs. In addition, the annual investment in the
satellite program has nearly tripled in the same time, while the
overall federal meteorological program has grown by only about 14%
in the same period, after adjustment for inflation.

In view of the current and planned substantial commitment of
resources to satellite centered earth observation systems, it is
appropriate to assess the impact of such systems to date and the
prospects for future impacts. This report attempts a concise
assessment with particular emphasis on past, present, and projected
future impacts of satellite systems on operational environmental
forecast operations.

Operational Environmental Satellite Accomplishments

Without question, the satellite has had a profound influence
on the sciences of Meteorology and Oceanography. Data provided
by the operational system thus far have been used in literally
thousands of scientific studies and have strongly influenced operational activities worldwide. In a recent review of meteorological
satellite activities in the U.S. between 1971 and 1974 Allison et
al(1974)2/ cite nearly 200 contributions from that four year period
alone. The pace of scientific investigation involving satellite
data has increased here since 1974 with a comparable volume of work
emanating from the international scientific community as well.

A survey of the literature in satellite accomplishments readily
reveals where most of the "action" has been. In order of quantity,
papers have dealt with sensor technology, the development of techniques to extract geophysical parameters from sensor measurements,
the use of derived geophysical data in basic scientific studies,
and the use of the data in global scale monitoring of the atmosphere

and oceans and finally the application of the data in environmental
synoptic analysis and forecasting. Particularly outstanding accomplishments in the area of monitoring include the operational implementation of global sea surface temperature analysis and sea ice
cover, and the early detection and continuous tracking of tropical
cyclones. Basic studies of cloud, snow and ice cover have revealed
new features of the global general circulation and of the earth-
atmosphere-ocean heat budget. Such knowledge should eventually
contribute to a better understanding of the earth's climate and its
changes.

Accomplishments in the application of meteorological satellite
data in analysis and forecasting are documented in the report by
Anderson et al(1969, with supplements 1971, 1973)3/ which is itself
based upon 75 individual contributions. That report and the many
papers published in this applications area since 1973 reveal that
the overwhelming activity in this area has been the development
of diagnostic relationships between satellite visible and infrared
imagery and meteorological and oceanographic phenomenon. Such
relationships are well developed for large scale features such as
extratropical cyclones and fronts, jet streams and related upper
atmospheric flow patterns, and for a wide range of tropical disturbances. High resolution satellite imagery has been shown to be
remarkably revealing of smaller scale local phenomena such as fog,
turbulence as revealed by lee waves, mesoscale vortices, sea breeze
circulations, soil moisture, snow depth, and a host of convective
phenomena ranging from cellular cloud patterns to thunderstorms,
local heavy rainfall, hail, squall lines, high wind gust zones,
and tornadoes.

There exist no quantitative measures of the impact of such
diagnostic relationships on improved weather forecast accuracy
but there is considerable reason to believe that considerable
impacts are possible in very short range forecasts, here defined
as the 1-24 hour range. The GOES system is providing, for the first
time, satellite data of quality and timeliness necessary to implement the well developed diagnostic relationships in a "nowcasting"
sense. Evidence is mounting rapidly that where the data is available
the GOES system will become an important if not dominant component
of the "man-machine mix" which consists of the very short range
operational environmental forecast interval. This promise is supported by several papers on the subject presented at the recent
conference in Albany, New York on Weather Analysis and Forecasting,
sponsored by the American Meteorological Society (AMS, 1976)1/,
and by the rapidity with which forecasters have accepted and come
to rely on GOES data in this country after installation of the GOES
picture receiving equipment. (This aspect of satellite data impact
will likely be discussed in more detail at this conference by
Wasserman and Stevenson.)

It should be stressed that the techniques involved in the use of satellite data for short range forecasting rely on pattern recognition and are thus quite subjective. The role of the man in the man-machine mix will therefore remain significant and possibly increase in this context. It is thus apparent that the benefits to be derived from operational earth-observation systems will depend critically upon the effectiveness of educational and training activities needed to transfer the necessary skills to present and future operational personnel who will be exposed to the GOES data as it becomes available. (Editor's note: Professor Cardone proposed that the ARI discuss, identify, and recommend the implementation of educational and training activities that can effectively transfer this new technology.)

The identification of a well defined technological benefit such as the above from a well defined component of the earth observation system (GOES) should be accompanied by a quantification of the economic and human benefits derived therefrom. This is admittedly a difficult task. In the U.S., the benefits will be associated with differential improvements in forecast accuracy rather than from cost savings made at the expense of other components of the earth-observation system. For example, the demonstrated ability of GOES satellite data to detect, track, and monitor tropical cyclones has not resulted in a decrease of alternate surveillance techniques. When a hurricane threatens a U.S. coast it is monitored almost continuously by reconnaissance aircraft and radar at considerable expense and some human risk. Indeed, the U.S. aircraft reconnaissance program and radar network are scheduled to expand further. However, in countries without such facilities the differential improvement in forecast accuracy should be larger and the benefit/cost ratio easier to estimate. (Editor's note: Professor Cardone proposed that the ARI should discuss this issue and recommend the initiation of benefit/cost studies related to the impact of the GOES system on very short range environmental forecasting on a regional basis.)

Remaining Data Deficiencies

While the proven accomplishments of the environmental satellite system probably justify its cost, the present global operational data base is still seriously deficient for many forecast applications. Satellite observation systems appear to have made little impact on environmental forecasts (weather and state of sea variables such as waves, currents, surges, temperature, sea ice, etc.) beyond the very short range, and have failed to extend the range of forecasts with useful skill beyond the present 3-5 day limit. Yet it is well established that considerable economic benefit can be associated with significant improvements in forecast accuracy in the 1-5 day time frame, particularly in areas of energy resource utilization, agriculture, and the expanding offshore activities associated with resource exploration, extraction and management and the maritime industry.

Modern environmental forecast systems for the 1 to 5 day time frame are built about complex dynamical models solved numerically in an objective automated procedure on large computers. The evidence suggests that forecasts made in such a system benefit little from the man-machine mix. Numerical environmental prediction is an *initial* *value* problem. For example, numerical weather prediction models require typically a specification of the atmospheric pressure, wind, temperature, and humidity from the earth's surface through the stratosphere such that horizontal scales of a few hundred of kilometers are resolved. Prior to the development of satellite sounders, satellite data were ill suited to this specification. Nevertheless, subjective techniques have been developed that transform satellite derived cloud data into a form useable by the models, usually by the creation of bogus pseudo-observations of wind or pressure. This technique is employed at some operational centers, particularly in the Southern Hemisphere, but its utility has not been well documented.

The satellite sounder measurements (VTPR) should be expected to provide useable temperature data of value in improving the initial value specification for numerical forecasting. However, objective tests of the role of NOAA-2 VTPR data in the NOAA National Meteorological Centre (NMC) forecast model (Bonner et al 1976)5/ showed that the data failed to measureably improve 1 to 3 day forecasts. Similarly discouraging results have been obtained at other national forecast centers that have attempted to use the data operationally.

Limitations in weather model forecast skill in this time frame, automatically limit skill attainable in the many forecast parameters derived from the model output. This is particularly true of "state of sea" forecasts which are produced typically from empirical, statistical or dynamical models that are driven by the output of weather forecast models.

A determination of the cause for those negative results is quite difficult because numerical forecasts are limited in accuracy not only by initial data deficiencies but also by misrepresentation of physical processes in the models and by restrictions on grid resolution that cause mathematical truncation errors and misrepresentation of sub-grid scale processes. Indeed, at the NOAA-NMC, which has used and evaluated numerical weather forecast models since the mid 1950's, the slow but steady improvement in model forecast skill evident over the U.S. and Canada in the past two decades can be attributed largely to improvements in the formulation and resolution of the models themselves rather than to improvements in initial data from satellites or otherwise. Experiments recently

conducted at NMC (Brown, 1976)6/ suggest strongly that the existing
data base over the U.S. could support further improvements in
forecast skill over the U.S. by the implementation of still higher
resolution models. One would suspect however that for areas such
as western U.S. and Europe errors in the initial state and model
deficiencies contribute about equally to forecast errors in 1-3
days while in the Southern Hemisphere forecast accuracy is limited
mainly by very large errors in the initial value specification.

The issue of sounder data utility is crucial to the future
role of operational satellite systems in environmental forecasting.
The sounders being considered for the TIROS N system are already
designed, candidate instruments are being flown on NIMBUS 6, and
no technological breakthroughs in remote atmospheric sensing appear
to be on the forseeable horizon. As a result, an intensive study
is underway at the Goddard Institute of Space Studies (GISS) in
New York. The study makes use of a general circulation model that
has been shown to possess skill in 1 to 5 day forecasts comparable
to operational weather models. In the study, specially enhanced
global data sets are used to make parallel sets of global forecasts
(with and without sounder data) with the forecasts verified object-
ively over the United States and Europe. The most recent results
reported by Halem (1976)8/ showed that when VTPR data from NOAA-4
were combined with data from the NIMBUS 6 sounders and assimilated
into the model in a "four dimensional" frame, there resulted only
a mean small net increase in skill in 48 to 72 hours forecasts and
a somewhat larger beneficial impact in about one third of the cases
studied. The effectiveness of the improvements shown in the skill
score appears to be slight as shown by Spar (1976)10/ in a synoptic
review of the results of the forecast runs. The consensus of current thinking is that full realization of the global satellite
system in short and medium range forecasting will require improvements in the predictive models, improvements in the methods of
data processing, and improvements in the accuracy of satellite
data.

One clear deficiency of sounder derived temperature profiles
relates to the so-called "reference level". Sounders provide estimates of the distribution of temperature as a function of pressure
but the computation of the vertical distribution of pressure as a
function of height, which is needed for the initial value specification in models, requires a knowledge of the pressure at one level.
Radiosonde stations use measured surface pressure as this reference
level. Over land areas, the surface pressure is usually known well
enough to serve as a reference level for VTPR data but over the
oceans uncertainties in surface pressure can cause considerable
errors in the vertical pressure-height specification.

It does not appear to be possible to sense surface pressure
remotely from a satellite; however, it can be retrieved from an

accurate knowledge of the surface winds over the oceans and a
small number of direct measurements of pressure. The analysis
of data from the combined microwave radar-radiometer on Skylab
by Cardone, Young, Pierson, Moore et al(1976)7/ established proof
of the concept of sensing marine surface winds from space, and
a microwave scatterometer and scanning five channel microwave
radiometer will be operated in an operational manner on the spacecraft SEASAT-A to be launched in 1978.

While SEASAT-A is considered an oceanographic satellite, it
may well have a significant impact on atmospheric forecasting
(and thereby enhance its value to oceanographic forecasting). The
next section outlines studies underway to test this concept.

SEASAT-A may also satisfy some additional outstanding data
needs. The international scientific community has also recently
identified meteorological and oceanographic measurement needs
within the context of the first GARP (Global Atmospheric Research
Program) Global Experiment (FGGE) to be conducted late in 1978
(after the launch of SEASAT-A). The prime objective of FGGE is
to define the three dimensional states of the atmosphere for studies
of the global circulation with general circulation models. The
FGGE requirements for surface variables are listed in Table 1.

A second objective of GARP is the investigation of the physical
basis of climate. Tentative measurement needs for this objective
have been formulated by the GARP Joint Organizing Committee for the
atmosphere, oceans and cryosphere. A partial list of tentative
specifications for Ocean Monitoring (including sea ice) for climate
study is also shown in Table 1. The promise of SEASAT-A to provide
the first wind stress, all weather sea surface temperature, sea ice
extent, and ocean current data makes SEASAT-A uniquely contributory
to both GARP objectives.

Data Utility Projections

This section discusses two of the more obvious and important
implications of a world wide capability in satellite surveillance
of ocean surface wind. The first concerns the improvements in
1 to 5 day forecasts made by numerical weather prediction models
that would result from an improved initial value specification
of the marine atmospheric boundary layer. The second concerns
the application of the new capability to the analysis and prediction
of tropical cyclones.

Extratropical Numerical Weather Predictions. One of the
readily apparent ways to use the global wind data involves the
use of the improved wind analysis in the planetary boundary layer
to define the surface pressure reference level over the oceans that

Table 1   A summary of GARP observing systems requirements relevant to SEASAT (McCandless and Cardone, 1976)

I. Surface Variables – For the first GARP Global Experiment (FGGE)

Variables	Space Resolution	Time Resolution	Period	Accuracy Desired
Surface Pressure	500 km	12 hr.	FGGE	± 3 mb
Sea Surface Temperature	500 km	12 hr.	FGGE	± 1°C
Wind Speed	500 km	12 hr.	FGGE	± 3m/sec
			FGGE	± 2m/sec (tropics)
Relative Humidity	500 km	12 hr.	FGGE	± 30%

II. Surface Variables – For Ocean (including sea ice) Monitoring and Climatic Studies (tentative)

Variables	Space Resolution	Time Resolution	Period	Accuracy Desired
Sea Surface Temperature	200 km	1-5 days	FGGE	± 0.5°C
Wind Stress	200 km	1-5 days	1-2 years	± .1 dynes/cm^2
Surface Currents	200 km	5-10 days	1-2 years	± 2cm/sec
Sea Ice Extent	50 km	5-15 days	FGGE + long term	yes/no
Sea Ice Thickness	200 km	15-30 days	FGGE + long term	10-20%
Sea Ice Melting	50 km	5 days	FGGE	yes/no
Sea Ice Drift	400 km	1 day	FGGE	5 km
Precipitation Over Oceans	500 km	5 days	FGGE	4 levels of discrimination

is required to use indirect sounding data accurately to specify the pressure distribution in the upper atmosphere.

Until real data becomes available from SEASAT to test ideas such as this, an effort is currently underway at various centers in this country to simulate the improvements in numerical weather prediction to be expected when SEASAT-A becomes operational. These studies typically use sophisticated numerical models of the atmosphere analogous to the operational numerical weather forecast models, to simulate the atmosphere over periods corresponding to weeks and months, to simulate the error characteristic of conventional meteorological observing systems and then to assess the impact of one or more alternative observation systems.

The results of one such study recently performed at the NASA Goddard Institute for Space Studies in New York is shown in Figure 1. In the study, a global general circulation model was used to simulate a 16 day period that served as the reference atmosphere state for this experiment. The reference state was used to simulate indirect soundings that would have been made by two polar orbiting satellites such as NIMBUS, with realistic errors imposed on the simulated soundings. The reference data were also used to simulate near-surface over-ocean winds that would have been measured by a fully operational SEASAT system consisting of three satellites providing global 12 hour repeat coverage. The root mean square errors imposed on the near surface winds were ±2 m/sec for the magnitude and ±15° for the direction.

The run labelled <u>control</u> in Figure 1 shows the ability of the indirect sounding data alone to control the departure of the model atmosphere generated from perturbed initial conditions from the reference state. The error shown is vertically averaged over the entire depth of the model atmosphere. The run labelled SEASAT in Figure 1 is the result of assimilation of both simulated sounding data and surface layer wind data. The reduction of error in this case is substantial not only over the ocean but also over downstream land areas. This is an encouraging result and more realistic simulation studies of this type are underway and are expected to lead to the development of optimum techniques for the assimilation of real satellite termed wind data into numerical forecast models.

<u>Analysis and Prediction of Tropical Cyclones</u>. The potential impact of SEASAT-A on tropical cyclone analysis is suggested by Figure 2, which compares radar backscatter measurements made from SKYLAB in tropical storm Christine and surface winds at the locations of the cells scanned. The winds were specified from a numerical model of the planetary boundary layer in hurricanes which is initialized from detailed data obtained by aircraft reconnaissance in the storm at the time of the SKYLAB pass.

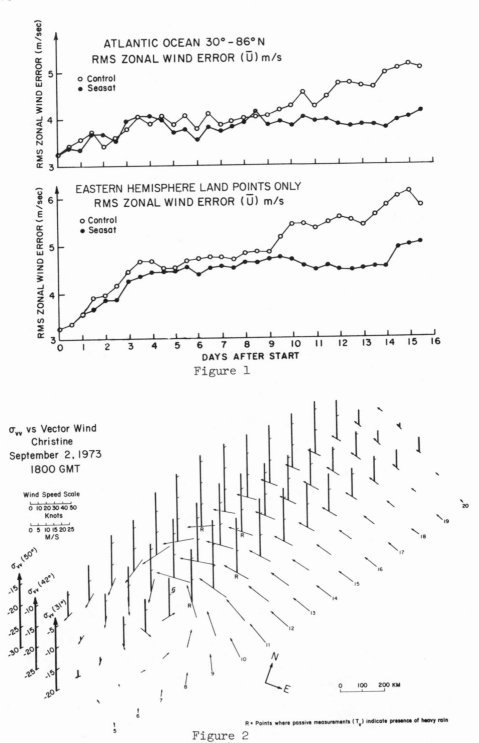

Figure 1

Figure 2

When SEASAT-A scans a tropical cyclone, many more radar cells than shown in Figure 2 would be available and the passive microwave and visible and infrared overlapping measurements could be used to locate the cells that were uncontaminated by heavy rain. With the surface vector wind distribution around the cyclone known, the same theoretical model could be used to derive basic characteristics of the cyclone sought by aircraft, such as eye pressure, radius to maximum wind, maximum sustained surface winds and the areal extent of the hurricane and gale force winds. Indeed, after a sufficient number of tests of this procedure with simultaneous aircraft data, the method may be used on a routine basis to define the character-istics of tropical cyclones world-wide. In countries with estab-lished reconnaissance programs it would then be possible to effect selective cost savings from the satellite system.

The spacecraft data should also impact tropical cyclone forecasting. Recent experiments in hurricane prediction with time dependent multi-level high resolution dynamical models applied to the tropical atmosphere show great promise and are already being tested on a quasi-operational basis by NOAA. The success of the tests to date suggest that the dynamical method may be able soon to replace current methods that rely mainly on climatology and statistics.

The major problem confronting the implementation of the numerical models is an adequate specification of the initial conditions from the typically very sparse data base characteristic of the tropical and sub-tropical oceans. However, recent simulation studies Anthes have demonstrated that wind observations in and around tropical cyclones from near the surface would be quite effective in controlling the growth of forecast errors, and these are precisely the kind of observations that SEASAT-A could provide.

Further Study. Simulation studies, such as the single SEASAT wind study reported above, can provide valuable insight into the potential of proposed earth-observation systems. They can also provide guidance as to how to best use data from existing earth-observation systems. The program of "SEASAT Simulation Studies" begun at GISS will be greatly expanded this year, and will lead to tests of data impact with actual SEASAT data obtained after launch.

(Editor's note: Professor Cardone recommended that the ARI endorse data utility studies which specifically focus on the impact of operational earth observation system on short to medium range environmental forecast systems). Inasmuch as the impacts will likely vary regionally, it is important that as many groups as possible engage in such studies. The development branches of the many national numerical forecasting centers are logical choices for the location of such studies. It is quite possible that such

studies can provide information needed to determine the fraction
of total available resources that should be assigned to model
development, development of in situ observation systems, and
development of satellite systems, to improve environmental forecasts.

REFERENCES

1/ American Meteorological Society, "Sixth Conference on
        Weather Analysis and Forecasting". Preprints, May 10-13,
        1976, Albany, New York.

2/ Allison, L.J., et al, "Meteorological Satellite Accomplish-
        ments". Reviews of Geophysics and Space Physics, U.S.
        National Report 1971-1974, July, 1975, Vol. 13: 737-746.

3/ Anderson, R.K., Ashman, J.P., Bittner, F., Farr, G.R.,
        Ferguson, E.W., Oliver, U.J., Smith, A.H., Skidmore, R.W.
        and Purdom, J.F.W., "Application of Meteorological
        Satellite Data in Analysis and Forecasting". ESSA TR-NESC
        51, NOAA, Washington, D.C., 1969, 1971, 1973, 232p.

4/ Anthes, R.A., "Data assimilation and initialization of hurri-
        cane prediction models". Journal of the Atmospheric
        Sciences, 1974, 31: 702-719.

5/ Bonner, W.D., Lemar, P.L., VanHaaren, R.J., Desmarais, A.J.,
        O'Neil, H.M., "A test of the impact of NOAA-2 V.T.P.R.
        Soundings on operational analyses and forecasts". NOAA
        Technical Memo N.W.S. - N.M.C. - 57, Washington, D.C.,
        1976, 43p.

6/ Brown, J.A., "Modelling and the man-machine mix at the National
        Meteorological Center". Wasington Papers presented at
        the WMO Symposium on the Interpretation of Broad Scale
        NWP Products for Local Forecasting Purposes, Warsaw,
        11-16 October, 1976, W.M.O. No. 450.

7/ Cardone, V.J., Young, J.D., Pierson, W.J., Moore, R.K., et al,
        "The measurement of the Winds Near the Ocean Surface with
        a Radiometer-Scatterometer on Skylab". Final Report on
        EPN 550, Contract No. NAS-9-13642, CUNY Institute of
        Marine and Atmospheric Sciences at the City College and
        The University of Kansas Remote Sensing Laboratory, 1976.

8/ Halem, M., "Report on a Four Dimensional DST Temperature Sound-
        ing Impact Test". Informal GISS Report (C.6 6.4),
        Goddard Institute for Space Studies, New York, 1976.

9/ McCandless, S.W. and Cardone, V.J., "SEASAT-A Oceanographic
      Data System and Users". Paper 76-061, 27th Congress of
      International Astronautical Federation, Anaheim, Cal.,
      U.S.A., 10-16 October 1976.

10/ Spar, J., "A Synoptic Review of Some Results from the DST-5
      Experiments". Informal GISS Report, Goddard Institute
      for Space Studies, New York, U.S.A., 1976.

SOME CONSIDERATIONS RELATIVE TO THE MAKING OF GLOBALLY

UNIFORM RESOURCE INVENTORIES BY REMOTE SENSING

Robert N. Colwell
Professor of Forestry and Associate Director of
Space Sciences Laboratory, University of California
Director, Berkeley Office, Earth Satellite Corporation

Introduction

Because the world's population is rapidly increasing, and because the demand for a higher standard of living likewise is increasing throughout the globe, the global demand for various natural resources also is rapidly increasing. This increasing demand is occurring at the very time when the supply of some of these resources is rapidly dwindling and the quality of others (and of the environment associated with them) is rapidly deteriorating, also on a global basis. As a result there is an urgent need to exercise, on each part of the globe, the wisest possible resource management and environmental control. An important first step toward the meeting of such a need is that of obtaining globally uniform inventories, of suitable accuracy and at suitably frequent intervals, so that the resource manager will know at all times how much of each kind of resource is present in each portion of the area for which he has management responsibility and also the condition of that resource. Within this context, all components of the "natural resources complex" are of interest (including timber, forage, agricultural crops, livestock, soils, water, minerals, oceanographic resources, and atmospheric resources) as are the environmental factors associated with them.

NASA-sponsored research that has been conducted by the present writer and others over the past several years strongly indicates the feasibility of our developing soon an operational satellite system which, through remote sensing of the surface of the Earth, will facilitate the making of globally uniform inventories of the Earth's resources and thus greatly facilitate our achieving wise resource management and effective environmental control on a global

183

basis. Despite the emphasis that will be given in this paper to
data acquisition from globe-circling satellites, however, the term
"remote sensing" will be used somewhat more broadly to include the
acquiring of information not only from spacecraft but also from
aircraft. Although the sensors used, whether they be conventional
cameras, optical mechanical scanners, or side-looking radar equipment, will, of course, be situated at a distance (i.e., remotely)
with respect to the area that is being sensed, possibilities will
be considered for augmenting these basic records with ancillary
data acquired either from direct field observation of selected
spots 'or through the use of instrumented field stations. The
property that we will consider as being sensed remotely will in all
cases be the amount of energy that is radiating from each of the
various resource-related features on Earth that are to be identified, measured, or otherwise analyzed. In most instances it will
be electromagnetic energy that is being sensed, i.e., energy
(including light and thermal energy) that is travelling at a
velocity of approximately 186,000 miles per second in a repetitive,
harmonic, wave-like motion.

As suggested by the title of this paper, its primary purpose
is to present some considerations relative to the use of remote
sensing as an aid in making globally uniform inventories of earth
resources. Reasons why the resource inventories should be made
globally are the same as those embodied in other aspects of the
"one world" concept, i.e., the concept that foresees the time in
the near future when man's very survival on this globe will depend
upon his pooling of the total supply of the Earth's resources in
order to meet the total human demand for such resources. This in
turn suggests the desirability of pooling the information pertaining to these resources. Most such information consists of inventories which are made periodically, through a remote sensing-aided
"monitoring" process, in order to provide current information as to
the type and condition of each resource component, area-by-area.
Reasons why the resource inventories should be uniform stem from
the need that is becoming of increasing importance to match the
total global supply of any given resource to the total global demand
for it. The fact that meaningful totals cannot be derived from
non-uniform components assumes great significance when the objective is to compile a global supply vs. demand "balance sheet" which
will facilitate our placing all components of the Earth's resources
complex under the wisest possible management.

Considerations With Respect to Timeliness of the Concept

It was the great French scientist and philosopher, Victor Hugo,
who said more than a century ago, "There is nothing in this world
as powerful as an idea whose time has come". There is steadily
increasing evidence that now is the time to implement one such

powerful idea, namely, that of using satellite-derived data, in conjunction with ancillary data from a variety of sources, to inventory periodically, and thus to monitor, the Earth's resources and its associated environmental state.

Realization of the timeliness of the concept has been rapidly developing in recent years as evidenced by the following events:

a) In 1969 at a symposium that was organized by Welch and Eakins, two papers were presented (Pecora, 1969; Colwell, 1969) which spoke of the increased need for, and increased feasibility of, remote sensing-derived uniform resource inventories, whether on a state, regional, national, or global level.

b) In 1971, NASA received and funded a Landsat (then called "ERTS") proposal for "The Uniform Mapping of Natural Ecosystems and Environmental Complexes from Space Acquired Imagery" (Poulton, Welch, and Colwell, 1971). Work done in conformity with this proposal led to a second award from NASA to the same investigators two years later for follow-on studies based primarily on imagery acquired by the SKYLAB astronauts.

c) During this same period ideas were being formulated jointly by three United States Government Agencies (Department of Agriculture, National Aeronautics and Space Administration, and National Oceanographic and Atmospheric Administration) for the conduct of "LACIE", the presently on-going Large Area Crop Inventory Experiment as recently summarized by MacDonald (1975).

d) In the inaugural lecture presented at the second biennial meeting of the FAO Committee on Forestry in Rome, Italy, on 22 May, 1974, the following specific comments were made on the topic "Remote Sensing as an Aid in Making a Globally Uniform Forest Inventory" (Colwell, 1974):

"In order for a global inventory of forest resources to be of value it should be made in a sufficiently short period of time to ensure the global comparability of inventory data. Otherwise serious inaccuracies could develop (especially with the passage of long periods of time while various parts of the inventory were being made) because of (1) the extensive depletion of forest resources in some geographic areas as a result of logging or the depredations of fire, insects, and diseases; and (2) the extensive accretion of forest resources in other geographic areas as a result of ingrowth and the establishment of new timber stands by either natural

or artificial means. Not only should the inventory
be made in a suitable short period of time, but also
to globally uniform standards. Otherwise problems
relative to the comparability of inventory data could
be so enormous as to negate any efforts we might make
in using such data to develop intelligent forest
management plans on a global basis. Until recently
it was virtually unthinkable that man would ever
have the means for making a globally uniform inventory
of forest resources. The first indication that foresters
were on the verge of an important breakthrough in this
regard came shortly after the dawning of the space
age when it was found possible by remote sensing to
obtain on a single space photograph an essentially
uniform look at an almost unbelievably large forest
area (approximately 3 million hectares) and to discern
thereon a very significant amount of detail (sufficient
for example to position timber boundaries to within
about 100 meters). This important capability improved
with successive space shots and hit a new plateau with
the launch of the world's first Earth Resources
Technology Satellite (ERTS-1) on 23 July, 1972. The
highly successful operations of that vehicle's 'cameras'
and other sensors has continued since then, even up
to the present time".

e) With reference to another category of resources, viz.,
agricultural resources, it was proposed at a recent United
Nations FAO-sponsored World Food Conference that the nations
of the world cooperate in a Global Agricultural Information
Project. At last report 45 countries had signed an agreement which would implement that proposal.

f) In his presentation earlier this year to another United
States Senate Group known as the Humphrey Panel, Dr.
Archibald Park, noted authority on applications of satellite-derived remote sensing data to resource inventory and
management, was asked the question, "Would it be possible
to make better estimates of crop production in other parts
of the world even without a cooperative agreement of the
countries involved?" His answer was (Park, 1976):
"Unequivocally, yes. The estimates could be even better
with cooperation, but the lack of such agreement does
not prohibit better estimates than are currently available
for many countries of the world".

g) Senator Frank Moss, Chairman of the Aeronautics and Space
Sciences Committee of the United States Senate, has recently stated (Moss, 1976):

"The new technologies available to us today as a
result of the space program make it apparent that
the interpretation of resource data and the management
of resources will vastly aid mankind in the preservation, through efficient use, of earth resources. . .
Few space efforts have the almost universal appeal of
NASA's earth resources surveys from space. The
first system dedicated to surveying the earth's
resources is the Landsat system initially launched
in 1972. The success of the Landsat mission has
exceeded the most optimistic predictions; it has
laid the groundwork for a global inventory of man's
resources. . . It would be sheer negligence to
disregard the potential available to us. We must
establish an entire Global Resource Information
System and on August 23 (1976) when the Congress
reconvenes , I intend to introduce legislation
which will establish such a Global Resources
Information System. . . Scientists have learned
in a short time how to aid diverse interests
including agriculture, forestry, fishing industries,
land use planners, geologists, hydrologists, oceanographers, and those concerned with pollution monitoring, highway planning, coastal zone monitoring
and demography".

h)   Senator Wendell Ford, Chairman of Moss's Subcommittee
     on Aerospace Technology and National Needs, has stated
     unequivocally, when speaking of an operational Landsat
     system (Ford, 1976): "Today the world is waiting for
     a commitment to such a system. . . I urge my colleagues
     to give their support to the general effort to create a
     Global Resources Information System".

Potential Users of Globally Uniform Resource Inventories

     One of the primary users of such inventories would be the
resource manager, himself. The usefulness to the resource manager
of accurate periodic information about the status of the "resource
complex" within the area for which he has management responsibility
is most clearly seen when we consider that the wise management of
earth resources usually requires the implementation of a three-step
process: inventory, analysis, and operations.

     In the inventory step a determination is made as to the
amount and quality of each type of earth resource that is present
in each portion of the area to be managed. In the analysis step,
certain management decisions are made with respect to these resources. This is accomplished for each portion of the area by considering, on the one hand, the nature of its resources (as previously

Table 1   Basic and applied data sought through multispectral
          remote sensing by workers in various disciplines.

I. Foresters and Agriculturists
   A. Basic

   1. Amount and distribution of the "biomass"
   2. Nature and extent of important "ecosystems"
   3. Amount and nature of energy exchange phenomena.
   B. Applied

   1. The species composition of vegetation in each area studied
   2. Vigor of the vegetation
   3. Where vegetation lacks vigor, the causal agent
   4. Probable yield per unit area and total yield in each
      vegetation type and vigor class
   5. Information similar to the above for livestock, wildlife
      and fish.

II. Geologists
    A. Basic

    1. Worldwide distribution of geomorphic features
    2. Energy exchanges associated with earthquakes and
       volcanic eruptions.
    B. Applied

    1. Location of certain or probable mineral deposits
    2. Location of certain or probable petroleum deposits
    3. Location of areas in which mineral and petroleum deposits
       of economic importance probably are lacking.

III. Oceanographers
     A. Basic

     1. Diurnal and seasonal variations in sea surface temperatures
        and subsurface temperatures
     2. Vertical and horizontal movements of ocean currents and
        individual waves
     3. Global, regional and subregional shoreline characteristics
        and the changes in these characteristics with time
     4. Diurnal and seasonal movements of fish, algae and other
        marine organisms.
     B. Applied

     1. The exact location, at a given time, of ships, icebergs,
        tsunamis, storms, schools of fish and concentrations of kelp
     2. The location of ocean beaches suitable for recreational
        development
     3. The rate of spread of water-pollutants and the kind and
        severity of damage caused by them.

IV.Meteorologists
  A. Basic

  1. Diurnal and seasonal variations in cloud cover, wind velocity
     and air temperature and humidity in relation to topography
     and geographic locality
  2. Accurate statistical data on the points of origin of storms,
     the paths followed by them, their intensities, and their
     periods of duration.
  B. Applied

  1. Early warning that a specific storm is developing
  2. Accurate tracking of the storm's course
  3. Accurate periodic data on air temperatures, humidity and
     wind velocity
  4. Accurate quantitative data on the response of the atmosphere
     to weather-modification efforts.

 V.Hydrologists
  A. Basic

  1. Quantitative data on factors involved in the hydrologic
     cycle (vegetation, snow cover, evaporation, transpiration,
     and energy balance)
  2. Quantitative data on factors governing climate (weather
     patterns, diurnal and seasonal cycles in weather-related
     phenomena).
  B. Applied

  1. The location of developable aquifers
  2. The location of suitable sites for impounding water
  3. The location of suitable routes for water transport
  4. The moisture content of soil and vegetation.

VI.Geographers
  A. Basic

  1. Global, regional and subregional land use patterns
  2. The nature and extent of changes in vegetation, animal
     populations, weather, and human settlement throughout
     the world.
  B. Applied

  1. The exact location, at any given time, of facilities for
     transportation and communication
  2. The interplay of climate, topography, vegetation, animal
     life and human inhabitants in specific areas
  3. The levels of economic activity and the purchasing habits
     of inhabitants in specific areas.

established in the inventory phase) and, on the other hand, the "cost-effectiveness" of each management alternative that might be exercised with respect to these resources. In the operations step, the resource manager implements each decision that has been made in the analysis step (e.g.,the decision to apply the appropriate fertilizer in certain mineral-deficient parts of an agricultural area or to cut only the over-mature trees in a certain portion of the forest area, or to practice "deferred rotation grazing" in certain parts of a rangeland area). The resource manager, especially when dealing with such renewable natural resources as agricultural crops, timber, and forage, is likely to find that these resources are highly dynamic rather than static and that he needs to obtain a new inventory periodically - a process known as "monitoring".

A second category of users would be those at the national and/or international level who, under auspices of FAO, NATO, or other appropriate international agency could suggest, or require compliance with, measures which were based on such inventories and which were designed to correct obvious imbalances in the production and/or consumption of certain critical resources.

Further insight with reference to these users and the actions which they might take will be provided by the sections which follow.

The Specific Information Requirements of These Users

In this portion of our discussion, reference to Table 1 should prove helpful. In Table 1 we have listed the kinds of information sought (and hopefully obtainable through the use of remote sensing derived information) by workers in various disciplines that entail the inventory and management of the Earth's natural resources. Listed in that table, under each discipline, are both the basic and applied data sought by the workers in that discipline*.

---

* Basic research by definition seeks to understand the fundamentals on which a particular science rests; the types of data derived therefrom are therefore called "basic data" in Table 1. On the other hand, applied research seeks to solve specific problems in a direct manner. The types of data derived from applied research (and subsequently from operational systems and procedures based on such research) are termed "applied data" in Table 1.

In considering the informational requirements of scientists in any of the disciplines listed in Table 1, we need to provide a level of detail beyond that given in the brief listing that is contained in that table. Realization of this fact has prompted the author to prepare three amplifying tables which pertain specifically to the information requirements of those who are concerned with the management of <u>vegetation</u> resources. The first of these (Table 2) dealing with the informational requirements, themselves, assumes its greatest significance when viewed in concert with Tables 3 and 4, which are discussed in succeeding sections.

Judging from Table 2, there are only four main categories of vegetation for which information is sought, viz., agricultural crops, timber stands, rangeland vegetation, and brushland vegetation. Starting with the left-hand column of that table we see that, by and large, the users of agricultural crop data need only six categories of information, viz., crop type, crop vigor, crop-damaging agents, crop yield per acre by type, crop acreage by type, and total yield. Proceeding to the other three columns in Table 2 we note that essentially these same six categories of information likewise are the ones sought by the managers of timber lands, rangelands, and brushlands, respectively.

Table 3 provides a listing of the various agencies and groups, within the United States that desire information about vegetation resources. The counterparts to these agencies are found in most of the countries of the world. Perhaps our first observation, upon examining that table, is that we can logically list these many users under the same four column headings as were used in the previous table, viz., the agencies and groups concerned primarily with agricultural crops, timber stands, rangeland forage, and brushland vegetation, respectively. This is, indeed, the case.

In Table 4 as in the two which preceded it, the same headings can be used for the four vertical columns. At the risk of some oversimplification, the table lists six time intervals that are indicative of the frequency with which various kinds of information about the vegetation resource are needed (10-20 minutes; 10-20 hours; 10-20 days; 10-20 months; 10-20 years; and 20-100 years).

In considering relationships between the frequency with which earth resource data should be collected and the rapidity with which the collected data should be processed, the writer has found it useful to employ the term "half-life" in much the same way as it has been employed by radiologists and atomic physicists. The shorter the isotope's half-life, the more quickly a scientist must work with it once a supply of the isotope has been issued to him. One half-life after he has acquired the material only half of the original amount is still useful; two half-lives after acquisition only one quarter of the original amount is still useful, etc.

Table 2 User requirements for vegetation resource data: type of information desired.

For Agricultural Crops: Crop Type (Species and variety)	For Timber Stands: Timber Type (Species composition)	For Rangeland Forage: Forage Type (Species composition)	For Brushland Vegetation (mainly shrubs): Vegetation Type (Species composition)
Present crop vigor, and state of maturity	Present tree and stand vigor by species and size class	Present "Range Readiness" (for grazing by domestic or wild animals)	Vegetation density
Prevalence of crop-damaging agents by type	Prevalence of tree-damaging agents by type	Prevalence of forage-damaging agents (weeds, rodents, diseases, etc.) by type	Other types of information desired will depend upon primary importance of the vegetation (whether for watershed protection, game habitat, aesthetics, etc.)
Prediction of time of maturity and eventual crop yield per acre by crop type and vigor class	Present animal-carrying capacity and probable future capacity per acre by species and range condition class in each forage type	Present animal-carrying capacity and probable future capacity per acre by species and range condition class in each forage type	
Total acreage within each crop type and vigor class	Total acreage within each stand type and vigor class	Total acreage within each forage type and condition class	Same as above
Total present yield by crop type	Total present and probable future yield by species and size class	Total present and probable future animal carrying capacity	Same as above

Table 3 User requirements for vegetation resource data:
U.S. examples of agencies and groups desiring information.

Agricultural Crops	Timber Stands	Rangeland Forage	Brushland Vegetation
Federal Agencies: Agricultural Stabilization and Conservation Service; Cropland Conservation Program; Conservation Reserve Program; Agr. Conservation Program; Emergency Conservation Measures Program; Commodity Credit Corp.; Agri. Marketing Service; Statistical Reporting Service; Economic Research Service; Soil Conservation Service; Federal Crop Insurance Corp.; Farmers Home Adm.; Rural Community Development Service; Foreign Agri. Service; Famine Relief Program; Foreign Economic Assistance Program Dept. of Commerce Agri. Census Program	Federal Agencies: U.S. Forest Service Bureau of Land Mgt. plus many federal agencies listed in column 1.	Federal Agencies: U.S. Forest Service Bureau of Land Mgt. plus many federal agencies listed in column 1.	Federal Agencies: Primarily U.S. Forest and Bureau of Land Management.
State and County Agencies: Agri. Extension Service State Tax Authority	State and County Agencies: Division of Forestry; Forest Extension Service; State Tax Authority.	State and County Agencies: Livestock Report- ing Service; Range Extension Service; State Tax Authority.	State and County Agencies: Division of Forestry; Division of Beaches and Parks; Water Resource Agency; State Tax Authority.
Private Agencies: Producers of Fertilizers and Pesticides; Crop Harvesting Industry; Food Processing and Packing Industry; Transportation Industry; Food and Fiber Advertising and Marketing Industry	Private Agencies: Producers of Fertilizers and Pesticides; Logging; Wood Processing; Transportation; Wood Products; Advertising and Marketing Industry.	Private Agencies: Producers of Fertilizers and Pesticides; Meat Packing; Tanning Industry; Transportation.	Private Agencies: Hunting and Fishing Clubs; Public Utilities Commissions; Local Irrigation Dist.

Note: Deal with the first column separately from the last three columns.

Table 4   User requirements for vegetation resource data

Frequency with which the information is needed (examples only).
To convert this Table to "rapidity with which information is needed"
use "Half-Life" concept (see text).

For Agricultural Crops

10-20 minutes
Observe the advancing waterline in croplands during disastrous
floods. Observe the start of locust flights in agricultural areas.

10-20 hours
Map perimeter of on-going floods and locust flight. Monitor the
Wheat Belt for outbreaks of Black Stem Rust due to spore showers.

10-20 days
Map progress of crops as an aid to crop identification using "crop
calendars" and to estimating date to begin harvesting operations.

10-20 months
Facilitate annual inspection of crop rotation and of compliance
with federal requirements for benefit payments.

10-20 years
Observe growth and mortality rates in orchards.

20-100 years
Observe shifting cultivation patterns.

For Timber Stands

10-20 minutes
Detect the start of forest fires during periods when there is a
high "Fire Danger Rating".

10-20 hours
Map perimeter of on-going forest fires.

10-20 days
Detect start of insect outbreaks in timber stands.

10-20 months
Facilitate annual inspection of fire-breaks.

10-20 years
Observe growth and mortality rates in timber stands.

20-100 years
Observe plant succession trends in the forest.

Table 4 (continued)

For Rangeland Forage

10-20 minutes
Detect the start of rangeland fires during periods when there is
a high·"Fire Danger Rating".

10-20 hours
Map perimeter of on-going rangeland fires.

10-20 days
Update information on "Range Readiness" for grazing.

10-20 months
Facilitate annual inspection of fire-breaks.

10-20 years
Observe signs of range deterioration and study the spread of noxious
weeds.

20-100 years
Observe plant succession trends on rangelands.

For Other Vegetation (mainly shrubs)

10-20 minutes
Detect the start of brushfield fires during periods when there is
a high "Fire Danger Rating".

10-20 hours
Map perimeter of on-going brushfield fires.

10-20 days
Update information on times of flowering and pollen production in
relation to the bee industry and to hay fever problems.

10-20 months
Facilitate inspection of fire-breaks.

10-20 years
Observe changes in "Edge Effect" of brushfields that affect suitabi-
lity as wildlife habitat.

20-100 years
Observe plant succession trends in brushfields.

By coincidence or otherwise, this half-life concept applies remarkably well to nearly every item listed in Table 4. Specifically, if the desired frequency of acquisition of any given type of information, as listed in that table, is divided by two, a figure is obtained indicating the time after data acquisition by which that particular item of information should have been extracted from the data. It is true that some value will accrue even if that item of information does not become known until somewhat later. But the rate at which the value of the information "decays" is in remarkably close conformity to the "half-life" concept.

Each item that is listed in Table 4 can be placed in one of the six categories of information listed in Table 2 and identified as a user requirement for one or more of the agencies listed in Table 3. Thus a certain unity can be found in these three tables. Consequently, by studying them, not just individually but in concert, we can better appreciate the true nature of the multifaceted user requirement that we seek to satisfy by the remote sensing of vegetation resources.

Remote Sensing Capabilities and Limitations of Landsat in Relation

To the Satisfying of These Information Requirements

A summary of the highly favorable spectral, spatial, and temporal characteristics of Landsat, together with certain other favorable characteristics of this truly.remarkable sensor system, will be found in Table 5.

During recent years progress in the field of remote sensing has been characterized by the development of multiple data acquisition capabilities and multiple data analysis techniques. This progress has, in turn, given rise to the need for more clearly understanding both the advantages and the limitations that are inherent in what tentatively might be termed the "multi" concept as applied to the acquisition and analysis of remote sensing data. This concept, by whatever name, is of great significance in relation to the potential usefulness (for the inventory of earth resources) of Landsat imagery augmented as necessary with aerial photography and ground observations. Consistent with the "multi" concept, remote sensing scientists are compiling an ever increasing and highly impressive list of examples which demonstrate, collectively, that:

a) More information usually is obtainable from multistation photography than from that obtained from only one station*.

*The term "multistation photography" (not to be confused with "multistage photography") pertains primarily to successive overlapping photographs, taken along any given flight line as flown by a photographic aircraft or spacecraft. When two such photographs are studied stereoscopically the photo interpreter is better able to perceive features than if a photo from only 1 of the 2 stations were available.

Table 5   Valuable characteristics of Landsat data in relation to
         the globally uniform inventory and monitoring of earth
         resources.
         (No other vehicle-sensor system provides this important
         combination of characteristics).

1.   Multispectral Capability
     A. Senses for the optimum wavelength bands for use in the
        inventory and monitoring of most types of earth resources
        (timber, forage, agricultural crops, minerals, water,
        atmospheric and oceanographic resources).
     B. Provides high spectral fidelity within each of these bands.

2.   Multi-Temporal Capability (Provides Multiple "Looks" for
     Monitoring Seasonal Changes in Vegetation, Rate and Direction
     of Plant Succession and the Accumulation or Receding of Snow
     or Flood Waters).

3.   Constant Repetitive Observation Point (Facilitates Change
     Detection by Matching of Multi-Temporal Images).

4.   Sun-Synchronous (Nearly Constant Sun Angle) Ensures Nearly
     Uniform Lighting and Uniform Image Tone or Color Characteris-
     tics for Use in Feature Identification.

5.   Narrow Angular Field of Sensors (570 Mile Altitude and Only
     115 Mile Swath Width Avoids Tone or Color "Fall Off" at Edges
     of Swath and Thus Increases Still Further the Uniformity of
     Image Tone or Color Characteristics).

6.   Provides Computer-Compatible Products Directly (Facilitates
     Automatic Data Processing).

7.   Potential Minimum Delay in Data Availability to User (Permits
     "Real-Time" Analysis and Facilitates Making Globally Uniform
     Resource Inventories, when Appropriate, or Analyzing Troubled
     Areas such as Sahel, in Africa).

8.   Systematic Coverage of Entire Earth Except for Near-Polar
     Regions.

9.   Capability for Receiving Data from Ground-Based Data Platforms
     (Facilitates Use of "Ground Truth" Data in the Inventory and
     Monitoring of Earth Resources).

10. Spatial Resolution is Optimum for "First Stage" Look and is
     Politically Palatable, Both Domestically and Internationally.

11. Data Routinely Placed in Public Domain for Benefit of All
     Mankind.

b) More information usually is obtainable from <u>multiband</u>
    photography than from that taken in only one wavelength
    band.

c) More information usually is obtainable from <u>multipolari-</u>
    <u>zation</u> photography than from that taken with only one
    polarization.

d) More information usually is obtainable from <u>multi-stage</u>
    photography than from that taken from only one stage or
    flight altitude.

e) More information usually is obtainable through the
    <u>multienhancement</u> of this photography than from only
    one enhancement.

f) More information usually is obtainable by the <u>multidis-</u>
    <u>ciplinary</u> analysis of this photography than if it is
    analyzed by experts from only one discipline.

g) The wealth of information usually derivable through
    intelligent use of these various means usually is
    better conveyed to the potential user of it through
    <u>multi thematic maps</u>, i.e., through a series of maps,
    each dedicated to the portraying of one particular
    theme, rather than through only one map.

All of the foregoing principles of the "multi" concept are
illustrated and more fully discussed in Chapter 1 of the recently
published "Manual of Remote Sensing". Suffice it here to state
that, if several of these components were to be used in proper
combination, the feasibility of using remote sensing techniques
as an aid to the global inventory and management of the Earth's
natural resources would be greatly increased.

A Tabular Presentation of Some Additional Considerations

Since this presentation deals with procedures for determining
the usefulness of various types of aerial and space photography in
relation to the inventory of natural resources, it is considered
appropriate to conclude with the summaries appearing in Tables 6
through 10. Table 6 seeks to differentiate between areas which
are simply structured and those which are complexly structured.
Tables 7 and 8 deal, respectively, with feasibility ratings that
might be applied in a typical case to simply structured and complexly structured areas. Table 9 recognizes that these ratings
may constitute oversimplifications, however, because the complexity
of the area is only one of at least 12 factors which govern the
potential usefulness of a particular type of aerial or space photography to those who wish to inventory, develop, and manage natural

resources. Table 10 provides details relative to still another
complicating factor, viz., whether the resources being inventoried
are renewable or nonrenewable.

The significance of these five tables in relation to the title
of this presentation is best appreciated as one attempts, in any
specific instance, to determine the usefulness of a given type of
aerial or space photography as an aid to the inventory of natural
resources within a specified geographic area.

Environmental Issues

In a soon-to-be-published report* to the International Council
of Scientific Unions the council's Scientific Committee on Problems
of the Environment (1976) notes that (1) a proposal for a global
environmental monitoring system was endorsed in 1972 by the United
Nations Conference on the Human Environment, and (2) little has
been done since to implement that proposal. The committee expresses
great concern over the prospect that increasing industrial activity
throughout the globe during the next two centuries would raise the
carbon dioxide content of the atmosphere at least 4-fold and perhaps
by as much as 8-fold. This in turn could be expected to bring about
major climatic changes. The report also analyzes the possible
long-term detrimental effects to the environment from human production of phosphorus, sulfur, mercury and nitrogen compounds.

Of interest with respect to the topic of the present paper is
the Committee's assertion that "it is difficult to see how best to
improve the environment without first establishing fundamental
facts. . .to offset the interminable speculation passing for knowledge". They specifically advocate that some kind of program be
instituted promptly for "monitoring levels of potentially harmful
or beneficial substances. . .in air, land and water", on a global
basis.

Some Unanswered Questions

Since one of the eventual objectives of this conference is
the developing of an <u>operational</u> satellite system for resource
management and environmental control, numerous considerations not
discussed in the preceding pages also should be mentioned. In
this section, through the raising of some unanswered questions,

---

* The probable citation that will be used for this highly significant
report is: Scientific Committee on Problems of the Environment,
"Environmental Issues - A Report to the International Council of
Scientific Unions". Paris, France, August, 1976.

Table 6   Characteristics of simply structured versus complexly
          structured areas in relation to natural resources.

Simply Structured Areas

A. Agricultural Vegetation

    1. Fields large, regularly shaped, usually homogeneous
        with respect to crop condition.

    2. Few competing crops and cultural practices.

    3. Little interspersion of cropland with noncropland.

    4. All fields of a given crop planted on about the same
        date and hence developing in essentially the same
        seasonal pattern.

B. Range and Forest Vegetation

    1. Blocks of rangeland and forestland are large and
        relatively homogeneous.

    2. Elevational range is low to moderate and hence vegetation
        of a given type tends to develop with essentially the
        same seasonal pattern.

    3. Few vegetation types present, all adapted to the
        same elevational and climatic range.

    4. Topography flat to gently rolling so that few
        vegetational differences are the result of differences
        in slope and aspect.

    5. Cultural practices with respect to range and timber
        resources are few and uniform.

C. Geology, Soils, and Hydrology

    1. Geologic, soil, and hydrologic formations are
        relatively large, simple, discrete, and homogeneous.

Table 6 (continued)

Complexly Structured Areas

A. Agricultural Vegetation

    1. Fields small, irregularly shaped, frequently
        heterogeneous with respect to crop condition.

    2. Many competing crops and cultural practices.

    3. Much interspersion of cropland with noncropland.

    4. Fields of a given crop planted on many different
        dates and hence developing with many different
        seasonal patterns.

B. Range and Forest Vegetation

    1. Blocks of rangeland and forestland are small and
        relatively heterogeneous.

    2. Elevational range is high to very high and hence
        vegetation of a given type tends to develop with
        many different seasonal patterns.

    3. Many vegetation types present, each adapted to a
        particular elevational and climatic range.

    4. Topography steep so that many vegetational differences
        are the result of differences in slope and aspect.

    5. Cultural practices with respect to range and timber
        resources are many and varied.

C. Geology, Soils, and Hydrology

    1. Geologic, soil, and hydrologic formations are relatively
        small, complex, intermingled, and heterogeneous.

Table 7 Hypothetical example of the tabular summary that has proved
to be useful in evaluating any given type of aerial or space photography in terms of its usefulness for vegetation analysis in <u>simply
structured</u> areas.

Information Desired	Feasibility
Agriculture	
Crop type	++
Crop vigor	++
Identity of crop-damaging agents	++
Crop yield per unit area, including forecasts	+
Area of ground occupied by each type and condition	++
Previsual evidence of plant stress	++
Slope of crop lands	+
Soil moisture	-
Soil nutrient effects	+
Range	
Forage type	++
Forage vigor	++
Identity of forage-damaging agents	++
Forage yield per unit area, including forecasts	+
Area of ground occupied by each type and condition	+
Previsual evidence of plant stress	+
Slope of range lands	++
Soil moisture	-
Soil nutrient effects	-
Forestry	
Timber type	++
Timber vigor	++
Identity of timber-damaging agents	++
Timber yield per unit area, including forecasts	+
Area of ground occupied by each type and condition	+
Previsual evidence of plant stress	-
Slope of forest lands	++
Soil moisture	-
Soil nutrient effects	-

++ Usually determined with ease from this type of photography after
limited on-the-ground calibration. May require that the photography be
obtained at proper seasonal state of the vegetation.

+   Determined often enough to make this type of photography useful as
the first element in a multistage sampling scheme. May require image
enhancement and large amounts of on-the-ground calibration, together with
proper season of photography.

-   Only rarely obtained from this type of photography. Rarely will such
photography be useful as an element in a multistage sampling scheme.

-- Information greatly desired by the managers of vegetation resources,
but unobtainable from this type of photography.

Table 8 Hypothetical example of the tabular summary that has proved
to be useful in evaluating any given type of aerial or space photography in terms of its usefulness for vegetation analysis in complexly
structured areas.

Information Desired	Feasibility
Agriculture	
Crop type	+
Crop vigor	+
Identity of crop-damaging agents	−
Crop yield per unit area, including forecasts	−
Area of ground occupied by each type and condition	++
Previsual evidence of plant stress	+
Slope of crop lands	+
Soil moisture	−
Soil nutrient effects	+
Range	
Forage type	+
Forage vigor	+
Identity of forage-damaging agents	−
Forage yield per unit area, including forecasts	−
Area of ground occupied by each type and condition	+
Previsual evidence of plant stress	+
Slope of range lands	++
Soil moisture	−
Soil nutrient effects	−
Forestry	
Timber type	+
Timber vigor	−
Identity of timber-damaging agents	−
Timber yield per unit area, including forecasts	−
Area of ground occupied by each type and condition	+
Previsual evidence of plant stress	−
Slope of forest lands	++
Soil moisture	−
Soil nutrient effects	−

++ Usually determined with ease from this type of photography after
limited on-the-ground calibration. May require that the photography be
obtained at proper seasonal state of the vegetation.

+   Determined often enough to make this type of photography useful as
the first element in a multistage sampling scheme. May require image
enhancement and large amounts of on-the-ground calibration, together with
proper season of photography.

−   Only rarely obtained from this type of photography. Rarely will such
photography be useful as an element in a multistage sampling scheme.

−− Information greatly desired by the managers of vegetation resources,
but unobtainable from this type of photography.

Table 9 For each of the factors listed in this table an entire
"spectrum" of conditions is theoretically possible, ranging from
very unfavorable to very favorable as regards its effect on the usefulness of the given type of photography in relation to the inventory, development, and management of natural resources. However, in
any given instance, the applicable situation is likely to be well
locatable and quantifiable. It follows that, in any given instance,
the overall usefulness of this type of photography for the stated
purpose will be determinable quantitatively by the aggregated effects

---

To the extent that the factors listed in this column pertain, there will be
minimum benefit derived from the use of this type of photography in relation
to the inventory, development, and management of natural resources.

1.   Area to be analyzed is very *complexly structured* in terms of
     the criteria appearing in Table 6.

2.   *Only photos having a GRD of, say, 10 feet* are available for
     use.

3.   Clouds *usually* obscure the area that is to be analyzed.

4.   Remote sensing can only be done on *one date* and at *one time of
     day*.

5.   There is a *very long delay* after the photos have been taken
     before they can be retrieved and placed in the hands of ana-
     lysts.

6.   Because of *rigid time constraints*, only *"quick look" analysis*
     can be made.

7.   Only *one data analyst is available* and he is inexperienced,
     poorly trained, poorly funded, poorly equipped, little appre-
     ciated, and poorly motivated.

8.   The analysis required is limited to only *one natural resource*
     and consists of a *one time inventory* of it in its static state.

9.   The resource classification scheme that is used is of *limited
     extensibility* because it is *locally specific*.

10. The derived inventory data must be *tightly held* because of
     sensitivities that relate to the economic or military security
     of the area under study.

11. The *sole purpose* of obtaining the inventory data is to faci-
     litate *resource preservation*.

12. *Few funds are available* with which to implement decisions de-
     rived from a study of the resource information that has been
     acquired; furthermore *the decisions, themselves, are suspect*
     because they were based on inadequate information as to the
     cost-effectiveness of each of several resource management al-
     ternatives.

Table 9 (continued)　　of these various factors. Usually, however, a weight will need to be assigned to each factor, in proportion to its estimated importance; hence the aggregated value normally will reflect these individual weights.

To the extent that the factors listed in this column pertain, there will be *maximum* benefit derived from the use of this type of photography in relation to the inventory, development, and management of natural resources.

1. Area to be analyzed is very *simply structured* in terms of criteria appearing in Table 6.

2. To the extent desired, *photos having a GRD of, say 10 feet plus* any or all other forms of remote sensing can be used.

3. Clouds *rarely* obscure the area that is to be analyzed.

4. Remote sensing can be done on each of *many dates* and at *many times of day*.

5. There is only a *very short delay* after the photos (and other remote sensing data) have been obtained before they are retrieved and placed in the hands of the analysts.

6. For all practical purposes there are *no time constraints;* hence the making of a *complete data analysis* is feasible.

7. *An entire multidisciplinary team of analysists* is available and each of them is well-experienced, well trained, well funded, well appreciated, well supported by consultants (when they are needed), and well motivated.

8. The analysis required is one which will integrate all components of the *entire "resource complex"*, including renewable resources, and will make *repeated inventories* to monitor them in their dynamic state.

9. The resource classification scheme that is used has *great extensibility* because it comprises one component of an overall scheme that is *globally* uniform.

10. The derived inventory data can be made *freely available* to all interested parties without fear of economic or military sensitivities.

11. The *multifaceted purpose* of obtaining the inventory data includes the facilitating of *resource development*.

12. *Very substantial funds are available,* and with them the necessary equipment, engineering knowledge, and local political stability, to ensure that both short-term and long-term benefits will derive from implementation of the resource management decisions; furthermore *the decisions themselves are sound* because they were based on reliable information as to the cost-effectiveness of each of several resource management alternatives.

Table 10 Considerations relative to the value of remote sensing-derived information about the non-renewable vs. the renewable natural resources of an area.

Non-Renewable Resources (e.g., Minerals and Fossil Fuels)	Renewable Resources (e.g., Agricultural Crops)
1. Short-term payoff is potentially very great because of prospects of discovering important resource deposits of great and almost immediate value to the developing country.	1. Short-term payoff is potentially less great because of the smaller immediate economic value to the developing country.
2. Because the resources are non-renewable, and in finite quantities , the long-term payoff steadily decreases and may become less than for renewable natural resources.	2. Because the resources are renewable, the long-term payoff steadily increases and may become greater than for non-renewable natural resources.
3. Desirability of fully exploiting the information through full resource development may be questionable because to do so may lead to early and permanent depletion of the resources. This, in turn, might diminish the birthright or inheritance of succeeding generations, both within the immediate area and throughout the world.	3. Desirability of fully exploiting the information through full resource development is rarely questionable because to do so may lead to steady increase of the resource. This in turn, may increase the birthright or inheritance of succeeding generations, both within the immediate area and throughout the world.

several such considerations will be briefly alluded to:

1. In each of the various resource-related decision models
   that need to be constructed, what are the relevant parameters, what are the "drivers" and to what extent can
   each be analyzed, mapped, monitored, and/or evaluated
   through the use of a suitable operational satellite
   system?

2. What can "sensitivity analysis" show with respect to the
   probable consequences of various incremental changes in
   the critical parameters that have been factored into
   these models? More specifically:

   a) How do such incremental changes affect the yield
      and quality of certain resource products (e.g.,
      how do they affect animal carrying capacity of
      the range or bushels of wheat produced, area-by-
      area)? and

   b) What impact do such incremental changes exert
      upon the environment?

3. In light of our experience to date with the total Landsat
   system, to what extent will data from stations on the
   ground that were instrumented to measure such factors
   as temperature, humidity, and wind velocity be usefully
   telemetered to an overflying Landsat vehicle, thereby
   facilitating the interpretation of objects and conditions
   as recorded by the satellite's own remote sensing device?

4. With specific reference to the production of agricultural
   crops and livestock, to what extent can "reciprocity"
   between the northern and southern hemispheres be developed
   through exploiting of the fact that the growing season
   in one of these hemispheres is six months out of phase
   with that in the other hemisphere? For example, presuming
   that at some time in the near future our capability had
   been developed for promptly making globally uniform and
   accurate resource inventories, might steps then be taken
   on rather short notice to increase the production of corn
   and pigs and decrease the production of oats and cattle
   in the southern hemisphere if a resource inventory com-
   pleted near the end of the immediately preceding growing
   season in the northern hemisphere clearly showed that
   there was going to be a very sizable overproduction there
   of oats and cattle and also a sizable underproduction
   there of corn and pigs?

5. Let us consider that, eventually, a globe-encompassing
    satellite system might enable us to locate, map, and
    evaluate virtually all of the world's available <u>non-
    renewable</u> natural resources (fossil fuels, minerals,
    etc.). The record to date shows a remarkable determination
    by man to exploit such resources as quickly as they are
    discovered. Since the supply of these resources is dwind-
    ling at an alarming rate and cannot be renewed (at least
    in this geologic era), will we have rendered a disservice
    through our very success in the making of such global
    inventories, unless we insist that parallel and vigorous
    measures be taken to safeguard the birthright of future
    generations, as represented by their "fair share" of these
    resources?

6. Similarly, let us consider that, eventually, a globe-
    encompassing satellite system might enable us to inven-
    tory and accurately monitor all of the world's <u>renewable</u>
    natural resources (agricultural crops, livestock, timber,
    etc.). The recent record shows a disturbing tendency
    by man to exploit, to his own personal benefit, the
    temporary regional or global imbalances of such resources
    as quickly as they are made known to him. In some instan-
    ces there could even be a tendency for certain nations,
    when quickly given such globally uniform and reliable
    information, to exploit it as an economic weapon. There-
    fore, we again need to ask whether we will have rendered
    a disservice through our very success in making such
    inventories, unless we insist that parallel and vigorous
    measures be taken to ensure that such information be used
    to the equal benefit of all mankind rather than to the
    preferential benefit of a few individuals or nations.

7. In light of these last two considerations, should a more
    careful look be given with regard to the organization
    that might most appropriately administer a Global Resources
    Information System based on satellite-acquired data?
    For example, would it not be timely for a multi-nation
    free-world group, such as NATO, to play a vigorous role
    not only in the establishment of such a system but also
    in the devising of safeguards that would ensure its use
    for the benefit of <u>all</u> mankind--now and in the future?

8. With our increasing awareness of problems that are being
    created by overpopulation and by the encroachment of man
    on our few remaining wilderness areas, to what extent
    should we be emboldened to revise the God-given and age-
    old mandate "Be fruitful - multiply - and subdue the Earth"?

Summary and Prospectus

In this paper several important considerations have been discussed relative to the making of globally uniform resource inventories with the aid of Landsat or other forms of space-acquired data. These considerations pertain, in most instances, to the technological feasibility of making such inventories, but their cost-effectiveness, political palatability, and susceptibility to various kinds of misuse also are considered. The paper also considers certain differences in the resource-oriented attitudes of people in developed countries versus those in developing countries. The differences found have led the author to conclude that the use of space photography and other modern forms of remote sensing data, by permitting a globally uniform inventory of natural resources to be made, would do much to reconcile these two opposing viewpoints. This, in turn, probably would lead to closer agreement as to how the world's natural resources might best be managed, area-by-area and country-by-country, for the benefit of all mankind.

The paper concludes with speculation on the possibility that, given such natural resource inventories of sufficient accuracy and made at sufficiently frequent intervals, some kind of free-world international organization such as NATO might constitute the ideal group for overseeing of the program to ensure that its potential benefits were shared equally by people throughout the world.

As a final thought, might we logically foresee the possibility that the techniques for remote sensing might evolve into a highly automatic operation, in which an unmanned satellite orbiting the Earth will carry multiband sensing equipment together with a computer? Thus equipped the satellite could, for any particular area, take inventory of the resources and produce a printout that would amount to a resource map of the area. The computer could then use the inventory data in conjunction with pre-programmed factors (such as what ratio of costs to benefits would be likely to result from various resource management practices) and could reach a decision for the optimum management of the resources in the area. The decision would be telemetered to the ground for whatever action seemed necessary.

As a simple example, the satellite's sensors might spot a fire in a large forest. Its computer might then derive information on the location and extent of the fire and could assess such factors as the type and value of the timber, the direction and speed of the wind and the means of access to the fire. On the basis of the assessment the computer would send to the ground a recommendation for combating the fire.

Capabilities of this kind might not need to be limited to

emergencies. Many routine housekeeping chores now done manually
by the resource manager conceivably could be made automatic by
electronic command signals. Examples might include turning on
an irrigation valve when remote sensing shows that a field is
becoming too dry and turning off the valve when, a few orbits
later, the satellite ascertains that the field has been sufficiently watered.

A satellite of such capabilities may seem now to be a rather
distant prospect. After a few more years of developing the techniques for remote sensing the prospect may well have become a
reality.

REFERENCES

Aviation Week and Space Technology, "Proposed Resources Corporation
     Could Resemble Comsat General". August 23, 1976, p. 52-53.

Colwell, Robert N., "Needs at the EROS Data Center for Rapidly
     Processing Landsat Data into High Quality Products for Resource
     Inventory". Statement before the Interior Subcommittee of the
     Senate Appropriations Committee, May, 1975.

Colwell, Robert N., "The Present Status and Potential Future
     Usefulness of Earth Resources Survey Programs Based on
     Landsat". Statement before the Subcommittee on Space Science
     and Applications of the Committee on Science and Astronautics,
     U.S. House of Representatives, Ninety-third Congress, Second
     Session, October, 1974.

Colwell, Robert N., "The Future for Remote Sensing of Agricultural,
     Forest, and Range Resources". Statement before the Subcommittee
     on Space Science and Applications of the Committee on Science
     and Astronautics, U.S. House of Representatives, Ninety-second
     Congress, Second Session, January, 1972.

Colwell, Robert N., "Gains to be Made in Natural Resource Manage-
     ment Through the Use of Remote Sensing Systems". In The Use
     of Remote Sensing in Conservation, Development and Management
     of the Natural Resources of the State of Alaska, Department of
     Economic Development, State of Alaska, p. 5-26, 1969.

Ford, Wendell, "Need for an Operational Global Resources Informa-
     tion System". Congressional Record, Senate, p. S-13816 and
     13817, August, 1976.

MacDonald, Robert B., Hall, F.G. and Erb, R.B., "Large Area Crop
     Inventory Experiment (LACIE). I. The Use of Landsat Data".
     Proceedings of the NASA Earth Resources Survey Symposium,
     Houston, Texas, June, 1975, Volume IIA: 43-62.

Merewitz, Leonard, "Special Study No. 3, On the Feasibility of
    Benefit-Cost Analysis Applied to Remote Sensing Projects".
    In Annual Progress Report, An Integrated Study of Earth Resour-
    ces in California, 1974, p. 5-40 to 5-49.

Moss, Frank E., "Meeting Future Needs of Man". Congressional
    Record, Senate, August, 1976, p. S-13815 and 13816.

Park, Archibald, "Presentation to Humphrey Panel Recommending that
    a Landsat Program be Made Operational". January, 1976.

Pecora, William T., "The Need for Mapping and Inventorying Natural
    Resources". In The Use of Remote Sensing in Conservation,
    Development, and Management of the Natural Resources of the
    State of Alaska, Department of Economic Development, State of
    Alaska, December, 1969, p. 30-32.

Poulton, C.E., Welch, R.I. and Colwell, R.N., "The Uniform Mapping
    of Natural Ecosystems and Environmental Complexes from Space
    Acquired Imagery". A Technical Report prepared for NASA by
    Earth Satellite Corporation, 1973.

Sattinger, I.J. and Polcyn, F.C., "Peaceful Uses of Earth
    Observation Spacecraft". Volume II, Survey of Applications
    and Benefits. Report No. 7219-F(11): 159. Institute of Sci-
    ence and Technology, University of Michigan, 1966.

DATA PROCESSING AND DISTRIBUTION

FOR EARTH OBSERVATION SATELLITES

John M. DeNoyer
Director, Earth Resources
Observation Systems (EROS) Program
U.S.G.S., Department of Interior
Reston, Virginia, U.S.A.

Introduction

Automated satellite data acquisition systems have reached a level of development where our ability to acquire remote sensing data exceeds our capability to use the data in terms of extracting information. The amount of data that can be acquired can also exceed the practical limitations for capital investments in equipment and associated operating costs.

Providing a smooth flow of data from acquisition through information extraction depends on several factors. These factors include realistic data acquisition strategies, screening of data for quality, data requirements that are consistent with recognized information needs, advances in technology that will reduce the costs of data processing and automated and semiautomated information extraction, recognition of the characteristics of related information gathering capabilities, and design of total systems that incorporate all of these factors.

Remote sensing systems often start with the capabilities of sensor technology rather than the requirements of the user. This is appropriate for research in sensor technology, methods of analysis and application techniques. Systems to serve operational programs, however, require a much more comprehensive approach that is based on results from experimental programs and considers user requirements.

Initial Processing

Image data acquired by spacecraft-borne electro-optical sensors require initial corrections before they can be portrayed in a consistent image format that is suitable for resource and environmental analyses. These initial corrections are complex.

Many of the corrections are not practical to attempt for data acquired by film cameras that might be placed on similar spacecraft. The fact that modern technology makes it practical to make these corrections to data acquired by electro-optical sensors is a major factor in determining the relative utility of these two types of imaging systems.

Specific initial corrections that are applied to data acquired by Landsat include Earth rotation, orbit height, spacecraft attitude, rates of change in spacecraft attitude, mirror velocity and radiometric calibrations. Some of these corrections are determined from the telemetry, from the spacecraft calibrations from flight and preflight data and orbit determinations from tracking stations. Other corrections depend on a combination of these sources and the image data such as the use of ground control points to accomplish precision geometric corrections.

The difficulty of carrying out these initial corrections depends on the accuracy of parameter determinations, the rates of change of the parameters, and the accuracy of the final product that is desired. For example, the geometric corrections would be relatively simple for a spacecraft sensor system with very precise attitude determination, slow rates of change of attitude relative to local vertical and very small or zero deviations from the desired scan velocity. Few ground control points would be required and the data processing could be optimized. Realistic solutions are in tradeoffs among the one-time costs of the space portion of the system and the one-time cost of purchasing ground processing equipment and the recurring costs of operating the ground portion of the system. These relative costs vary with changes in the state-of-the-art technology that are available at any particular time.

Data Management

Management of the data acquired by remote sensing systems may be one of the most formidable challenges in the future. Examinations of the data that could be acquired by the six-channel Thematic Mapper that NASA is developing for operation in the early 1980's shows that complete coverage over land areas of the world on a 9-day repeat cycle would yield 10^{12} bits per day. The annual costs for high density digital tape and the associated processing would be very high. These high costs for operations on the ground have led to

planning that includes selective acquisition for particular coverage where the uses can be well defined. This approach reduces the
estimate of the frames to be acquired by a factor of 5. An additional factor of 2 reduction in data processed can be achieved by
rejecting frames that have excessive cloud cover or low quality
introduced by noise. These two considerations provide nearly an
order of magnitude reduction in the amount of data to be processed
and archived. The data collection capabilities could provide additional coverage, and the ground processing capabilities could be
augmented, if justified by future requirements.

The selection of the areas to be covered will be made on the
basis of a priori knowledge of discipline requirements and by observing the data acquired by a lower resolution Multispectral
Scanner that can produce repetitive, lower data rate, coverage of
the land areas of the world.

An argument against such preselections is that some important
observations will not be made either because the á priori knowledge
is not available or because the tradeoffs in priorities for acquisition cannot be accommodated. Removal of this deficiency, using
present technology, will require larger investments for equipment
and operating costs. Substantial improvements in accessible mass
storage systems and in data processing efficiency could introduce
efficiencies for future systems.

Another important factor in data management is accessibility
of stored data. Satellite systems with stable, repetitive orbital
parameters provide systematic repeat times for specific geographic
areas. The indexing, archiving, and retrieval of these data by
geographic location and date of acquisition is relatively easy for
general areas such as a frame of Landsat coverage. Capabilities
such as change detection have been demonstrated. Automatic change
detection requires precise overlay or registration of images acquired at different dates. The analysis requirements can be very
large if the images are not precisely corrected for geometry and
associated with specific geographic locations. The capability to
achieve automated registration between multidate images on a regular
basis appears to be a realistic objective and will be implemented
for Landsat-C and later Landsat-type spacecraft. There will be
more difficult problems in obtaining this registration capability
when attempts are made to combine data from Landsat, Seasat, Heat
Capacity Mapping Mission, and other satellites.

Data archived on a geographic and temporal basis can be retrieved from archives relatively easily as long as an adequate data base
is maintained. A computer is essential when large amounts of data
are in the data base, data of different types are present, and access must be rapid. Inability to find out what data are available
in an archive makes the archive useless for the general user.

Quality criteria of the data in an archive should also be readily available. The current data bases containing Landsat data are inadequate in this respect. Improvements are possible in the format of catalogs and the quality of browse film. More reliable and useful information on the percent of cloud cover, location of clouds and the type of clouds would be desirable. In the future it may be possible to provide histograms of the radiance values for the spectral bands and key information concerning quality of electronic recordings such as dropouts and parity errors.

The rapid retrieval, reproduction and distribution of archived data to users is probably more important for satellite acquired data than for data acquired by aircraft. Aircraft data are generally acquired for a specific purpose, such as mapping, that it serves first. The data may then be placed in an archive and distributed as appropriate. These aircraft data have served their initial purpose. Subsequent use adds to their value, but the acquisition objectives have already been satisfied. In contrast, most satellite data are acquired and then wait to be used for a large variety of applications. The data will not be applied if they cannot be identified and provided to the user.

Data Processing For Distribution

Methods used to reproduce data for distribution to users should be designed to minimize the risk of degrading the quality of the archived material. Primary archives on film do not satisfy the minimum risk requirement. First, the dynamic range of electro-optical sensors is often greater than the dynamic range of the film; second, scratches, dirt and dust on film degrade the quality of products that are produced from the archival masters; and, third, even under ideal conditions, there is loss of quality in making photographic copies from film archives. During the Landsat-C time-frame the primary archive in the United States will be changed from film to high density magnetic tape to make it possible to retain and retrieve the maximum amount of the data acquired by the sensor system.

The use of high density digital tape as the archival medium makes it possible to provide for a number of improvements in the data products. Choices in map projections in which the images are presented are possible. Resampling of the data will be used to estimate the appropriate radiometric values for new picture element locations to produce geometrically corrected images in specific map projections. There are several resampling strategies such as nearest neighbor, bilinear interpolation and cubic convolution. The choice of the resampling method can affect the results of the analysis that is planned. Choices between resampling methods can be used to meet specific user needs. Unfortunately, the digital

processing systems that will be in use by the United States at
the time of Landsat-C will provide only one map projection that
is produced by using one resampling method. This will probably
satisfy most users, but there will be deficiencies that cannot
be satisfied by the system that is now being implemented. It
may be possible to augment the system to provide greater flexibility
in the future.

The appearance or cosmetic characteristics of Landsat-C photographic products will be improved. These improvements will be
accomplished by using processing methods to remove the "striping"
characteristic of many images, removal of haze or radiance bias
that tends to reduce image contrast, and adjusting the radiance
values to fit the maximum number of gray levels within the dynamic
range of the film. These cosmetic improvements will be applied on
a scene-by-scene basis.

The use of high density digital tape as the archival medium
also makes it practical to provide some specialized products applicable to specific disciplines such as edge enhancement to aid in
geologic and geographic interpretations and contrast stretch techniques that emphasize water features, vegetation, or other land
cover types. The increased variety of image products that will
become available in the near future will increase the user's responsibility to specify precisely what product is required. There
will have to be a period of learning by both the users and the
data producers before an effective communication can be established.
This communication problem will be repeated many times as new types
of data become available and combinations of data from different
acquisition systems are attempted. Explanatory material concerning
new or improved products will help, but experience and patience
will be essential.

Summary

Imaging systems to be used on Earth-orbiting satellites to
operational programs can provide a very efficient method of data
acquisition for a number of purposes. Tradeoffs between "space
segment" and "ground segment" capabilities need to be planned as
a total system to keep initial costs and operating costs in proper
balance.

Two areas where technological advances are needed are:

(1) development of large, random access, mass storage
     for digital data, and

(2) development of lower cost systems to process image
     data rapidly.

SOIL DIAGNOSIS THROUGH REMOTE SENSING

J.L. D'Hoore
Professor, Faculteit Der Landbouwwetenschappen
Katholieke Universiteit Te Leuven
Heverlee, België

Abstract

The deduction of soil profile properties from landscape aspects bears some resemblance to the recognition of deep seated illnesses by means of exterior body signs. For both purposes signs of the existence of hidden properties are sought out whose indicator quality was previously established either by rational deduction or by experience. In both cases it is admitted that external signs alone are not yet able to provide unambiguous identifications - there is some doubt as to whether this could ever be the case - but constant efforts are being made in order to render symptom recording, data retrieval and correlation, as precise and as objective as possible. Non-destructive systems capable of probing in depth, and methods of pertinent deductive reasoning utilizing any new form of reliable input data, are welcome.

Theoretical and Practical Approaches to Soil Cover

Situated at the conjunction of atmosphere and land wherever life is able to subsist, soils integrate the cumulative effect of external influences, variable in kind, intensity and continuance, on a variety of parent materials. Together with the landscape within which they are situated, they may provide reliable testimony on the successive soil forming conditions that have prevailed there during the pleistocene or even since earlier times. Historical or genetic pedology must therefore take into consideration all materials covering the land surface that bear the marks of pedogenetic intervention. These may be weathering products in situ, either shallow or tens of meters thick, derived from hardrock or from loose

219

alluvial or eolian deposits. It may also be weathered material redistributed over a landscape through erosion and colluviation. Such materials may have been leached in the pedogenetic process and the extracts may have accumulated in adjoining areas. They may also have been enriched by water or wind-borne materials, the quality, quantity and origin of which it may be worthwhile to investigate. The soil layers to be studied for pedogenetic purposes may therefore be considerably thicker than the part now occupied by living plant roots, and the extent of the landscape necessary for the reconstruction of their evolution may be large and equal subcontinental dimensions.

Soils can also be considered as obligate substrata for land-based biocenoses, many of which are essential for the sustenance of mankind. Such an ecological approach comprises all land based biotic complexes, all forms of open air agriculture, taken in its broadest sense. All parts of the soil cover are not equally suited for biomass production, current climatic averages and extremes being taken into account. Soil fertility must be assessed on and in the soil body from the structural, biochemical and physico-chemical properties that regulate gas exchange, nutrient and water availability. Landscape features may furnish additional information concerning the hazards of leaching, erosion, enrichment by, or burial under, lateral influxes.

Soil layers primarily involved in ecological studies are those densely occupied by living plant roots: three meters underneath the land surface root density generally becomes of little significance. Additional attention goes to the deeper layers which may be genetically related to the overlying pedon or may be the source of ascending capillary ground water. On the whole, soil layers to be studied for ecological or agricultural purposes are shallower and laterally less extensive than these required by the genetic, historical approach earlier described.

A third approach, which is even more pragmatic, considers the soil as a foundation, a cover or a material for human constructions with an emphasis on subsoil layers. Main soil properties retained are mechanical and textural, and landscape aspects connected with drainage potential receive major attention.

All these approaches are interdisciplinary and interlock to a considerable degree. Their purposes are not limited to the recording of present situations nor to the analysis of their development in the course of time. They also provide the basic information required for active intervention in matters of environmental care, of sustained production of biomass, of land use, transportation planning, etc.

Landscape Aspects and Soil Profile Properties

In the present state of its art, remote sensing can only record surface properties of landscapes down to a skin depth, that is considerably shallower than that of soil profiles or layers of pedogenetically influenced parent materials. Notwithstanding its synoptic, multispectral and multitemporal capability, remotely sensed information remains limited to bidimensional surfaces. Soils and parent materials however are tridimensional bodies, whose essential properties are seated in depth, in the successive subhorizontal layers towards the differentiation of which pedogenetic development normally tends. With the exception of cracking clays, of typical microreliefs and of very striking colours, surface properties have not been retained as differentiating soil classification characteristics.

Besides, spectral signatures of soil surfaces are not very characteristic, at least in the visible range of the electromagnetic spectrum, and overall emission and reflection are strongly influenced by water content and surface structure. Moreover, most soils are more or less overgrown by plant communities, natural and cultivated; and both bare soil surfaces and plant canopies may be covered by dew, rain, ash, and dust. These factors render direct soil recognition by remote sensing methods highly aleatory or downright impossible save for a few exceptional cases.

Nevertheless, skeptical field pedologists have been repeatedly impressed by the strikingly apparent conformities or discrepancies that existed between the patterns recorded on their ground made maps and those on multiband colour composites (LANDSAT), especially when available in the multitemporal mode. Systematic comparisons between soil maps and overall ground truth on the one hand and RS imagery on the other, suggest that this cannot be due to mere coincidence, that certain landscape aspects can reveal subjacent soil quality, that soil diagnosis could eventually be approximated by indirect methods making use of extrinsic properties instead of the mainly intrinsic criteria of traditional soil classifications.

How this could be realized is predetermined by the way in which exploratory and reconnaissance soil maps often originate. Such maps are never computed from point by point measurements and may contain a considerable part of conjecture. Reconnaissance surveys generally start from working hypotheses, based on experience acquired in landscapes that are lithologically, geomorphologically and climatologically similar. Field work involves the checking and often the amendment of the initial hypotheses, and the identification and correlation of the units to be mapped. The boundaries that separate the latter are then recorded by traversing or more often by exact location of a limited number of transition points which are then interconnected by the least improbable interpolation lines.

This rational interpolation is based on presumed similarities of
parent material, landform, climate, vegetation, development time,
of prevailing soil forming conditions such as drainage potential
or other concomitant symptoms characteristic of given situations.

Soil exploration and reconnaissance are important components
of modern integrated surveys, they have developed into an increasingly multidisciplinary operation with its own methods and tactics
and also its provisions against inconsiderate deduction. Up to
now the additional information provided by LANDSAT imagery before,
during, and even after campaigns, has been widely used mainly for
the correction of base maps, the amendment of working hypotheses,
the outlay of itineraries, the correction of intrapolations, and
extrapolation towards adjoining areas.

The purpose of indirect soil diagnosis through remote sensing
goes further than that and aims at the construction of most probable soil distribution models, eventually to be checked by less extensive field inspection. In the general context of the exploratory
and reconnaissance soil survey, of which it is to become an essential part, remote sensing can be seen as verifying the initial hypotheses without the benefit of ground control. It consists of
the transformation of multispectral multitemporal landscape aspects
first into the pedogenetic conditions that most probably have been
active in distinguishable areas of the scenes, and second into the
actual soil properties prevailing there.

Soil recognition and delimitation will have to be done by inference, the premises being landscape features whose spectral,
spatial and temporal properties, measurable by means of remote sensing, faithfully reflect soil forming factors, processes or their
effects. Such properties have been tentatively called <u>soil
indicators</u>.

Soil Indicators

According to Dokuchaev's definition, soils are the unique products of unique interactions between factors such as parent materials, land forms, climates, life forms and development times. Cog-
nition of the factors should allow deductions about the products.

In addition, vegetation behaviour as reflected by canopy radiation may provide information about subsurface soil properties
affecting normal plant function, on resistance met by roots in
providing aerial parts with water and nutrients and in satisfying
gaseous exchange needs. Assessment of the stratification of water
distribution in soil and plant cover, by means of microwaves and
heat capacity measurements, may give further information in this
respect.

From the following not exclusive schema it will appear that most indicators do not result from single measurements but must be derived from several. Some of the most effective indicators, both for rapidly and slowly changing properties, rely for their identification on the monitoring of their interplay with climate; contemporaneous information from meteorological satellites is therefore desirable.

Factor: Parent Material

Indicators:

Landform.

Colour difference between wet and dry soils.

Aspect, origin of deposit, means of transportation.

Kind and seasonal behaviour of vegetation.

Heat capacity.

Possible distinctions:

Weathering rock in situ; lithological differences.

Deposits of slightly weathered materials -
    alluvial, colluvial, eolian, fluviolacustrine,
    fluviomarine....

Organic deposits.

Ferruginous, calcareous crusts.

Deeply weathered materials in situ.

Deeply weathered materials transported.

Catenary sequences.

Factor: Landform

Indicators:

Drainage, aggradation patterns.

Absence, presence, turbidity of open water and their seasonal variations.

Projected shadows of protruding features and of clouds.

Rain shadows and vegetation patterns induced by them.

Vegetation enhancement in depressions -
    prolonged turgescence into dry season, delayed combustion
    and, after burning, darker coloration due to greater
    amounts of fuel.

In humid forested areas, different reflection from forest on periodically flooded and on dry grounds.

Possible distinctions:

Mountains, hills and plains.

Floodplains, pediments and cliffs.

Peneplain fragments especially when bordered by laterite edges.

Dunes.

Alignments of structural origin.

Factor: Climate

Indicators:

Specific information from meteorological satellites.

Seasonal and diurnal variation of the spectral signatures of vegetation.

Seasonal and diurnal variation of the opacity of the atmosphere, cloudiness, kind, time and height of cloud development.

Colour change of bare soil and rock after the first rains.

River regimes, flooding of river plains.

Thermal infrared emission.

Possible distinctions:

Arid, humid, seasonally dry climate.

Precipitation reliability.

Factor: Vegetation

Indicators:

Spectral signatures.

Reaction on evapotranspirational stress.

Vertical profiling (active microwave).

Thermal infrared emission.

Burning patterns.

Possible distinctions:

Desert, steppe, savanna, forest, marsh (see also: human intervention).

Factor: Soil developing time

Indicators:

    Landform.

    Vegetation pattern.

Possible distinctions:

    Young deposits.

    Old deposits.

Factor: Human intervention

Indicators:

    Changes in vegetation pattern.

    Alignments different from natural forms.

    Burning effects, smoke.

    Modified drainage patterns.

    Effects, positive or negative, on erosion, salinization.

Possible distinctions:

    Roads, towns, canals, high tension lines....

    Different types of land use.

    Major agricultural developments.

Optimation of Soil Recognition Capabilities

It is readily admitted that soil diagnosis in itself is not sufficiently important to justify dedicated remote sensing systems. It mainly concerns marginal underdeveloped areas of uncertain agricultural value where surveys will have to be made only once, up to now reconnaissance surveys have been more reliable, and the field effort they require may be confidently replaced by remote sensing (e.g. LANDSAT).

Yet, pedology could be considered as a privileged user; its indirect recognition procedures can take advantage of almost the entire remote sensing arsenal developed up to now, and not much dedication will ever be required. Moreover, its findings and its methods may be useful to other disciplines from geology to agriculture.

Optimation can be attempted along at least four avenues:

- by concentrating research on the most significant indicators

- by recommending sensors, orbits, observation times and observation frequencies most suitable for the monitoring of the selected indicators

- by improving the quality and the accessibility of the data products

- by developing semi-automatic or optical procedures, allowing objective features recognition, which do not require equipment too complicated or too costly for soil research laboratories.

The most significant indicators appear to be those that will allow the monitoring of the interplay of climate, plant and soil. Plant communities, cultivated and wild, the variable climatic environment in which they grow and the soil in which they are rooted interact, but of the three, the vegetation reacts most rapidly and spectacularly to the variation of the others. Plant response to seasonal climatic change - phenological behaviour - is in part determined by genetic programming, but also by soil characteristics such as water retention capability, thermal properties, and maybe others, that either accelerate or retard spectral signature modification of the canopies.

It should be possible to elaborate models of the spectral behaviour of definable plant communities subjected to variable climatic and edaphic conditions. Attention should be given to root system geometry, indicative of the location of the soil layers on which the aerial parts inform.

Under uniform climates, uniform plant communities may be expected to behave equally throughout the year, if edaphic conditions remain constant or change equally over the entire area studied. Multispectral scenes recorded at any time of the year should all display overall uniformity, though variation between scenes may be observed in the multitemporal mode.

Under uniform plant communities over uniform soil, unequal plant behaviour both in the mono- and the multitemporal mode must be due to climate heterogeneity.

In the case of uniform communities under uniform climates edaphic differences tend to be reflected by vegetation behaviour, visible in each of the successive images, but also by significant differences in the time lags separating climatic change and plant response which are observable in the multitemporal mode.

A realistic model however will have to take into account:

- climate variability, overall, localized, differing from year to year

- plant cover heterogeneity, edaphically determined or induced by man

- soil cover heterogeneity.

Systematic research will certainly improve the detection capability of plant-climate couplings, some of which may be specific, i.e. particularly suitable for soil diagnosis or for agricultural monitoring for geomorphological or geological mapping, etc. Its scope may be suggested by the following tentative titles:

- Spectral identification of plant communities and their normal phenological behaviour.

- Spectral identification of aberrations of such normal behaviours.

- Spatial and temporal localization and measurement of aberrations.

- Correlation of aberrations with particulars of climate and underlying soil.

Sensor and orbit parameters, now in use or planned, are certainly not optimal for the observation of plant behaviour. In fact how could they be, as optima may differ from plant to plant, from season to season, and also with latitude? The search for better combinations, eventually making use of geosynchronous orbits and new sensors, offers an interesting challenge.

The advantage of product improvement, especially of imagery, is self evident. For a long time to come imagery will remain the format preferred by the natural scientist, whatever the real advantages of CCTs (Computer compatible tapes) for spectral identification. Because of their field training, their frequent use of maps and aerial photographs, imagery comes to researchers as something familiar, stirs their imaginations and encourages their pattern recognition (correlation) efforts. The broad synoptic overviews, eventually of several scenes simultaneously projected, the conveyance of clues as to causal relationships, the possibility of approximate adjustments of multitemporal overlays, of shape recognition even when the objects are deformed, all constitute distinct advantages of optical over automatic interpretation, and the former should always precede the latter.

Accessibility of data products to the research worker may be understood in terms of permission to use them, of time lags between reception, processing and distribution, and of acquisition and interpretation costs as compared to the working budget of soil science labs. The two first aspects are political and organizational. The third one is the concern of system owners and soil research organizations, private or governmental.

First order deductions, based exclusively on data measured by remote sensing, can be entrusted to computers. The production of system-corrected and enhanced imagery from crude signals from space could be considered as a first order deduction; but what is meant here is the computer identification of signatures and shapes. Deductions of higher orders result from interaction of first order deductions, either with exact information obtained by other means than remote sensing or with working hypotheses. Such higher order deductions, especially the latter, can only be made by the trained human mind.

Remotely sensed imagery in its present form has already considerably improved our understanding of landscape elements and of their evolution, without the benefit of first order deductions generated by computers. The potential superiority of the latter over the human capability is readily admitted.

Unfortunately, the costly equipment and specialized personnel required by computer methods may well push the practice of remote sensing out of the reach of many researchers on whom we must rely to extricate the complicated relationships between plant canopy signatures, climate and soil conditions. Many potential collaborators are natural scientists employed in universities that favour individual or small team research on a part-time basis, with the limited funding that normally goes with such a situation. Many are doing it just for scientific interest, their countries being adequately mapped and thus providing almost ideal ground truth bases, without much outlook for practical benefits for the community that supports them. As a compensation for the services they are bound to render, and as a further encouragement, such researchers should be provided at low cost with the data products they may require, including the thematic enhancements to which some of them may have contributed.

DEVELOPMENT AND PROSPECTS OF REMOTE SENSING

APPLIED TO EUROPEAN RENEWABLE EARTH RESOURCES

S. Galli de Paratesi
EURATOM Joint Research Centre - Ispra Establishment (Italy)
Commission of the European Communities

EEC's Requirements for Remote Sensing in Agriculture

The Commissions' Initiatives

Programmes in agriculture and in agronomic development are
heavily dependent on the availability of timely information for
wise decision making. This involves extensive data gathering and
information processing operations.

Renewable resource-consuming continents like Europe are continually seeking refinements and means of improving survey and analytical methods. In this respect the development of Remote Sensing
techniques can certainly contribute to accelerating and expanding
functions of data collection and analysis. The broad mission objectives of such systems in agriculture, from the Economic European
Community's (EEC) point of view, are ultimately to enable more food
to be produced by increasing yield and food quality from land under
cultivation, by decreasing losses in production due to disease,
weed and/or insect infestations and forest fires, and by increasing
the quantity and quality of cultivated soils. Each of these objectives requires accurate, timely and synoptic survey information in
order to manage the resources of the continent efficiently and to
bring about the desired increase in production of renewable
resources.

On this basis the Commission of the European Communities (CEC)
has promoted some initiatives aimed at evaluating the potential
of Remote Sensing methods as a support to agronomic research and
to agricultural management, applied in a context of typical
European conditions. The ultimate goal is to identify present and

229

future requirements for operational data to be gathered by means of
various observation platforms and for information processing systems.

Some pilot Remote Sensing experiments have been proposed for
a collaboration between various CEC Directorates General in Brussels
and some groups of national Institutes and Organizations, under the
overall leadership of the Joint Research Centre, Ispra (JRC-Ispra).

This idea of regrouping national laboratories to work together
with the JRC was chosen for the following reasons: because it
provides:

a)   a basis for short-term work efficiency, since it does in
     fact allow the immediate concentration of diversified
     competences and support facilities which could not other-
     wise be easy to regroup in a suitable European R & D
     frame-work;

b)   a basis for long-term creation of operational Remote Sensing
     infrastructures within the European Community.

The above current initiatives have demonstrated that the success of such an operation depends upon the following:

a)   the programme objectives, satisfying certain specific in-
     formation requirements of the central organization (i.e.
     the CEC in matters of policy for agriculture development,
     aid to developing countries, and environmental protection),
     must correspond in detail to the research objectives of
     the institutes and/or to some specific need for information
     on resources on the part of the national organizations;

b)   the various Remote Sensing platforms (satellite, aircraft,
     balloon, helicopter, etc.) are chosen because they provide
     data offering some significant operational advantage over
     those gathered by conventional methods.

As a matter of fact this form of infrastructure is effectively
demonstrating the feasibility of applying Remote Sensing techniques
to specific sectors of public interest in agriculture. This applies
particularly to comparative evaluation of the potential of various
space-and-air platforms, on-ground data acquisition systems, and
different processing algorithms for classification and inventory.

This form of supranational cooperation is likely to be preferred also from the view-point of the major space organizations
already providing users with satellite data. In fact such
"consortia" would offer them a single point of contact rather than
multiple contacts. The operation of distributing to users photo-
imagery, maps, CCTs and ancillary data can be organized more

rationally. Results gathered by the institutes collaborating
with the CEC on this research are standardized, compared and syn-
thesized by the JRC-Ispra before being reported. This service
enhances the value of the results themselves.

The JRC's Pilot Experiments

The Commission has put the JRC-Ispra in charge of two pilot
experiments involving renewable resources. Both utilize NASA
satellite data.

The AGRESTE Programme, Importance of Inventory

The objectives of the first pilot experiment - the AGRESTE
Project using LANDSAT data - were chosen to cover some typical
Southern European ecological conditions ranging from artificial
closed ecosystems (irrigated rice fields, poplar plantations,
conifer afforestations) to natural ecosystems (beech forests).
The main subjects of investigation are:

1/ rice: area inventory, yield forecasting,
                early detection and monitoring of
                diseases, varietal classification

2/ poplars: classification into age categories,
                inventory

3/ conifers: identification

4/ beeches: identification, inventory.

This selection was made following some specific recommendations
by the Directorate General for Agriculture (DG-VI). All the subjects are of considerable economic importance.

Rice is the most widely grown crop after wheat and its
consumption is increasing steadily. This choice was also supported
by the Directorate General for Aid to Development (DG-VIII) with
a view to some future global inventory in African countries, such
as Madagascar. Poplar and conifer grow rapidly and are in demand
as industrial wood; poplar cultivation is intensifying in some
EEC countries, where the present shortage of poplars is partly
responsible for paper import. Conifers are intensively used in
re-forestation projects. A rational exploitation of natural beech
wood as a substitute for poplar wood is now being carefully considered in Italy.

The choice of the above areas of investigation was purposely
made:

a) owing to the availability of sufficient ground truth and the certainty of providing a definitive check on the expected results;

b) with a view to obtain quantitative information for rapid assessments of inventory operations, costs and benefits.

To supplement the LANDSAT data, the programme employs an efficient combination of various platforms and equipment (MSS airborne sensors, spectrometers and radiometers installed on helicopters and cherry-pickers, ground measurements and observations in lysimeters and training fields, complementary laboratory investigations, conventional aerial photo-coverages etc.). This is in order to provide an exhaustive comparison basis for recommendations for improved Earth observation systems.

The TELLUS Programme - Social Impact

The second experiment is the TELLUS Project approved within the frame-work of the NASA/HCMM-A mission utilization. Specific recommendation by the DG-VI: investigating the feasibility of applying Remote Sensing techniques for monitoring soil moisture and vegetation stress in various European vegetative cover conditions.

Up to now conventional ground measuring instrumentation for evaluating moisture content in soil has been applied only on small soil patches. Networks composed of a great number of ground stations are needed for extensive hydrological data acquisition. In the event that the TELLUS investigations are successful, the DG-VI would consider it to be very important to have in the future a repetitive, comprehensive and synoptic facility (satellite and/or aircraft systems)for monitoring the status of soil and vegetation. This would also give the CEC an opportunity to put into effect certain directives of the Council, aimed at assisting the European zones classified as less-favoured from an agricultural point of view, owing to permanent natural handicaps (poor soil quality and/or a short growing season). These less-favoured zones consist of farming areas that are homogeneous from the point of view of natural production conditions and composed of rather infertile land, difficult for cultivation or intensification purposes, and suitable mainly for extensive livestock farming. The CEC effort is aimed at preventing further migration of local population from these areas, which are predominantly dependent on agricultural activity and the accelerated decline of which would jeopardize the viability of the areas concerned and their continuous habitation. These less-favoured zones also include areas in which farming is necessary to conserve the country-side, particularly for reasons of protection against erosion, to preserve the tourist potential, or to protect the coast-line.

Another main concern of the DG-VI is the surveillance of abandoned regions in order to assess their degree of degradation and the potential influence of various factors, among which soil moisture is important. This is particularly important in two cases:

a) The case of hydrologic deficiency, such as in hot semi-arid zones (South of Italy and EEC-associated African countries). In this case lack of water in the soil narrows the range of residual cultural possibilities. Synoptic monitoring of soil moisture content evolution seems to be essential;

b) The case of hydrologic excess, such as in saturated zones. In this case soil saturation is an important parameter, both from an agricultural view-point and for monitoring floods (Italy and the United Kingdom).

Recent JRC Experience: Some Applicability Considerations

Some positive results have been gained from the current investigations. Various difficulties are also coming to light during the R & D effort of the national partners of the JRC-Ispra. These may all contribute to a better assessment of requirements and benefits of global observation systems applied to renewable Earth resources.

Vegetation Acreage Estimates: Examination of the Most Influential Parameters and Improvement Techniques

Vegetation identification, mapping and acreage determination is an important step towards the most rational management of European renewable resources. Agricultural areas in use (93.4×10^6 hA) represent more than 60% of the total area of the 9 EEC countries. About 21% of this is represented by woodland.

A correct acreage determination is of great importance. Since total crop production is given as productivity (yield per unit area) multiplied by crop acreage, errors in acreage determination give rise to corresponding production estimation errors. This occurs mainly together with a significant deviation of productivity from the standard values (years of abnormal climatic conditions).

Spaceborne Sensor Spatial Resolution

Following the advice of the EEC Statistical Office in Luxemburg, the requested acreage determination accuracies should in some cases be higher than 90%. Agriculture holdings in Europe

can be divided into two main classes: holdings equal to or greater
than 20 hA of cultivated area (54%), and holdings less than 20 hA
(46%). Since several crops divide the latter into parcels smaller
than 1 hA, in order to achieve the desired accuracies, it turns
out that at least 50% of holdings of this class must be correctly
crop-classified. To satisfy such requirements, resolution performance for a satellite must be far better than that of the present
LANDSAT generation, and in any case less than 20 or 30 metres.

A consistent accuracy improvement is expected to be obtained
with the next earth resource satellite generation (40 x 40 m).
For fields of the same size as the satellite ground-resolution
element, accuracy will probably be improved little. But for
classification of crops which are grown on larger fields, the
improvement will probably be decisive. Field shape can, in this
respect, have an important influence. Using an automatic acreage
inventory technique, the computer time will be increased by four
unless the improved resolution justifies the use of some simpler
and/or faster classification algorithms.

Multidate imagery allows differentiation of some neighbouring
vegetal species and increases the probability of identification of
small fields when they have been wrongly classified using one single
scene only. Using multitemporal data analysis of the same scene, a
corresponding accuracy improvement can be obtained proportional
to the square root of the number of scenes. In this case the computer time is proportional to the number itself.

Given equal computer time, the small size of European fields
favours the use of only one scene corresponding to a certain
ground resolution rather than four scenes of a double-ground resolution. This may not apply to very homogeneous cultivated areas.

Masking Atmospheric Effects

A decisive improvement in spatial resolution towards future
operational applications can be obtained by correcting multispectral
data for atmospheric effects. Transformations may be successfully
applied to satellite data following correlation of count rate to
reflectance. Target signature may become independent of both
transmittance and path radiance and sun position.

These techniques are promising because contrast is increased
(particularly in bands corresponding to LANDSAT channels 4 and 5)
and data become compatible with ground reflectance measurements.
Systematic determination of atmospheric parameters (experimentally
and theoretically) for cloudfree satellite passages, can furnish
reflectance catalogues which may help in classification.

Some experiments performed at the JRC-Ispra have led to the
conclusion that ratioing techniques may be substantially improved

by correction of atmospheric effects. Some reflectance ratio
($\rho7/\rho5$)mappings of a training zone showed that correspondences
between different reflectance ratio ranges (i.e. grey tones, densities) and ground-cover type classes (i.e. grassland, open vegetation, closed vegetation, dead vegetation or town with some vegetation, etc.) are not so representative as those obtained by supervised classification.

The important point is that $\rho7/\rho5$ may be considered as an
indicator of standing biomass and therefore related to the so-
called "primary production" (chlorophyll content per unit area).
In fact the above class division of ground objects applies to
multitemporal imagery of the same scene, independent of the sun's
position. Therefore the ground irradiance and count rates are
quite different. Objects which change their reflectance ratio
class can be distinguished from objects which retain their class
during a part of the season. Inhabited regions, forests, unused
land and water fall within the first categories, whereas intensively cultivated land (irrigated and non-irrigated crops) belong
to the second one.

Texture/Spatial Feature Influence

Texture analysis techniques are in progressive evolution
and are expected to make a valuable contribution. A preprocessing
technique has been applied to space data at the JRC-Ispra using
textural measurements. Mixels (mixed picture elements) have been
identified. They are the inevitable result of the spatial quanti-
zation of low resolution images. Pixels which differ greatly from
their neighbours have a greater likelihood of being mixels than
pixels, which are similar to their neighbours. Mixels are shown
to be representative of types of land-use involving high texture.
Texture measurements are useful in determining land usage. By
screening data to eliminate mixels, improved results are obtained
in test-site selection, classification and cluster analysis. Thus,
in the computer processing and classification of low resolution
data, the role of mixels may be very important. Ignorance of their
presence can lead to poorer classification results and increased
processing costs. Constructive uses of mixels, however, can increase the usefulness of many common classification algorithms and
increase the interpretability of results.

Using Multidate Scene Merging

One way to improve accuracy and to limit dispersion when processing separately some scenes of the same geographic area, is to
merge the data. Significant results, for a possible operational
application of poplar inventory in the Po Valley (Italy), were
obtained using three 1975 LANDSAT -2 scenes. The evaluation
of the percentage of poplar-afforested areas appeared to be

difficult owing to the particularly fragmented field situation.
The study showed that an acreage estimation of poplar groves,
based on LANDSAT satellite data, in European land situations with
limited planted areas, is able to give useful results under nearly
operational conditions, when the input parameters for computer-aided
classification are evaluated over reduced training samples of the
area studied or even apart from it.

Processing separately any scene, the accuracy achieved for
poplars covering more than 25% of the ground is approximately 80%,
whereas processing the three scenes together as a unique 12-channel
assembly brings the accuracy up to nearer 95%. Dimensionality
data reduction from 12 to 6 was tested by principal component
analysis and appears to be promising. Since the above mentioned
poplars refer to "intermediate and adult" classes containing
approximately 90% of the overall timber volume, this result can be
profitably extended to an operational context area.

Level Slicing for Rice Area Inventory

Some encouraging results have been gathered as far as computer-
aided classification and area inventory of the irrigated rice crop
are concerned. A comparison has been made of the most current
automatic classification algorithm. From this it appears that level
slicing of LANDSAT channel 7 is the most effective and simple
method for rice identification. Rice fields are flooded in both
Southern Europe (Camargue and Po Valley) and Madagascar. In this
period (after sowing and before the complete crop emergence) the
water contrast versus surrounding landscape is very strong. Results
obtained using both LANDSAT 1 (1973) and LANDSAT 2 (1975) scenes of
the same area, confirm that the overall acreage of rice fields can
be assessed by satellite within an error of less than 2%. This
high quality performance is related to the relatively regular size
of the European rice fields. Some strongly absorbent targets such
as water bodies, streams, and towns were classified as rice. But
they can easily be deducted from the global inventory by comparison
with the processing of the scenes of the same area taken from time
to time, when the rice fields are not flooded. Repetitivity in this
particular case is a major concern. Considering the 50 percent
probability of cloud coverage and the fact that the most favourable
rice identification period lasts only one month (May in the Po
Valley), a nine-day cycle satellite is needed in order to obtain
at least one single scene.

Processing of rice scenes referring to a later phenological
situation (when spectral differences are introduced by the emergence
of the rice) seem to give a poorer result. A combination of the
clustering method and a uniformity mapping procedure was used. High
variability of reflectance data makes discrimination between rice
and the surrounding vegetal species difficult (20% error).

Space – and – Airborne Sensor Spectral Resolution

Under European agricultural conditions, the excessive width of the LANDSAT channels creates some difficulties in picking up the most peculiar feature differences among clonal (i.e. poplars) and varietal (i.e. rice) classes. Although the advantage provided by the 10 channels of an airborne MSS cannot be fully exploited in agricultural investigations, it seems to apply well enough to these particular cases. This acreage inventory may become a "parcel level" operation and has to be classed as an inventory of type(b). A narrower channel-width (50 µm) is probably needed in selected channels for phenological stage identification.

A method has been proposed for determining timber volume for poplars within the context of the AGRESTE Project. It is based on the possibility of developing a model which, starting from a discrimination between two or more age categories, correlates the poplar dimension with the crown area. For such a determination, aircraft MSS measurements would have to be associated with LANDSAT in order to take into account the many parameters (plantation density and mode, clones etc.),that are inaccessible with satellite measurements alone. A model was developed by French co-investigators to assess the accuracy of the measurement of crown surface percentage (scale 1/10,000). This model was digitized and computer-processed. The good agreement between the measurement and calculation (3%) seems very promising for the correct interpretation of aircraft MSS data.

Main Inventory Characteristics

Some considerations of the main characteristics of renewable resource inventory in EEC countries are presented here. Two broad categories can be determined:

a) Inventory aimed at obtaining statistical data on a broad basis to organize more rational management of the EEC resources and/or a better evaluation of foreign trade exchanges between the Community and external countries. Space Remote Sensing seems to be most promising in some cases, owing to its synoptic characteristics and capability to meet accuracy requirements at a global level.

b) Inventories aimed at precise (parcel level) determination. Aero Remote Sensing is able to provide basic crop identification, the essential element remaining aerophotography (high resolution, possibility of ortho-photos and ortho-maps, etc.). This category of inventory is generally rather expensive. The role of Remote Multispectral Sensing in assisting conventional photo-interpretation could probably lead to cost reduction. Some experiments now in

progress in the EEC countries will soon produce results offering a partial, but interesting answer to this question.

Yield Forecasting

Improvement Using Remote Sensing

In addition to the advantage of a better acreage estimate,
accuracy obtained from the examination of different scenes of the
same geographical area, repetitively, provides a basis for yield
forecasting. Yield prediction by means of Remote Sensing methods
is in fact based on the spectral observation of the main crop
phenologic characterization stages, in addition to the most important climatological parameter analysis.

The reliable making of yield predictions for some important
cereals (wheat, barley, rice, etc.), is considered to be a fundamental requirement in EEC policy concerning foreign trade. As mentioned earlier, significant annual yield variations may cause important production estimate errors when acreage determination is
incorrect. Conventional rice forecasts in Italy, based on statistical procedures, are normally affected by an error at least of
10-15%. In exceptional years the estimation error exceeds 25-30%
resulting from a considerable acreage variation and/or a yield
change owing to exceptional climatic conditions. Yield prediction
based on temporal signature variation detectability by Remote Sensing could lead to improved resource management in the case of
these exceptional situations. The cost of census-taking and
enquiries on acreage and crop production estimates in the EEC
countries is a sufficient justification for establishing agrometeorological models, mainly for the most important cereals
(i.e. barley in the United Kingdom and rice in Italy). (Editor's
note: see the paper by Baier.).

Rice Biomass Influence on Reflectance

As far as rice is concerned, some systematic studies are
being carried out within the framework of AGRESTE. Lysimeter
experiments on the influence of fertilizers (Nitrogen) on the
biomass spectral characteristics have shown that a linear correlation exists between total biomass and the ratio $\rho 7/ \rho 5$. Controlled
field campaigns confirmed that the function linking the values of
$\rho 7/\rho 5$ observed in different periods of the vegetative cycle to the
final total biomass can be assumed to be linear. Thus the above
ratio has been confirmed as a good physiological indicator for rice
when biomass variations are induced by different amounts of fertilizer. In order to take into account the great openfield variability of rice cultivation, a statistical approach has been adopted
to correlate variability characters and phenologic behaviour. It
was possible to note that the sequence of all characteristic

phenologic stages connected with the plant development causes
either a change in sign or an inversion, or at least a slope
change in some statistical curves (Pearson coefficient). This
demonstrates that variability is probably a very useful tool for
discrimination of rice phenological stages (like the full flowering
stage, singular point of the ratio $\rho7/\rho5$ curve).

For satellite identification of the most decisive rice stages
it appears mandatory to have 9 days repetitivity (cloud coverage
permitting). Making some air flights on selected training zones
sometimes seems to be the most effective and reliable solution.
Helicopters could also be used as reflectance measurement platforms
in coincidence with systematic ground morphological observations.

In concluding this section, one must stress the importance
of a further R & D effort to improve the present situation.

Crop Diseases: Early Detection and Monitoring

The intensity of diffusion and economic impact of some crop
diseases are increasing progressively. In order to limit both
disease propagation and the use of fungicides and pesticides
(which have negative effects on the environment and on the
fertility of soils), early detection methods could be exploited.
Remote Sensing could play an important role in this respect.

Experiments on diseased rice within the framework of the
AGRESTE Project have demonstrated that thermal-IR air surveys
provide accurate identification and mapping of a yellowing-type
disease called "giallume", several days before it can be detected
by conventional phytopathological measurements.

Some interesting results have also been obtained by investigating the rice reflectance behaviour in lysimeter-controlled
conditions. In this context Suit's multi-layer canopy model can
be applied to extrapolate the lysimeter situation (only partially
attacked by disease) to a 100% infected open-field hypothetical
operational situation. It has been found that "giallume" could
be detected by channel 5 of the LANDSAT satellite.

The above examples confirm the benefits of applying Remote
Sensing techniques in the fight against crop diseases. Following
recommendations by the DG-VI a specific invenstigation has been
proposed to be carried out within the framework of the next JRC
pluriannual programme.

Conclusions

The initiatives promoted by the Commission of the European
Communities, aimed at evaluating the potential of Remote Sensing
methods as a support to agronomic research and to agricultural management, are confirming its usefulness in identifying present and
future requirements for operational data.

Spectral, spatial and temporal characteristics of the synoptic
space imagery given by the present LANDSAT satellite generation do
not readily fit into present typical agricultural situations in the
EEC member countries. Nevertheless, in some cases of crop-and-
forestry classification and inventory, Remote Sensing is demonstrating that it is an effective tool fitting requirements of a future
global operational observation system applied to renewable resources.

REFERENCES

Agazzi, A. and Franzetti, G., "Effects of Rice Biomass and Yield
    on Reflectance in the LANDSAT Channels". AGRESTE Project
    Report, No. 15, 1975.

Cassinis, R., Galli de Paratesi, S. and Guyader, M., "Agricultural
    Resource Investigations in Northern Italy, Southern France and
    Madagascar". Paper presented at the Symposium on European Earth
    Resources Satellite Experiments, Frascati, January 28 - Febru-
    ary 1, 1974.

Colwell, J.E., "Grass Canopy Bidirectional Spectral Reflectance".
    9th International Symposium on Remote Sensing of Environment,
    Ann Arbor, April 15 - 19, 1974.

de Carolis, C., Baldi, G., Galli de Paratesi, S. and Lechi, G.,
    "Thermal Behaviour of some Rice Fields affected by a Yellowing
    Type Disease". 9th International Symposium on Remote Sensing
    of Environment, Ann Arbor, April 15 - 19, 1974.

Fraysse, G., "Perspectives offered by Remote Sensing in Agricultu-
    ral Resources Management". The Colston Symposium, April 5 - 9,
    1976.

Gatelli, E., "Development of a 4-channel Radiometer for Paralaxe-free
    LANDSAT Ground-truth Measurements with Variable Field of View".
    AGRESTE Project Report, No. 14, 1975.

Herzog, J.H. and Sturm, B., "Preprocessing Algorithms for Radio-
    metric Corrections and Texture Spatial Features in Automatic
    Land-use Classification". 10th International Symposium on
    Remote Sensing of Environment, Ann Arbor, October 6 - 10, 1975.

LANDSAT - 2 Satellite Follow-on Investigation No. 28790, AGRESTE
Project, First, Second and Third Progress Reports, 1976.

La Pietra, G. and Megier, J., "Acreage Estimation of Poplar planted
Areas from LANDSAT Satellite Data in Northern Italy". XVI
International IUFRO World Congress, Oslo, June 20 - July 2,
1976.

Maracci, G. and Sturm, B., "Measurements of Beam Transmittance and
Path Radiance for Correcting LANDSAT Data for Solar and
Atmospheric Effects". Paper presented at the "Cinquièmes
Journées d'Optique Spatiale". Marseille, October 14 - 17,
1975.

Megier, J., "Automatic Classification of LANDSAT Satellite Data
for Agriculture and Forestry". Proceedings of the 5th Conference on Space Optics, Marseille, October 14 - 17, 1975.

Rogers, R.H. and Peacock, K., "Machine Processing of ERTS and
Ground-truth Data", LARS Conference on Machine Processing
of Remotely Sensed Data, October 16 - 18, 1973 at Purdue
University BSR 4034,July, 1973.

Sturm, B. and Mehl, W., "Ein Programmsystem zur Verarbeitung
und Auswertung von ERTS Daten". Erderkundung Symposium of
DFVLRDPG, Porz-Wahn, 7 - 11 April, 1974.

PLANNING FOR AN EARTH RESOURCES

INFORMATION SYSTEM *

James G. Gehrig
Member of the Professional Staff
Committee on Aeronautical and Space Sciences
United States Senate
Washington, D.C., 20510, U.S.A.

United States Landsat satellites for remote sensing have been so successful in their experimental phase that there is a demand for an operational system. 1/ This demand is not confined to the United States. It is international and arises from the fact that remote sensing technology has already been used to assist in the solution of numerous practical resource and environmental problems. This particular application of space technology became available just when needed to deal with global problems of air, land and sea pollution, the adjustment of world food supplies to population growth, and the more effective management of resources. Worldwide appreciation of Landsat data is evident from the fact that some 50 nations and five international organizations have entered into co-operative agreements with NASA to use Landsat data. Some nations - Argentina, Brazil, Canada, Chile, Italy, Iran and Zaire - have made agreements with NASA to build ground stations at their own expense. Other countries have asked to participate actively in the acquisition and processing of remote sensing data.

United States management of the acquisition and dissemination of Landsat data is in accordance with domestic and international law. The United States National Aeronautics and Space Act of 1958 provides that:

> "...it is the policy of the United States
> that activities in space should be devoted
> to peaceful purposes for the benefit of all
> mankind."

* (The views expressed in this paper are those of the author and are not necessarily related to those of any organization with which he is associated.)

Furthermore, in establishing NASA as a civilian agency responsible for directing and exercising control of aeronautical and space
activities, the law provides that:

> "...activities peculiar to or primarily
> associated with the development of weapons
> systems, military operations, or the defense
> of the United States (including the research
> and development necessary to make effective
> provision for the defense of the United
> States) shall be the responsibility of, and
> shall be directed by, the Department of
> Defense; and that determination as to which
> such agency has responsibility for and direc-
> tion of any such activity shall be made by
> the President...."

In delineating the civilian responsibility of NASA, as distinct
from that of the Department of Defense, Congress provided that the
United States shall cooperate with other nations and groups of nations and authorized NASA to engage in a program of international
cooperation. During the past 18 years, NASA has developed hundreds
of projects that have furthered international cooperation through
bilateral and multinational agreements.

United States foreign policy concerning international space
activities is conducted in accordance with the NASA Act as well as
with international law, particularly with regard to the four space
treaties already in force. These treaties must be observed for any
application of space technology. Major provisions in the 1967 Treaty on Outer Space are that the exploration and use of outer space
"shall be carried out for the benefit and in the interest of all
countries" and that there shall be freedom of scientific investigation and encouragement of international cooperation. In addition,
"States Parties to the Treaty undertake not to place in orbit around
the Earth any objects carrying nuclear weapons or any other kinds of
weapons of mass destruction, install such weapons on celestial bodies, or station such weapons in outer space in any other manner."
Furthermore, "States Parties to the Treaty shall bear responsibility for national activities in outer space, including the moon and
other celestial bodies, whether such activities are carried on by
governmental agencies or by non-governmental entities, and for assuring that national activities are carried out in conformity with
the provisions set forth in the present Treaty."

In a number of cases when it has proved desirable to elaborate
on the fundamental principles contained in the 1967 Treaty on Outer
Space, additional treaties have been negotiated to deal with these
matters in more detail. Thus nations achieved the "Agreement on
the Rescue of Astronauts, the Return of Astronauts and the Return

of Objects Launched into Outer Space" (1968); the "Convention on
International Liability for Damage Caused by Space Objects" (1972);
and the "Convention on Registration of Objects Launched into Outer
Space" (1976) - all negotiated by consensus within the Legal Sub-
committee of the United Nations Committee on the Peaceful Uses of
Outer Space.

The subject of remote sensing of the earth by satellite has
been under consideration by the Subcommittees of the United Nations
Committee for several years and is on the agenda of the Legal Subcommittee for its next session in New York from March 14 to April 8,
1977. Proposals for an operational system for remote sensing of the
earth by satellites must necessarily consider not only the international law that already applies to this activity but also that currently being negotiated. In addition, proposals should take into
account the deliberations of the Scientific and Technical Subcommittee of the United Nations Committee on the Peaceful Uses of Outer
Space on such factors as predictable costs and benefits, organizational and financial requirements, and the feasibility of a coordinating function for the United Nations in a future operational remote sensing system.

Remote Sensing Issues Raised in the United Nations

To identify the issues that have been raised in the Legal Subcommittee concerning remote sensing is to reveal the complexity of
the task of obtaining a consensus on international guidelines for
the conduct of States engaged in this activity. The principal issues are embodied in the following questions:

Should there be a requirement that States must give their consent prior to remote sensing by satellites of their territories?

Should access to remote sensing data and its dissemination be
restricted to the sensed State? Or should it be unlimited and
available to all persons and nations?

What methods could be adopted to ensure that the benefits are
enjoyed by all States?

Under what conditions and terms should States consult with each
other concerning remote sensing of the earth by satellites?

Is it technically feasible to separate the remotely sensed data
and information on natural resources 2/ so that they could be made
available to the sensed country before or at the same time access
is granted to third parties?

What methods does a State have for exercising sovereignty over

its natural resources in order to control exploitation by other
nations and persons? Does remote sensing affect such control?

If an international organization with jurisdiction over remote
sensing is established, how would it be funded?

To what extent can countries share regional facilities engaged
in remote sensing acquisition, analysis and dissemination?

Concerning the present status of international law in this
area, Ronald F. Stowe, Assistant Legal Advisor for United Nations
Affairs, Department of State, has said that

> The view of the United States was, and remains,
> that there is no provision of applicable inter-
> national law which restricts or inhibits remote
> sensing of the Earth from outer space. Quite to
> the contrary, the 1967 Treaty on Principles Go-
> verning the Activities of States in the Explora-
> tion and Use of Outer Space, Including the Moon
> and Other Celestial Bodies expressly proclaims
> in Article I that "Outer space, including the
> moon and other celestial bodies, shall be free
> for exploration and use by all States without
> discrimination of any kind..." "There shall be
> freedom of scientific investigation in outer
> space..." and "States shall facilitate and en-
> courage international cooperation in such in-
> vestigation." 3/

There is a long history of uncontested remote sensing of the
Earth, including sensing related to natural resources. The earliest
meteorological satellites and the early manned space flights engaged
in remote sensing, prior to the adoption of the 1967 Space Treaty.
It is the view of the United States that neither the advancement
of the technology nor a wider practical application of remote sensing data constitutes any reason to distinguish Landsat-type data
or Seasat-type data or other future remote sensing systems data from
the universally accepted policies of the peaceful exploration and
use of outer space. Fortunately, it appears from reports of the
activities in the Legal Subcommittee that there is little or no continuing support in the Subcommittee for the idea that remote sensing
is an activity outside the scope of the Outer Space Treaty or that
acquisition of data by remote sensing should be subject to the prior
consent of the sensed countries. This feeling, of course, does not
mean that the anxieties of all countries that gave rise to the discussion in the first place have been quieted.

The dissemination of the remote sensed data and information
derived from that data is also of great legal, political, economic

and technical interest. A wide diversity of opinion has been expressed as to how remote sensing data and the information derived
from that data should be handled and disseminated. There seems to
be some agreement that international law does not impose any re-
striction on open dissemination; however, the Legal Subcommittee
is examining carefully whether restrictions should be applied. As
with the issue of acquiring the remotely sensed data, those States
advocating restrictions on the dissemination of the data are concerned with the protection and control of activities within their
national boundaries. These activities are related to the development and exploitation of natural resources.

The position of the United States on this issue is that open
dissemination is more likely to enhance than diminish the ability
of individual countries to control their natural resources. The
adoption of a restricted data dissemination policy would most likely
establish two classes of countries, one technologically advanced
enough to have its own remote sensing programs and therefore capable
of obtaining data on other countries directly; and the other (consisting of most other countries) that could obtain data only about
their own territories indirectly. Further, neither the dissemination nor the analysis of remote sensing data can affect the control
of natural resources. It is only at the point that someone attempts
to apply that information to an actual development plan or exploitation of a natural resource that the question of State control
arises.

It appears that the Legal Subcommittee's examination of the legal implications of remote sensing is thorough and detailed and will
help clarify the law and reassure those States concerned for their
natural resources. Some draft principles have been formulated in a
Working Group of the Legal Subcommittee. Of particular interest to
this NATO Advanced Research Institute is Principle IV which reads
as follows:

> Remote sensing /of the natural resources of the
> earth/ /and its environment/ from outer space
> /should/ /shall/ promote the protection of the
> natural environment of the earth. To this end
> States participating in remote sensing /should/
> /shall/ identify and make available information
> useful for the prevention of phenomena detrimental
> to the natural environment of the earth.

The bracketed language has not yet been agreed to in the Working
Group, but this paragraph gives the flavor of the Group's thinking.

Congressional Interest

The interest of the United States Congress in remote sensing is based upon its responsibility for authorizing and appropriating funds for such programs. The legislative process whereby such decisions are made requires initial consideration by Congressional committees that have been delegated jurisdiction over these matters. In the House of Representatives this responsibility rests with the Committee on Science and Technology. In the United States Senate, the Committee on Aeronautical and Space Sciences holds hearings and reports to the Senate on authorizing funds and other legislative matters for outer space projects such as Landsat. Thereafter, the Appropriations Committees of the House and Senate consider and recommend to the House and Senate, respectively, the necessary funds for authorized activities.

Thus far, Congress has approved the experimental system launched by NASA. It is evident from testimony taken by the Committees during the past few years that results from the acquisition and dissemination of Landsat data have far exceeded expectations. The studies and reports made by the Senate Committee on Aeronautical and Space Sciences reveal that Landsat information has enabled many countries to improve the management of their resources. In line with U.S. policy on the distribution of information from the United States Geological Survey established in the last century, earth resources data produced by satellites are available for purchase by any person, industry or nation.

We have now come to the point where a decision must be made about the future of Landsat. The issues are whether Landsat should be continued on an experimental basis and, if so, for how much longer? Should Landsat be made operational and, if so, how and when? If a commitment is made to fund an operational system, how should it be organized and managed? What should be the relationship between government and private industry? How should we provide for coordination between national and international requirements?

Planning for the funding, management and operation of a continuing earth resources information system involves consideration of some unique features. The problem is unlike that of establishing the U.S. Communications Satellite Corporation and the global commercial International Telecommunications Satellite Organization, which could be built upon the strong foundation of an existing communications system of private enterprise and governmental involvement.

The problem of designing an earth resources information system is also different from that involving meteorological data, which is traditionally organized and operated by governments on the basis of the best available technology at any given time. This function is now well organized and operated in the United States by the National

Oceanic and Atmospheric Administration (NOAA), coordination being at the international level with the United Nations specialized agency, the World Meteorological Organization (WMO).

Unlike communications and meteorology, which are so specific as to be easily identifiable, earth resources survey information can be used for a great variety of purposes. This raises the question of how such a multi-purpose activity should be organized and managed. It seems that something almost entirely different may be devised to move from the experimental to the operational phase. The chairmen of Congressional committees have taken the initiative in introducing legislation designed to address these problems.

Major Features of the Proposed

"Earth Resources Information System"

On August 24, 1976, Senator Frank E. Moss, Chairman of the Senate Committee on Aeronautical and Space Sciences, introduced S. 3759, 94th Congress, 2d session, a bill "to develop, establish, and validate an Earth Resources Information System, to direct the National Aeronautics and Space Administration to continue a research and development program, to provide, if necessary, for establishment of a corporation to operate the domestic ground segment of such a system, /and/ to establish an Office of Earth Resources Policy."4/

The bill states it is the policy of the United States that advanced technology shall meet public needs for information on food, water, air, minerals, materials, and may include other matters; that the Earth Resources Information System serve the needs of the United States and other countries and contribute to world peace and understanding; that data be furnished to developing as well as highly developed countries; that all users have equal access to products of the system; and that competition shall be promoted in acquiring equipment and services. The System will be developed in two phases: validation and operation.

The bill calls for two primary management segments. First, there is to be the "space segment", which is defined as a portion of the System which includes satellites or other observation sources and the associated ground equipment for command and control of the satellites. Second, there is to be a "data handling segment", that portion of the System "which includes receiving data from the space segment, archiving, retrieval, processing, and duplication and dissemination of the products of the system on demand to users or subscribers."

The private sector is to be encouraged "on a commercially viable basis" to take responsibility for establishing and operating the U.S.

data handling segment and for disseminating U.S. gathered earth resources data. This activity is designed to facilitate the validation and development of the System. Qualified private entities are
to have access on an equal basis to the data if they are prepared
to handle the data handling segment during the validation and operational phases. When the private sector has established its capability of delivering processed earth resources data, the Federal Government shall use this source for its requirements.

Under this plan, NASA has responsibility for managing all government research and development related to the System, establishing and operating the space segment, and replacing satellites and
other equipment to ensure that data are continuously available.
Priority is to be given to the continuity of the space segment.

In Title III of the bill, "Federal Coordination, Planning, and
Regulation", an Office of Earth Resources Policy is to be established in the Executive Office of the President. The Director of this
Office is to aid in the planning and development of the System and
review all activities, including those of the Earth Resources Information Corporation. The Director is given a number of coordinating
and supervising functions within the government and between the government and the private sector. NASA is also given the role of ad-
vising and consulting with the Corporation or other private entities
on technical characteristics, and to provide reimbursable services
to the Corporation to the extent feasible.

An Earth Resources Information Corporation is to be created by
the President if he determines that one is required. The Corporation is to be established for profit and is not to be a government
agency. If the Corporation is created, it is to operate a commercial domestic data handling segment of the Earth Resources Information System. The Corporation is authorized to "plan, initiate,
construct, own, manage, and operate itself or in conjunction with
foreign governments or business entities a commercial Earth Resources Information System." The Department of State is to be notified
of foreign business negotiations so that relevant foreign policy
considerations may be taken into consideration.

After this bill (S. 3759) was introduced in the Senate, Senator
Moss asked for comments from a broad spectrum of users. The matter
is still in the discussion stage and no hearings have been held. 5/

The second session of the current (94th) Congress adjourned sine
die without passing S. 3759. Bills not passed during one Congress
do not carry over to the next. They "die". However, a revised bill
to establish an "Earth Resources Information System" most likely
will be introduced in the Senate soon after the new Congress con-
venes on January 4, 1977. The proposed bill will be reviewed and
commented on by all interested departments and agencies of the

United States Government and by any interested parties outside the Government. Any comments, suggestions or recommendations on S. 3759 from this Advanced Research Institute or from any of you as individuals would be gratefully received by the Senate Committee on Aeronautical and Space Sciences.

FOOTNOTES

1/   Landsat is an experimental system whose prime purpose is the testing of new technology and new procedures. While the Information from it has many applications, it does not guarantee a flow of data to the users. Here <u>operational</u> system means a system that would provide an assured continuing service based on user requirements.

2/   There is <u>no</u> agreement on the definition of "natural resources".

3/   Stowe, Ronald F., The Development of International Law Relating to Remote Sensing of the Earth from Outer Space. A paper given at the 19th Colloquium on the Law of Outer Space of the International Institute of Space Law. Anaheim, California, Oct. 12, 1976.

4/   An identical bill, H.R. 15736, was introduced in the United States House of Representatives on September 28, 1976, by Representative Olin Teague, Chairman, House Committee on Science and Technology. Copies have been distributed to members of this NATO Advanced Research Institute, as an appendix to this paper.

5/   Background information is available in "An Analysis of the Future Landsat Effort", Staff Report prepared for the use of the Committee on Aeronautical and Space Sciences, United States Senate, August 10, 1976, (for sale by the U.S. Government Printing Office, Washington, D.C. 20402). Copies of the Committee Print have been distributed to members of this NATO ARI, as an appendix to this paper.

APPLICATION OF REMOTE SENSING FOR POLICY, PLANNING

AND MANAGEMENT IN FORESTRY AND AGRICULTURE

Gerd Hildebrandt
Professor and Director of the Department of
Photogrammetry and Photo Interpretation
Institute of Forestry and Forest Management
University of Freiburg, Germany

Introduction

In a world of limited natural resources, increasing population, increasing demands on food supplies, goods, timber, recreation areas, energy, water, etc., the rational use and conservation of resources and improvement of the management of agricultural land, forests and water, as well as environmental protection, are indispensable.

All these require knowledge of land capability and long term planning in land use, forestry, and agriculture. Therefore there exists a distinct need for timely and reliable information and data concerning the resource base and the land to be developed and managed. For this purpose, it is absolutely necessary

- to inventory and map the status quo of land use, resources and conditions as well as the accessibility of the land and its resources,

- to analyse and research the relationships between natural or cultivated vegetation and the geo-conditions in the region, with the aim of defining land capability for the different kinds of possible sustained agricultural use,

- to permanently monitor the natural or man made changes in environmental conditions and the stock of resources, including monitoring the land and its resources for prompt detection of damage, diseases, fires, landslips, flooding etc.,

- to forecast the more important agricultural crops as well as environmental developments.

For a realistic discussion of the role of remote sensing in
the fields of operational natural resources, especially forest
inventory, one has to distinguish between:

- reconnaissance surveys

- preinvestment surveys

- inventories for operational management.

The above mentioned purposes require

- spatial / area information and data about the land, its
  natural condition, occurence and acreage of forest types,
  land use classes, crop types, etc.,

- qualitative information about growing crop species, tree
  species, species composition in natural or cultivated
  stands of vegetation, as well as about such stand quality
  aspects as vigor, healthiness, increment rates, timber
  quality, etc.,

- quantitative information and data about cultivated stands
  of vegetation, or natural but usable vegetation, crops,
  timber volume and age, plantation density, grazing capacity
  of pasture grounds, figures about the loss on any kind of
  resource after disasters, etc.

Information and data for decision-makers and managers in agriculture
and forest services are useful only

- if they meet specified, required precision or reliability

- if they are available soon after acquisition

- if relatively low cost and simple evaluation and interpretation techniques are available

- for certain tasks if repeated coverage is possible.

Reconnaissance Surveys

In many parts of the world knowledge and information of land
capability and natural resources in terms of quantities and quali-
ties are not available or not complete. In this stage "reconnaissance" inventories and mapping are of greatest importance especially for land use policy and development. This type of survey
can be related to global, regional, or country-wide inventory
areas, or in very large countries - even to parts of a country.

The ongoing FAO/UNEP Tropical Forest Cover Monitoring Project
with its objectives of collecting and evaluating data on a small

scale on present forest cover in the tropics, and serving as a
basis for subsequent monitoring of the changes in that cover, can
be mentioned as an example of a global reconnaissance survey of
resources (see Baltaxe and Langley,1976). This well-known RADAM
Project, designed for the reconnaissance of natural conditions of
approximately 5,000,000 square kilometers of the Amazon area in
Brazil, exemplifies a reconnaissance survey related to a part of
a large country.

Informational needs are related to the main geomorphological
features, the definition of broad bioclimatic zones, broad classes
of vegetation, soils, land use types, etc., and degraded or destroyed tracts. The area to be inventoried is usually very large,
but resurveys are either not necessary or are required only years
later. The extent of the inventory area and the synoptic type of
data required suggest the use of satellite or high altitude airborne remote sensing. The survey, however, can benefit from multitemporal observations that take advantage of phenological changes.
Observation once each "season" may improve the interpretation and
the classification results in many cases. It seems unimaginable
to perform the reconnaissance survey without remote sensing; in
some cases, as in perennially cloudy regions, the use of imaging
systems with all weather capability like the Synthetic Aperture
Radar is an essential requirement.

In considering satellite imagery we have to realize that in
spite of impressive research results, the development of operational
applications is still in its beginning. For most tasks, operational
trials are still very necessary. Only for a few purposes are methods of spaceborne remote sensing really ready to put into practice.

Considering Landsat MSS data for natural resource surveys,
forest inventories and mapping, some present possibilities, as
well as limitations in interpretation, can be described as follows:
It has proved possible to separate forest land from non-forest
land when the data acquisition takes place in the proper season.
In some cases it is possible to separate main forest types. Kalensky
and others (1975, 1976), for instance, were able to consistently
discriminate and map coniferous and deciduous stands in Canadian
forests. But they could not separate individual tree species or
associations of species. Our experience with the interpretation
of Landsat imagery in Central Europe is very similar. Also the
Anderson Report (Anderson et al, 1976) indicates that the more
successful Landsat vegetation discrimination studies attempted
only limited or simple cover type identification, while the less
successful inventories were aimed at multiple-category, detailed
classification or simple classifications in areas of heterogeneous
cover types.

In a comprehensive and critical review Heller (1976) concluded
that "Although it is evident that detailed forest type mapping by
species cannot be accomplished consistently with Landsat data,
stratification of forest stands into strata related to density
class, condition class, age class, etc., may be possible with
controlled clustering techniques". Furthermore, he pointed out,
in full conformity with the author's opinion, that the correlation
of Landsat MSS spectral signatures with parameters related to stand
conditions, varies as a function of phenology, aspect, slope, time
of year, etc.

Kalensky et al, the Anderson Report, and Heller, refer to
experiences and results that are more or less limited to forest
and land use conditions within the temporal zones. For the tropics
and subtropics some similar, but also some different and more
complicated problems, must be considered; there is a huge number
of tree species, and sometimes only a loose correlation between
canopy characteristics as seen from above and bole dimensions
and defects. Experiences with Landsat imagery have shown that
wavebands of about $0.04\mu m$ (against the $0.10\mu m$ bands of the Landsat
sensors) would probably help improve the inventories (see Heller,
1976).

Realizing that the spectral signature of the same vegetation
type as seen from space can vary greatly for many reasons, and
that many species and types have very similar average reflectance
properties, one can expect only limited improvements with better
sensor sensibility.

A major present limitation of Landsat imagery relative to
natural resources, especially forest inventories, is the relatively
low ground resolution of the MSS data. For detection of small
objects and for the identification of features with similar spectral
reflectance properties but typically small texture patterns, much
better ground resolution is necessary. Therefore many users are
focussing their attention on improvements of multispectral scanners,
on the concept of the charge coupled detectors (CCD) and other
developments of high resolution sensor systems, as well as on
promising high altitude aerial photography with high resolution
films. An additional remark is necessary with regard to operational
global or regional crop forecast systems, as well as crop production
alarm systems (see A.B. Park's paper in this Proceedings): a high
resolution sensor system with all weather capability is needed.
The latter is necessary because systematic observation with a high
repetition rate during the season is required in order to identify
and to distinguish the various crop types, to estimate their acreage
and their vigor, state of health, etc. In regions with permanent
cloud cover, but also in many other parts of the world with "only"
more or less temporary heavy clouding, the necessary interpretation
and assessment cannot be done successfully with "good-weather"
sensors alone. On the other hand, we must not close our minds to

the fact that the interpretation of microwave images or data is still in an early stage of development. Nobody knows at present to what extent the identification of crops, the detection of diseases or damage, etc., can be done through the use of a high resolution imaging radar such as the Seasat-A Synthetic Aperture Radar (see the paper by D.J. Clough and L.W. Morley in this Proceedings).

In conclusion: At present Landsat MSS imagery meets the requirements of operational reconnaissance surveys and mapping in agriculture and forestry only in some cases. But, together with Sayn-Wittgenstein (1976), I don't believe that any future large area surveys "should go ahead without considering the possible contributions of satellites". More detailed and therefore more useful inventories and mapping may be expected by using sensors with selected, narrower wavebands in conjunction with improved ground resolution and classification systems which use spectral as well as textural information. For crop forecasting on a global or regional level, specific requirements must be considered. All weather imaging is needed, and strong efforts in research are necessary to develop successful and reliable interpretation methods for this purpose.

Preinvestment Surveys

After the initial reconnaissance surveys, more specific detailed surveys and investigations are required. Local conditions related to more or less large areas become objects of the "preinvestment" survey. Quantitative data receive more emphasis than in reconnaissance surveys,for example merchantable timber volume, average yield per unit area, grazing capacity, etc. For economic decisions, planning and development in agriculture and forestry, detailed and reliable statistical data and qualitative information are needed in addition to maps.

Spaceborne remote sensing data with ground resolutions like those obtained from Landsat 1 and 2 are only of interest for preinvestment surveys in regions with rather uniform conditions and large homogeneous units. But even in such regions preinvestment inventories cannot be based on satellite imagery alone. This is especially true of forest inventories where merchantable timber volumes classified according to species and bole size classes or, in many cases according to periodic increment, must be inventoried.

Therefore one has to consider inventory methods using remote sensing data from several altitudes in combination with ground measurements. In practice multistage inventory using high and/or low altitude aerial photography in combination with ground measurements within sample plots has proven to be very useful. Such methods profit by the increasing ground resolution of each subsequent

stage. Information can be provided by interpretation of remote
sensing data which are useful

- either to achieve higher precision of the data
  measured on ground sample plots for given costs,

- or to decrease the costs to reach a given
  level of precision for the ground measured data.

The statistical approach has been developed from the conventional multistage sampling technique using replacement with equal
probabilities. According to Langley (1969) a three stage estimator
v with fixed or variable probabilities can be described as follows:

$$v = \frac{1}{m} \sum_{i=1}^{m} \frac{1}{p_i n_i} \sum_{j=1}^{n_i} \frac{1}{p_{ij} t_{ij}} \sum_{k=1}^{t_{ij}} \frac{v_{ijk}}{p_{ijk}}$$

v_{ijk} = the measured attribute in stage three

p_i = the probability of drawing the ith first stage unit

p_{ij} = the conditional probability of drawing in the jth
second stage unit given the ith first stage unit

p_{ijk} = the conditional probability of drawing the kth third
stage unit given the first-and second-stage units
which have been drawn

m, n_i, t_{ij} = the sample sizes in the first, second, and
third stages, respectively.

"While we have found the above form of multistage sampling to be
effective in several instances" Langley (1976) said "it should be
used with discretion. The (proportional sampling form) of the model is capable of achieving the highest gains in precision of supplementary variables obtained from remote sensors. On the other
hand, it is also sensitive to low correlations between the measurement and auxiliary variables".

In 1969 Langley, Aldrich and Heller applied the multistage
sampling technique including space photography for a forest resources inventory for the first time. In a pilot timber inventory
of 4.8 million hectares in the southeastern United States they carried out a five-stage probability sampling using Apollo 9 space
photography, 1:60,000 and 1:12,000 and 1:2,000 scale aerial photography, in combination with ground sampling. The results of this
inventory met requirements of the preinvestment survey in countries
with low intensity forestry.

On certain premises, especially in regions with relatively homogeneous forest conditions with large silvicultural units, thematic mapping using Landsat 1 or 2 data with computerized classification may be applicable and useful as a part of preinvestment surveys.

Considerable improvements are needed to develop an operational earth observation satellite system which could be more useful on a worldwide scale, especially for the very important "preinvestment" type of survey. Multistage inventory approaches, the analysis of data for land capability studies and special purposes in agriculture and forestry, as well as thematic mapping using multispectral satellite remote sensing with computerized classification, would all profit from improvements in ground resolution of satellite imagery (up to 10 m). Furthermore, the proper choice of the season of observation and observation frequency, and the proper selection of bands with narrow spectral resolution will result in more useful information. Relative to the latter, two remarkable considerations should be mentioned. Both are based on many experiences with the Landsat 1 and 2 data. For purposes of agriculture and forestry Heller (1976) and the Anderson Report (1976) point to the usefulness of the following bands:

Heller, 1976		Anderson Report, 1976	
Band µm	Bandwidth µm	Band µm	Bandwidth µm
0.54 – 0.58	0.04	0.53 – 0.59	0.06
0.58 – 0.62	0.04	0.58 – 0.63	0.06
0.66 – 0.70	0.04	0.62 – 0.68	0.06
0.80 – 1.10	0.30	0.76 – 0.90	0.14
2.00 – 2.60	0.60	2.00 – 2.60	0.60

Inventories for Operational Management

For sustained and careful management of forests, agricultural land and rangeland, as well as for improvement and protection of well-developed regions, a third type of inventory is necessary. A system of periodic and very detailed inventories and mapping must be established. Referring to European and partly also to North American experiences in intensive land management, a resurvey is typically necessary in intervals of 5 to 10 years. The resurvey and the corresponding map adjustments must provide, of course, comparable data and measurements. The role of remote sensing in such detailed and repeated inventory systems is very important. However spaceborne data seem to be less important than airborne imagery. It should be remembered that since the 1920's aerial photography has proven to be an essential and effective tool for important thematic and topographic mapping, as well as for detailed land use surveys.

Table 1:   Information Sources

Main Information Sources

A = Area mapping/classification
P = Properties, quantitative and qualitative.

SURVEY TYPE	remote sensing			Ground level field work
	satellites	high altitude aircraft	medium altitude aircraft	
Reconnaissance: very large areas, broad classes, less details.	A P	A P	A P	A P
Pre-investment: large areas, refined classes, many details.	A	A P	A P	A P
Inventories for Operational Management: small areas, detailed classes, very many details.			A P	A P

Another important task in well-developed agriculture and forestry areas is the permanent monitoring and control of crops or forests to detect diseases and stand damage as early as possible. In Europe and North America the importance and applicability of airborne remote sensing for these purposes have been clearly demonstrated. In developing countries with their relatively less developed infrastructure and large areas, this monitoring problem can be solved in the near future only by remote sensing.

In North America and Europe infrared colour aerial photography has proven to be a very useful tool for monitoring purposes of such a kind. Early visual detection of diseases is possible, however, only in certain cases. Airborne infrared fire detection and monitoring is another outstanding example of a successful and important operational application of remote sensing in forestry.

Conclusion

Spaceborne and airborne remote sensing can be employed successfully in the surveying and monitoring of agriculture, rangeland, and forest lands. As Table 1 shows, the more detailed and precise the information requirements are, and the smaller the area to be inventoried or to be mapped is, the more the importance shifts from satellite to medium and low altitude aircraft data and imagery, and to ground-level field observations.

Some fundamentals concerning the methodology and the planning of a general or specific natural resources survey should also be considered, as follows:

1. Successful and effective inventory work require a project-related concept of the survey method. This means that no unique "best solution" or "best information source" exists for any particular project.

2. Therefore, the selection and balance of inventory components for the project must be adjusted to the inventory goal, and must consider all local conditions. (These may be climate, accessibility of the land, financial funds, existing information, available facilities, educational status of manpower, etc.).

3. Therefore, planning of a survey implies a comprehensive knowledge of -

   (a) inventory methodology,
   (b) the possibilities and limitations of the feasible tools
   (c) all relevant local conditions and
   (d) a definition of the goals, which must set in advance.

Within this frame of reference for remote sensing one can state:

1. that remote sensing techniques can play a very important
    role in the survey and monitoring of agricultural, range
    and forest lands, if those who handle these techniques
    know how to handle them in the proper way;

2. that remote sensing from space platforms has many potential
    applications in natural resource inventory and monitoring
    of very large areas, and in situations where dynamic
    processes must be observed in short periods; however,oper-
    ational use is still in its infancy and successful appli-
    cations in one area cannot be transferred to other
    areas without new experiments and research;

3. that conventional aerial photography, which has in the
    past been the only remote sensing tool, will certainly
    remain a most important and essential tool in the future;

4. and that promising results of multispectral scanning,
    imaging radar, thermal remote sensing, multi-stage samp-
    ling, and computer-aided image analysis and classification
    may lead to the solution of the still unsolved problems;
    but in many cases these remain to be tested before they
    can be practically applied.

REFERENCES

Baltaxe, R. and Langley, J.P. "The UNEP/FAO Pilot Project on
    Tropical Forest Cover Monitoring". Proc. Remote Sensing in
    Forestry, Symposium. IUFRO Sub. Group, Remote Sensing, Oslo,
    1976, publ. Freiburg 1976 pp.

Hansen M.H., and Hurwtiz,W.N. "On the theory of sampling from finite
    population." Ann. Math. Statist., 1943, 14: 330-362

Heller, R.C., "Natural resources surveys". Invited paper XIII
    Congress of the ISP, Comm. Interpretation of Data, Helsinki
    1976.

Hildebrandt, G., "Forstliche Grossrauminventuren,Allg. Forstzeit-
    schrift"1964, p. 100 -107.

Hoffer, R.M., "Techniques and applications for computer - aided
    analysis of multispectral scanner data." Proc. Remote Sensing
    in Forestry, Symp. IUFRO - Subj. Group Remote Sensing, Oslo,
    1976, publ. Freiburg 1976, p.103-114.

Howard, J.A., "Remote Sensing in tropical forests with special reference to satellite imagery". <u>Proc. Remote Sensing in Forestry, Symp. IUFRO Subject Group Remote Sensing</u>, Oslo 1976, publ. Freiburg 1976, p. 211-226.

Kalensky, Z. and Scherk, L.R., "Accuracy of forest mapping from Landsat computer compatible tapes". <u>Proc. X. International Symposium on Remote Sensing of Environment</u>, Ann Arbor, Michigan.

Kalensky, Z. and Wightman, J.M., "Automatic forest mapping using remotely sensed data". <u>Proc. Remote Sensing in Forestry. Symposium IUFRO Subj. Group Remote Sensing</u>, Oslo, 1976 p. 115-136, publ. Freiburg, 1976.

Langley, P.G., "New multistage sampling techniques using space and aircraft imagery for forest inventory". <u>Proc. VI. International Symposium on Remote Sensing of Environment</u>, Ann Arbor, Michigan, 1969, (2) p. 1179-1192.

Langley, P.G., Aldrich, and Heller, R.C., "Multistage sampling of forest resources by using space photography - an Apollo 9 case study". <u>Proc. 2nd Ann. Earth Aircraft Progr. Review</u>, NASA, MSC. Houston, 1969, (2) 19: 1-21.

Langley, P.G., "Sampling methods useful to forest inventory when using data from remote sensors". 1976.

Sayn-Wittgenstein, L., "The role of remote sensing in thematic mapping with special reference to the tropics ". <u>In Remote Sensing International Training Seminar on Remote Sensing Applications</u>, Lenggries, Fed. Republic of Germany, 1976, p. 57-62.

CONCEPTS OF REMOTE SENSING APPLICATIONS FOR FOOD

PRODUCTION IN DEVELOPING COUNTRIES

J.A. Howard
Senior Officer, Remote Sensing Unit
Food and Agriculture Organization of the United Nations
Rome, Italy

INTRODUCTION

In recent years, advances in remote sensing methodology have outstripped the useful application of the technology to man's needs. This may be partly due to the heavy reliance of remote sensing applications on the relatively undeveloped methods of deductive reasoning rather than the customary inductive approach to problems.

A perusal of technical publications will indicate that leadership in applying remote sensing to the Earth's resources, commencing with aerial photography over 50 years ago (Howard, 1970), has rested mainly with either forestry or geology. Now, with the emergence of a wide variety of problems associated with the rapidly increasing world population and the increasing capability of remote sensing for continuous inventory (monitoring) including repetitive daily coverage of the Earth by satellite, (albeit at present at very low resolution and often with a cloud-cover limitation), conditions favour agricultural leadership. However, this leadership to harness remote sensing to food production is not readily forthcoming, although the potential role of remote sensing applications to the labyrinth of food production problems was recognized by the World Food Conference in 1974.

The reasons for the lag in agricultural leadership are complicated, but must be overcome if human society is to benefit from remote sensing in combating the emerging worldwide traumas. With a world population growth of 2.5 percent to 3 percent, the present world population of 4,000 million (March, 1976) is expected to exceed 7,000 million by A.D. 2000 and the population density will

be acute in some regions (e.g. an estimated 146 million in Java!).
Also resulting primarily from the population explosion (more serious
than the future threats of pollution) will be the accelerating
threats to agriculture caused by deforestation, soil erosion,
declining soil fertility including salinity and in some regions
increased desertification (Howard, 1976). For example, in North
Africa alone, the loss of land to desert encroachment has been
estimated as exceeding 100,000 ha. per year (Le Houerou, 1970).

Reasons for the lag in agricultural initiative appear to be
political, economic and technical. It should be appreciated that
politically many countries are restrictive in the availability of
aerial imagery and international opinion is sharply divided on the
future free access to space imagery (see UN debates). It is
unfortunate that there is not a greater realization that, based on
published information, the resolution provided by sensors on military satellites is many times greater than Landsat. Further many
developing countries cannot afford to use the sophisticated remote
sensing techniques developed in countries with advanced economies
and therefore new efforts are needed in improvising inexpensive
techniques. As pointed out by McNamara (1975) at least 1000 million
people, or about a quarter of the world's present population, have
a per capita GNP of only US$ 105 which is expected to be US$ 108
by 1980.

The need for the transfer of technical know-how on remote
sensing applications through educational training programmes is
not confined only to the developing countries. It seems that in
most countries graduate agriculturists have received little or no
formal training in remote sensing technology (cf. forestry, geology)
and therefore at senior technical and administrative levels in
agriculture there is a lack of understanding of the potential
technical role of remote sensing applications. For example in
Australia, an agriculturally advanced country, there are no undergraduate or graduate courses in agriculture remote sensing. On
information available, between 1965/75 less than 2% of Australian
university students, who received formal training in remote sensing,
were agriculturists.

State-of-the-Art of Remote Sensing Technology in Agriculture

As we know, satellites have the capability of providing man
with repetitive imagery of the same ground areas. However, different field disciplines have markedly different time-frequency
requirements. For example, the geologist and soil surveyor require
few repeated satellite coverages of the same ground area. Amongst
other natural resource managers, the hydrologist and forester will
probably be satisfied by annual seasonal coverage, other than in
periods of natural disasters (e.g. drought, flood, fire, biotic

damage). Thus agriculturists, including agricultural economists,
fishery officers and rangeland managers, are potentially the major
natural resources users of repetitive satellite imagery.

It is unfortunate that cloud cover resolution of the imagery
(see Table A) and the global availability of ground receiving
stations restrict the immediate wider-use of satellite data. It
is also unfortunate that at the present time only one of the four
tape recorders on the two Landsat polar orbiting satellites is
functioning, since regions without receiving stations cannot benefit
continuously from the 9-days coverage of Landsat-1 and Landsat-2.
The launching of Landsat-C in 1977, Seasat in 1978 and the GOES
programme with its five geostationary satellites (to be in orbit
by 1980 to provide primarily global low resolution weather data
at ½ hour intervals) may partly alleviate the situation. It should
be noted that weather satellite data receiving stations are much
cheaper to establish than Landsat stations and at commercial prices
Landsat imagery and tapes are relatively expensive, being many
times the NASA/EROS Data Center prices. These are obviously important decision making factors in planning a satellite application
system for food production.

The operational role of aerial imagery for improving food
production seems in the short-term primarily confined to aerial
photography. It is over 5 years since agricultural research papers
were published on thermal sensing and side-looking airborne radar
(SLAR); but, with the exception of thermal sensing in hydrology,
the agricultural operational stage has not been achieved. SLAR,
at normal scale, obviously has the limitation of low resolution
(10 - 20 or 40 - 60 metres); but, with its cloud penetrating capability, is the only system suited to agricultural purposes in much
of the tropical rainforest areas of the world. A television system
for assessing coffee production in Brazil is being developed but no
up-to-date information is available. Such a system is designed to
favour a computer analysis of spatial data and not spectral data.
Probably, recent developments in high-flight photography, and low-
level aerial photography, have been eclipsed by the development of
satellite remote sensing. It is possible to identify several major
information thresholds dependent on the imagery resolution/photogra-
phic scale; but, what is not well determined is the resolution
threshold between high-flight photography and the best available
satellite imagery. Possibly this resolution threshold is in the
vicinity of 5 - 20 metres. Reference to U.S. publications on high-
flight aerial photography, mostly about the time of the launching
of Landsat-1 (e.g. NASA Symposium on Significant Results, Maryland,
1973) indicate the type of information obtainable. In comparison
with the ground resolution of SLAR and of Landsat (theoretically
about 80 m) and the higher resolution of imagery on manned spacecraft (e.g. U.S. Skylab; U.S.S.R. Salut), high-flight photography
with a potential ground coverage of 10,000 km^2 per hour has the

Table A. Satellites suited to providing input to food production

Satellite	Spectrum	Resolution (metres)	Time Interval between repetitive coverage of the same ground area.
* Landsat-1	Visible and near infrared	80 (0.4 ha.)	Every 18 days(since 1972)
* Landsat-2	Visible/near IR	80 (0.4 ha.)	Every 18 days(since 1975) (combined with Landsat-1 every 9 days)
DMSP (VHRR)	Visible/thermal IR	600	12 hours
** NOAA (VHRR)	Visible/thermal IR	900	12 hours
NOAA (SR)	Visible/thermal	4,000	12 hours
Landsat-C (1977)	Visible/IR/thermal	0.2 ha; 0.4 ha	18 days
GOES (by 1980)	Visible/thermal	2,500/5,000	1/2 hour
Nimbus-G (1978)	Visible 5 Narrow bands (± 10 nm)	800	---
SEASAT (1978)	Microwave (Radar L-band)	25	72,151 days

* Used in agricultural studies. Data receiving stations are located in United States, Canada, Brazil and Italy with a range of up to 3,000 km. Satellite has telemetry capability for transmitting data from ground control platforms. Stations are planned in Iran, Japan, Zaire and Chile.

** Receiving stations are located in United States, Brazil, Canada, France, West Germany, Japan, China and are planned in Belgium, Italy, Sweden and United Kingdom.

advantage of providing a practical ground resolution better than
3 metres.

Low-level strip and pin-point aerial photography, commonly
taken with panchromatic black-and-white or colour film and using
35 mm or 70 mm cameras, is not even as well documented as high-
flight photography. This type of photography, with a resolution
often better than about 15 cm, has the advantage of being less
politically restricted in use, relatively inexpensive and suited to
amateur operation to supplement other imagery. Helicopter twin-
camera fix-beam aerial photography with its stereoscopic coverage
but higher cost is viewed as of little operational potential in
agriculture. However, helicopters are ideal for collection of
ground data in difficult terrain and possibly could be cost/benefit-
wise employed more often in multi-stage sampling systems.

State-of-the-Art of Remote Sensing Applications in Agriculture

A distinction must be made between the state of development
of the methodology and techniques of remote sensing applications,
particularly those related to satellite systems, and their operational testing and operational use. Whilst Landsat is classed as
an experimental system in the United States, Landsat data is increasingly being used operationally in developing countries (e.g. for
thematic mapping at 1:250,000). It is also considered useful to
recognize that remote sensing as an aid to food production is
applied either directly or indirectly to agricultural crops or to
other Earth's resources which influence food production. Concerning
the latter, remote sensing applications related to forestry, hydrology, geomorphology etc., must be included in the writer's opinion
in the overall application of remote sensing to aid food production;
but sensing other resources will not be pursued. In this context,
however, it needs noting that there has been a marked lack of action
in applying the synoptic approach provided by the remote sensing
of 'other resources' to the improvement of food production.

The direct application of remote sensing to food production
depends on assessing from imagery the crop areas and the state and
condition of crops and pasture. The aerial photographic techniques
used nowadays have been developed and proved technically and economically sound over several decades. Careful choice of the film-
filter combination, season of photography in relation to plant
growth, and photographic resolution (expressed as scale in terms
of the focal length of the camera lens and flying height) enables
crops including cereals to be identified (cf. Brunnschweiler, 1957;
Bomberger et al, 1960; Aunuta and Macdonald, 1971), their areas
measured, their conditions and yields assessed (e.g. Howard and
Price, 1973); insect infestations located (Spurr, 1960, Wenderoth
et al, 1975) and the presence, absence and development of disease

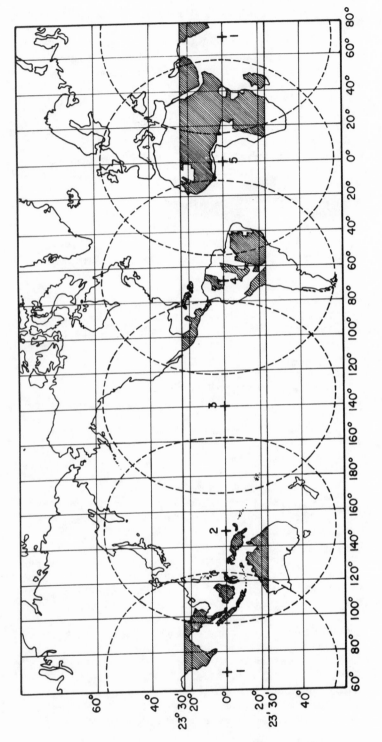

Fig. 1 – Grey areas indicate the approximate ground coverage with 10% cloud cover or less of tropical countries by the polar orbiting Earth resources satellites, Landsat-1 and Landsat-2 (July 1972 to April 1975). The broken lines indicate approximate coverage of the GOES geostationary weather satellites to be in operation before 1980: (2) GMS, Japan. (3) SMS-B USA. (4) SMS-A USA. (5) METEOSAT, Western Europe). These geostationary satellites may have an important impact on agricultural crop forecasting.

determined (e.g. Brenchley and Dadd, 1962). Frequently photographic scales between about 1/8,000 and 1/20,000 are used, (i.e.
resolution in practice better than about 1½ metres). Much smaller
scales are satisfactory for land-use and land capability mapping
(e.g. 1/80,000 in Australia); and infra-red colour photography at
a scale of 1/75,000 (resolution better than 2 metres) is being used
by FAO in a national land-use and land capability study in Sierra
Leone. Infrared colour photography at 1/120,000 has been used in
several studies including the study of maize blight (under 1m
resolution) and in comparison, low-level strip, pin-point and
oblique photography is favoured in wild-life and livestock studies
(e.g. Jolly, 1969).

As indicated by the increasing range of published works on
the applications of satellite imagery (see Manual of Remote
Sensing; NASA Proceedings, 1973, 1974; Abstracts NASA Survey Symposium, 1975; Proceedings, Remote Sensing of Environment, 1973, 1974;
Canadian Remote Sensing Symposia), Landsat imagery is proving
increasingly useful in different parts of the world in providing
direct and indirect inputs in programmes associated with food
production. For example, in Nebraska, biomass estimates covering
52,000 km^2 has been shown to be a viable method for making decisions
on rangeland management, including stocking. In part of North
Dakota, the areas under small grains have been inventoried with an
accuracy of 96.5%. The Large Area Crop Yield (LACY) programme in
the U.S.A. has given 10% error estimate experimentally for areas
under cereals. Temporal imagery composites, combining two or more
bands in time sequences, are proving useful in the study of seasonal
changes occurring over large areas (e.g. pastoral savanna burning
in Africa). Spectral band combinations have been found useful in
showing up hitherto unknown large-scale natural patterns on the
Earth's surface, the significance of which are yet to be determined.
In Thailand, Landsat-1 imagery was used to determine the rate of
de-afforestation (between 1961 and 1973 the forest cover decreased
from 50% to 37%), and in the Sudan the seasonal inundation of the
Sudd is being examined using a temporal sequence of Landsat imagery.

Various satellite remote sensing programmes illustrate an
indirect approach to improving food production. Thematic maps at
scales up to 1/250,000 are being prepared or revised in an increasing number of developing countries using Landsat imagery (i.e.
related to landforms, hydrology, soil, forest cover, erosion, land-
use and land capability). Landsat imagery has also been reported
as useful for up-dating existing thematic maps at scales as large
as 1/50,000 (UN, 1976). Both the FAO Desert Locust Commissions
pilot study on monitoring factors influencing the development of
the migratory desert locust (Hielkema and Howard, 1976) and the
USA-Mexican study on monitoring the screw-worm (NASA, 1974) illustrate that low resolution weather satellite imagery can be applied
beneficially in the management of Earth resources. The screw-worm

study showed that NOAA imagery can be used to detect temperature
differences (± 1½°C) close to the Earth's surface with sufficient
accuracy to locate areas with temperatures above the minimum
required for the breeding of screw-worm. Similarly imagery is
being used in Florida to monitor for frost-damage of citrus crops.

In the pilot study of the desert locust, NOAA VHRR imagery
was used in 1976 to locate desert areas in the Ahaggar region of
Algeria having recently received rainfall sufficient for the breeding of the desert locust and then these sites were monitored by
Landsat imagery for the development of ephemeral vegetation and
the flush of perennials. Rainfall was estimated from the cloud
types recorded on the imagery. Based on the results, it is now
planned to monitor rainfall in a semi-operational mode throughout
the region of the N.W. Africa in 1977 and to continue development
work with Landsat imagery, since with the latter ephemeral vegetation has only been detected so far under very favourable conditions.
The experience gained with NOAA imagery in Algeria indicates the
method could be applied to a wide range of problems including crop
forecasting and for supplementing rainfall data from sparsely
located ground meteorological stations or stations poorly located
for agricultural purposes.

Organization and Planning of Global Systems Using Satellites

Four approaches can be identified in the problem of monitoring
the Earth's food production. The most obvious consists of a single
global system, including a single read-out and analysis centre.
Remote sensing information, including information collected by
ground data control platforms (DCPs), would be provided either from
orbiting satellite(s) with data tape-storage and transmission to
a single satellite data receiving station or data from geo-station-
ary satellites would be transmitted directly by inter-satellite
communication to the central station. If required, local/national
aerial survey information could be used to supplement the satellite
data. This satellite approach is viewed as technically the quickest
to implement and probably the least costly (although still very
expensive); but there could be political objections and the very
basis of the method precludes the world-wide input of local
expertise.

Alternatively regional read-out and analysis centres could be
established, as recently recommended by ECA for Africa and/or a
central organization can be modelled along the lines of WMO. What
at first appears to be a global system may in effect be a single
(or extended) regional system. For example, one may speak loosely
of global monitoring of factors influencing the migratory desert
locust, but as the extensive area to be monitored is mainly in
northern and central west Africa and the southern, Near and Middle

East, it requires essentially a regional or inter-regional approach.
Monitoring regional specific problems may be the best approach to
developing global monitoring.

Fourthly, a global satellite system may be structured at the
country level. This will be possibly the most attractive at the
present time to many developing countries concerned with improving
or obtaining estimates of their crop production and crop yields.
It has the advantage of a step-wise build up, country by country;
but on a global basis is likely to be the most expensive to operate.
The system could commence in cooperation with a few developing countries selected either on a global basis or within a key region.
Country level systems have the further advantage of easy "tailoring"
the satellite data inputs to established statistical crop reporting
services, which at the present time do not use remotely sensed
information.

When appraising future crop forecasting in developing countries
using satellite imagery, it is suggested that the role of Landsat -
type imagery, probably with multi-date imagery, is to provide a
seasonal check on the planted area of identifiable crops within
a region, but bearing in mind that in some countries the field-
size is very small and that clouds are often a limiting factor to
good image quality. The role of weather satellite imagery is seen
firstly to help provide a meteorological data bank related to crop
yields and crop production (i.e. yield x area) and then in association with crop calendars/ecozone maps for individual crops to
provide during the growing season, information on temperature,
precipitation, etc.,(and their transformations) and to monitor
extreme weather conditions. For countries with well-developed
statistics reporting services, the main advantage of satellite
sensing is seen to lie in reducing overall costs, in reducing the
field staff component, in reducing the error estimates of yield
and production, in reducing the sampling errors, in reducing the
time period of the crop forecast and in providing information
quickly of extreme conditions (e.g. drought, flood). In some
developed countries an estimate error within 5% is achieved 5 or 6
weeks before the harvesting of major crops, but in some developing
countries an error as great as 20% may be acceptable, although the
long term aim may be to achieve an error of about 5%.

We must bear in mind that statistics reporting services using
only ground sampling, and occasionally supplemented by postal information provided by farmers, have been developed in many countries
(e.g.,FAO, 1964), that visual field forecasts of yields have been
made for over 200 years (e.g. in Hungary; Simon, 1975) and that
field techniques being developed today can be applied to estimates
of grassland production (cf. Vinczeffy, 1975; Budyko, 1968).
Obviously, information on actual yields can only be obtained after
the harvest during which further losses of several percent occur;

but often overlooked is the fact that a gap of several months may
elapse between the collection of the sample field data and completion of the data processing and that during this period a number of
important decisions may need to be made (e.g. adjustment of future
crop subsidies).

In passing, it is important to recognize that yield forecasts
using ground data are based on either the yield of individual plants
or on samples of the whole crop on large farms or in a county/
district or on samples within a region, and that it is the crop
production/yield forecasts for a region which may be best correlated
with macro-climate data of the type observable by satellites. Crop
yields by region/country vary yearly less than by districts or farms
(e.g. for the GDR co-efficients of variation for 1962-72 were 20%,
24% and 30% respectively; Strumptel, 1975). In Italy, where only
data for temperature and precipitation were available, close correlation was achieved for wheat for the period 1953 to 1969 except
for 1960 following extreme weather conditions (Haunus, 1975).
Landsat seems well suited to providing a check on re-sowing following extreme conditions, and weather satellite data may well be
essential for countries lacking adequate historical agro-meteoro-
logical data or a sufficient number of well sited meteorological
stations or well educated farming communities.

Discussions and Conclusions

A widening gap exists between the developed and developing
countries in the applications of remote sensing, which is emphasised by the problem of increasing food production. Steps need
to be taken through international cooperation to improve the know-
how of the developing countries and to associate the applied
research of the developed countries with the problems of developing
countries, particularly those related to the survey of agricultural
resources, crop forecasting, crop census and the monitoring of
pests and disease. Often what appears to be an identical applications problem in developed and developing countries assumes a
different priority in developing countries, and, also, different
methods and techniques are needed if the infra-structure and know-
how are to be built up and retained in the developing countries.
Too frequently, a problem existing in a developing country, which
requires the application of remote sensing (either aerial or
satellite) has been tackled with external expertise and little or
no local involvement including training. Thus, no permanent expertise is retained by the developing country.

At the present time satellite sensing is still very much at
the applied research stage of development as compared with aerial
photography, but its operational uses are rapidly increasing.
Based on the experience of the life of Landsat and the recent U.S.
announcement on Landsat-C (Frutkin, 1975), global coverage by these

satellites seem secure until nearly 1980. Further, Landsat-C
is likely to be equipped with a thermal scanner and the RBV sensor
will have a ground resolution of twice that of Landsat-1. Seasat,
to be launched in 1978, should provide experimental data useful
to the management of marine resources and possibly some land sources.
The GOES programme, with its complete low resolution coverage of
the Earth at ½ hour intervals by 1980 for weather forecasting, will
no doubt be beneficial to Earth resources management related to
food production. The future role of high spectral resolution (as
distinct from high spatial resolution) will be more fully evaluated
through the Nimbus-G satellite with its narrow-band coastal zone
scanner (see Table A). The results of research into the variations
of spectral signatures of species and crops, according to their age,
vigour, leaf area index, physiognomy, season of the year, weather
conditions, incidence of disease, background, noise, etc., indicate
the complexity of and potential importance of sensing simultaneously
in several key narrow bands.

For the future development of remote sensing applications
to food production at least three basic systems need recognizing.
These are training systems; national/local application systems
and transnational (global) applications systems. As suggested
earlier, training systems are viewed by the writer as needed by
both the developed and developing countries, since available evidence indicates that agriculturists have not received adequate
training in remote sensing. Training in the developing countries
is needed urgently not only at the country-level but also at the
local level, so that hard-worked local officers can achieve results
more quickly by associating information from imagery with local
management and planning at the grass-roots level.

For maximum net benefits in a wide spectrum of agricultural
activities, more care needs to be exercised in the correct timely
integration into local and national rural development of remote
sensing applications, including aerial photography, photogrammetry,
photo-interpretation and satellite imagery analysis. Too often in
the past, field programmes which could benefit from remote sensing,
were initiated either without using remote sensing or the imagery
was acquired or interpreted too late in the programme for maximum
benefit. It is suggested that if network analysis is being applied
to agricultural field activities, which could benefit from remote
sensing, the network specialist should be trained in or at least
be conversant with remote sensing techniques, since remote sensing
can have an important influence on overall costs. With the ready
availability of Landsat imagery at low cost, there is now little
justification in not using it in the planning and the initial
execution of many national and regional field programmes and when
appropriate in local field programmes of a country. Transnational
systems, whether on a global basis or synthesised from national
level are likely to receive major impetus from the emerging

satellite programmes of the late 1970's. Further, it is suggested
that by the 1980's hybrid systems will emerge, which will require
meteorological data transformations not available at the present
time. It is viewed timely therefore to determine now the phasing
and types of remote sensing inputs needed to help establish or
improve/crop yields and crop production in the 1980's; but these
inputs should be closely integrated with ground-based techniques
and should take into consideration the need for information to be
collected in the field.

BIBLIOGRAPHY

Anuta, P.E. and Macdonald, R.B., "Crop surveys from multiband
    satellite photography using digital techniques". Remote
    Sensing of Environment, 1971, 2: 53-67.

Bomberger, E.H., Dill, H.W., et al, "Photo-interpretation in
    agriculture". Manual of Photo-interpretation, 1960, ch.
    11: 561-666.

Brenchley, G.H. and Dadd, C.V., "Potato blight recording on aerial
    photographs". N.A.S.S. Q, Rev., London, 1962, p. 21-25.

Brunnschweiler, D.H., "Seasonal changes of the agricultural
    pattern". Photogramm. Engng., 1957, vol. 23, p.131-139.

Budyko, M.I., "Solar radiation and the use of it by plants".
    Agron. Methods, UNESCO, VII, 1968, p. 39-51.

FAO, "National methods of collecting crop statistics". Rome,
    1974, p.609.

Frutkin, A., "U.N. Committee on the Peaceful Uses of Outer Space",
    U.N., New York, 1975.

Hanus, H.,   "Crop forecasting based on meteorological data of
    larger areas". International Symposium on Crop Forecasting,
    Kompolt, Hungary, June 1975.

Howard, J.A., Aerial Photo-Ecology. Faber & Faber, London, 1970,
    p. 325.

Howard, J.A.,"Satellite remote sensing of agricultural resources
    for developing countries - present and future - an interna-
    tional perspective". AGD(RS)2/76, FAO, 1976, Rome.

Howard, J.A.,"Remote sensing of tropical forests with special
    reference to satellite imagery". AGD(RS)3/76, FAO, 1976, Rome.

Howard, J.A. and Price, R.D., "Using aerial photography to estimate yield loss in oats due to barley yellow dwarf virus". Revue Photo-Interpretation, Paris, 1974.

Hielkema, J.U. and Howard, J.A., "Pilot project on the application of remote sensing techniques for improving desert locust survey and control". AGD(RS)4/76, FAO, Rome, 1976.

Jolly, G.M., "The treatment of errors in aerial counts of wildlife populations". East African Agricultural and Forestry Journal, Vol. XXXIV - July 1969.

Le Houérou, H., "North Africa: past, present and future. Arid lands in transition". (ed. H.E. Dregre), AM. ASSOC. ADV. Science, Washington, 1970, p. 227-278.

McNamara, R.S., Address to Board of Governors, IBRD, Washington, D.C., 1975.

NASA, "Project plan - remote sensing as an aid to screw-worm eradication". (LEG 3752). Lyndon B. Johnson Space Center, Houston, 1974, p. 32.

Simon, B., "System and methods of crop forecasting in Hungary". International Symposium on Crop Forecasting, Kompolt, Hungary, June 1975.

Spurr, S.H., Photogrammetry and Photo-Interpretation, Ronald Press, 1960, NYS p. 433.

Strümpfel, J., "Results of early crop forecasting of cereals under German Demoncratic Republic conditions". International Symposium on Crop Forecasting, Kompolt, Hungary, June 1975.

United Nations, Report on the United Nations Interregional Seminar on Remote Sensing Applications in cooperation with Canada and the United Nations Educational Scientific and Cultural Organization, Guelph and Ottawa, Canada, 1975.

Vinczeffy, U., "Crop forecasting of sward". International Symposium on Crop Forecasting, Kompolt, Hungary, June 1975.

SPECIFICATIONS FOR SOME SATELLITE APPLICATIONS OF

DATA TELEMETRY AND PLATFORM LOCATION

A. D. Kirwan, Jr.
Research Scientist, Department of Oceanography
Texas A & M University, College Station, Texas

Introduction

The technical capabilities of satellite systems traditionally are divided into two categories:

1) Remote sensing.

2) Data transmission and platform location.

Most of the emphasis in resource management and environmental monitoring has been on the remote sensing capability. Consequently, this has been an area of rapidly evolving technology.

The data transmission and platform location (DTPL) capability of satellites has a much less glamorous background. However, it is a proven technique and has a history of solid, if unspectacular, accomplishments in environmental research and monitoring.

The DTPL capability is particularly well suited for environmental monitoring and research in the world's oceans. It is noteworthy that over this two-thirds of the earth's surface, remote sensing has not been as spectacularly successful as it has been over land. There are two reasons for this. First, many of the areas of interest are normally covered by clouds; thus, remote sensing techniques are severely limited as a systematic monitoring tool. Secondly, many of the oceanic processes which are important to resource management and environmental research take place well below the sea surface. This, of course, is inaccessible to remote sensing.

Actually, monitoring the ocean environment has proved to be an extremely difficult technical problem. It is apparent that no single technique such as remote sensing or regular hydrographic surveys or buoys can meet all requirements. What is required is a rational mix of various techniques. In such a mix, a DTPL capability and remote sensing would be highly complementary.

This paper addresses the application of the DTPL capability to monitoring environmental conditions in the upper layers of the ocean. The analysis is especially applicable to in situ observations routinely made from free-drifting ocean platforms.

Two basic problems are considered. The first is concerned with the accuracy of in situ observations received via a satellite telemetry link. This is discussed in the next section. The third section analyzes platform location errors and fix rates. Such an analysis is important in estimating the accuracy of ocean currents determined from the drift of the buoys and in evaluating a search and rescue application of this satellite capability. The last section summarizes the findings and presents some recommendations for future utilization and development of the DTPL capability.

Data Transmission Errors

Proper utilization of in situ observations telemetered through a satellite requires knowledge of both instrumentation errors and transmission errors. These errors are independent in that the accuracy and/or precision of a particular sensor is determined solely by hardware constraints completely unrelated to the satellite telemetry link. The transmission error from platform to satellite is typically ± 1 bit. For example, for an 8-bit data word, the accuracy of the transmission is about ± 0.5%. If more accuracy is required than is available in a single data word, then it is necessary to transmit the particular observation in multiple data words. Thus, a single observation transmitted as two 8-bit words has an error of less than 0.25%. From this it is seen that the telemetry link will introduce no additional error, provided an adequate number of data words are available in the system.

A DTPL capability currently being utilized is TWERLE/RAMS on Nimbus 6. This system has the capability of transmitting 8 data words each of 8 bits. I have used this system extensively since its launch in June, 1975. Table 1 summarizes the accuracies I require for transmission through this system.

Inspection of this table indicates that the 8 data word, 8 bit/word telemetry capability of TWERLE/RAMS can easily meet these requirements. It is noteworty that existing barometric pressure sensors, suitable for deployment, may not be able to

Table 1   Example of Data Accuracy Requirements for TWERLE/RAMS

Variable	Accuracy	Range
Barometric pressure	± 1 mb	900-1100 mb
Sea surface temperature	± 0.1°C	0-25°C
Air temperature	± 0.2°C	-25 to 25°C
Battery voltage	± 0.1 volt	10-12 volts
Drogue indicator (strain gauge)	± 10 lbs.	25-150 lbs.

meet the design accuracy. Generally, the telemetry link is not
the limiting factor in measurement accuracy in oceanography; it
is usually the in situ sensor.

An upgraded telemetry capability is scheduled for the Tiros N
series. In this system at least 32 data words will be available.
Most projected data requirements in oceanography will hardly tax
the capability of that system.

Position Determination

There are at least three uses that can be made of precise
platform location by satellite. The first use is location determination for in situ measurements made from free-drifting ocean
buoys. Such observations would be incorporated in numerical prediction models of the atmosphere and/or ocean environments much
along the lines of the First Garp Global Experiment (FGGE). The
second use would be for mapping the characteristics of major ocean
current systems. A preliminary example is given in Kirwan et al,
(1976). The third use would be a practical and humanitarian
application of this capability. This is search and rescue operations for survivors of maritime disasters. This would be a very
practical spin-off of a technology which heretofore has been used
solely for research purposes.

The paragraphs below outline the technical requirements for
each of these uses.

Location for In Situ Measurements

Here the requirement for platform location is to provide a
position for environmental data being incorporated in atmospheric

and/or ocean prediction models. Most of these models require data
every synoptic period (every six hours) at the model grid points.
Typically, these grid spacings are of the order of 1° or approximately 100 kilometers. Current data requirements are for sea
surface temperature, air temperature, and barometric pressure.
When available, the air-sea fluxes of sensible and latent heat
and momentum should be included in these models.

As stated before, a DTPL capability currently exists in
Nimbus 6. The most conservative position error for this system
is of the order of 5 kilometers with positions being available
normally only during daylight hours. I have found that a more
realistic estimate for the position error is of the order of 2
kilometers.   Furthermore, even though the satellite does not fix
the platform on every fly-by, it typically receives platform signals
on as many as 8 fly-by's during a twenty-four period. During each
fly-by, reliable transmissions of the in situ observations are
nearly always received.

During the course of twenty-four hours, a drifting buoy may
move up to 100 kilometers. This distance is characteristic of one
caught in a strong current such as the Gulf Stream or the Kuroshio.
More typically, 10 to 20 kilometers is a typical displacement for
this period.

From this it is clear that the satellite normally does not
receive observations from the buoy at each grid point exactly at
each synoptic period. In practice, the forecast models take data
at a synoptic time anywhere within the model grid. This data is
then interpolated in both space and time to the required grid point
and period by the process of "four-dimensional data assimilation".
It appears from this, then, that the position error with existing
DTPL capability is an unimportant consideration as far as accuracy
of the model is concerned. Even one position a day, along with the
data from 8 fly-by's, is adequate unless the platform is in a strong
current. In such a case, at least two fixes would be required.

Ocean Currents

The problem of the effect of position error and platform fix
rate on estimates of ocean currents has been investigated in detail
by Kirwan and Chang (1976). The pertinent conclusions from that
analysis are:

1. A random position error of 5 kilometers at a fix rate of
     once per day produces a 10% error for current estimates
     typical of mesoscale eddies.

2. Increasing the fix rate while leaving the position error the same does not significantly improve the estimates.

3. In order to detect inertial and tidal cycles in ocean currents, it would be necessary to go to a fix rate of at least ten per day along with a decrease of position error to the order of 100 meters.

It is concluded from this that the existing DTPL on Nimbus 6 and the one proposed for Tiros N is adequate for studying mid-ocean general circulation and mesoscale eddies. It can also be used for studying strong currents such as the Gulf Stream as long as the phenomenon of interest has a time scale of the order of several days.

Search and Rescue

The capability for monitoring routinely platform location to within 5 kilometers on a worldwide basis suggests an immediate application to search and rescue (SAR) operations. Such an application is now feasible because of the rapid advance in transmit terminal and antenna technology that has taken place in the past few years. Using a coil antenna, it is now possible to reach DTPL satellites routinely with less than 2 watts. This makes it feasible to have a transmit terminal and antenna which could be clasped on a life jacket or readily attached to a lifeboat. The transmission could be activated by either contact with water or by a switch operated by the survivor.

In order to be applicable to SAR operations, the DTPL capability on a satellite should encompass:

1. Rapid data turn-around -- This means a minimal interval between the time the platform is activated and the time the signal is received at the appropriate SAR station. At the least, the satellite would have to broadcast once an orbit (approximately every two hours) the time and locations of any distress ID's. The distress signals would have to be flagged so that they would have highest priority in processing at the receiving stations. The time interval could also be shortened by a network which monitors on a real time basis the transmissions received by the satellite. Of course these stations must be directly tied in with national SAR centers.

2. Frequent monitoring of position -- After the distress signal is first received, it is highly desirable to monitor the platform location as continuously as possible. With one satellite, there may be as much as twelve hours between reliable satellite fixes. Any reduction of this time would require multiple SAR satellites.

3. Duration of transmissions -- In mid-ocean regions it may take twelve to twenty-four hours to get search and rescue equipment into the SAR area . Also, the SAR operations may be hampered by darkness or bad weather. It is essential, then, that enough power be available to keep the transmit terminal in operation continuously for at least one week, preferably two. Such a requirement may preclude transmitting other data concerning the survivor or his/her environment.

The DTPL capability in existence on Nimbus 6 is certainly adequate to initiate such a program. However, major administrative changes would be required. A special group responsible for continuous monitoring of the satellite data for distress ID's must be identified, as the project managers for Nimbus 6 have neither the mission responsibility nor the personnel to perform this service. Similar changes would have to be made for the DTPL capability on the Tiros N series.

Note also that there may be as much as a twelve hour turn around time for distress signals received through Nimbus 6 as this satellite does not dump its data on every orbit. With the Tiros N series this time may be cut to a few hours. The Global Position System satellites scheduled for launch in 1978 would be ideal for SAR as they could meet the requirement of continuous monitoring of positions.

Summary

A) Satellite data telemetry links for in situ observations have proved to be extremely efficient and reliable. Virtually all problems and data errors have been due to failures of the sensor system and not the telemetry system.

B) The capability of existing satellite data telemetry systems is adequate for present applications. The next generation telemetry systems will have considerably more capability than required.

C) Platform position error and fix rate of the current DTPL system are not significant problems for data used in environmental prediction models. Moreover, it is not expected that future model requirements will be any different.

D) Present platform position error and fix rates are also adequate for measuring large-scale ocean currents. They are not sufficient for measuring tidal oscillations.

E)   The simplicity, efficiency, and reliability of present
     platform location capability by satellite are such that
     they could be readily applied to search and rescue
     operations. The major obstacle to this application is
     organizational, not technical. A dedicated operational
     group must be identified to perform the function of
     continuously monitoring for distress signals.

F)   The Nimbus 6 and Tiros N satellites could be easily
     adapted for a search and rescue mission. There might,
     however, be as much as a twelve hour delay before a
     distress signal was processed. The Global Position
     System would not have this shortcoming; moreover, it
     could provide continuous monitoring of platform position.

REFERENCES

Kirwan, A.D. Jr., McNally, G. and Coehlo, J., "Gulf Stream kinema-
    tics inferred from a satellite-tracked drifter". Journal of
    Phys. Oceanography, 1976, 6: 750-755.

Kirwan, A.D., and Chang, M.S., "Effect of sampling rate and random
    position error on analysis of drifter data". Submitted for
    publication.

Acknowledgements

The research reported here has been sponsored by the Office of
Naval Research and the National Science Foundation.

INFORMATION VALUES OF REMOTE SENSING

A.K. McQuillan
Research Scientist, Canada Centre for Remote Sensing
Department of Energy, Mines and Resources, Ottawa, Canada

Introduction

Many of the thoughts presented in this paper are well known to thoses involved with the study and applications of remote sensing. There is merit sometimes though in going back over simple ideas that are basic and important. There is merit too in constantly groping to understand the role of a technology in the context of a world of changing lifestyles and values. This paper attempts to do that, and at the same time suggest a set of values which stresses quality of life, harmony with nature, international respon- sibility and the principle of self limitation.

Remote Sensing as a Measurement Technique

An act of measurement by its very interference with the medium to be measured may perturb and change that medium. Remote sensing is a form of measurement by which contact is made with the target through electromagnetic radiation. In the case of passive sensors such as cameras, spectrometers or line scanners, the measurement involves observation at a distance of reflected or radiated electro- magnetic energy and <u>no interference</u> with the target. In the case of active sensors such as radar, interference with the measured object is small. Information provided by remote sensing, therefore, shows the measured target "as it is", in an <u>unmodified</u> form, and may provide more accurate information than that obtained through direct mechanical contact.

Information about the target is transmitted by electromagnetic radiation. In electromagnetic wave theory, waves are represented

by terms which contain information on wave frequency, amplitude,
polarization and phase. Multispectral scanners or aerial photographs, for example, take advantage of the amplitude (i.e.intensity)
information and the manner in which that varies in selected frequency channels. Some sensors however, such as synthetic aperture
radar, take advantage of total information transfer including phase
and polarization information.

Probing the properties of a target such as a hydrogen nucleus
by protons, electrons or other radiation has been done for many
years in elementary particle physics. The indirectness of the
method together with target complexities have stimulated construction of a multitude of different accelerator sources. Although
the sun has provided a powerful source of electromagnetic radiation,
the requirement for more selective probing of the earth's environment combined with a more complete information transfer will result
in a greater use in future of coherent sources operating at various
frequencies throughout the electromagnetic spectrum.

Understanding the earth's environment requires three dimensional space measurements of the earth's atmosphere, water and
land and the nature of their change with the fourth dimension, time.
Use of the radiation of the sun to probe the atmosphere permits
absorption or emission measurements integrated spatially through
columns of atmosphere. Radar systems present the capability of a
more refined spatial diagnostic probe with range resolution. Requirements for selective probing and information transfer ensure
that coherent radar systems will become an increasingly important
measurement tool.

Measurements of phenomena on the earth's surface require
passing the electromagnetic radiation through an atmosphere which
will modify it. If absolute radiance measurements are required
to facilitate detection of environmental change from one time to
another, calibration techniques are needed. The synergistic use
of both atmospheric sensors and sensors to monitor the earth's
surface, together with selected surface reflectance measuring
calibration standards (e.g. groundplots with calibrated reflectance)
seem essential for optimum performance of satellite earth observation systems. While the qualitative environmental relationships
provided by a satellite multispectral image, for example, are very
important, future research must concentrate more on quantification
of remote sensing measurements.

Satellite earth observation systems should greatly improve
the sampling in space and in time of enviromental information
systems. The crucial question is, can we obtain the type of
information or the accuracy of information that one really wants.
Are the data requirements we now perceive to need really the correct ones? The high accuracy requirements we perceive now may be

due to the fact that the parameters we are measuring are not the best ones. Because they are correlated, it may take considerable accuracy improvement in the parameter measurements to give an accuracy improvement in a particular type of forecast. The question is, "What are really the orthogonal parameters which need to be measured and to what accuracy must we measure them?"

Users frequently have great difficulty in establishing their requirements. Their perception is conditioned by their experience with traditional forms of data. Operational systems, for example for crop forecasting, can be expected to be evolutionary in nature with some iterative steps involved as more appropriate acquisition, processing and interpretation techniques evolve.

It is a feature of surveillance either of the environment or of human activity that as one requires coverage of greater and greater areas, the techniques which were attractive for small areas from both a cost and effectiveness point of view become less suitable. In general, satellite methods become more attractive as the requirement for larger area coverage increases.

The Character of a Remote Sensing Observation

In assessing the quality of a remote sensing observation, reference is usually made to "ground TRUTH" information. Such information is usually taken to be correct or nearly so. Frequently, the intrinsic properties of a remote sensing observation are not fully appreciated. A remote sensing observation is an objective observation which can stand on its own as a valid observation. Although of a different nature than "ground truth" measurements which man has used for centuries and which have conditioned his way of looking at the world, a remote sensing observation provides a new but equally valid observation. This does not mean to imply that "ground truth" observations are not required for resource and environmental monitoring. Ground and remote sensing measurements provide complementary information to assist in the decision making process. An example may be drawn from oceanographic environmental monitoring. Ships and data buoys can provide frequent, ocean surface point measurements over a limited area but complementary airborne and satellite systems are required to give broad area coverage. Ships and data buoys can, therefore, provide frequent sampling in time but limited spatial sampling while satellites and aircraft provide frequent spatial sampling but limited sampling in time. An optimum system mix will involve a combination of these platforms.

The value of an objective observation can be appreciated by considering the distortions which the human mind can unconsciously introduce in analyzing a sensory perception. Consider Koestler's example 1/ of holding the index finger of the right hand ten inches,

and the same finger of the left hand twenty inches in front of
the eyes. A person "sees" them as being of equal size because
we <u>know</u> their size remains constant even though the image on the
retina of one is twice as large as the other. The photographic
lens will <u>honestly</u> show the right index finger twice as large as
the left. For the human, the visual input is interfered with and
<u>falsified</u> at some level of the nervous system. The implications
are profound in all areas of human experience from science to reli-
gion. We see the world as we <u>want</u> <u>to</u> <u>see</u> <u>it</u> or believe it should
be and adjustments are made automatically and <u>unconsciously</u>. Thus,
for example, visual reports of wave heights from large ships may
be smaller than from small vessels and visual classification of
forest disease damage from aircraft has been frequently found in
error by many tens of percent. Man's experience conditions his
behaviour. His brain has been developed to do automatic processing
involving <u>data</u> <u>integration</u> which may result in erroneous inferential
constructs. The <u>objective</u> <u>honesty</u> of remote sensing observations
cannot be questioned. Another kind of bias error can arise in
"traditional" measurements. Considering the ocean wave height
example again, climatological data on wave heights is usually biased
on the low side because ships avoid high sea states whenever possible. Satellite derived wave heights should not show such biases.

Point Versus Field Measurements

Traditional environmental measurements have of necessity been
point measurements, sometimes of a vertical profile nature, of
land, oceanographic or atmospheric phenomena. These data inputs
have been used in operational systems for many years and have become
ingrained in the models and decision making processes. A whole
modelling and analysis machinery in meteorology, hydrology and
oceanography has been developed over many years to accept these
data inputs. While modest success has been achieved in most
operational systems such as stream flow forecasting, weather prediction, crop yield forecasting, etc., there has always been a
demand for more data points. This requirement can be satisfied
fairly well in populated areas but becomes difficult in some circumstances such as mountainous regions or oceanic areas, particularly of the southern hemisphere.

Because of data gathering costs, sampling techniques have been
used which generally yield a high information content for a particular sample at a particular point. The tendency, therefore, has
been to make point measurements of water temperature or vertical atmospheric profiles, for example, sometimes to an ultra high degree
of accuracy.

Of more fundamental importance than the number of data points
to the fulfilment of user requirements is the answer to the

fundamental question "How representative of an area or volume of atmosphere, water, or soil are those point measurements?" Environmental phenomena are generally distributed and interrelated in nature but those relationships may be first order, second order ....nth order....in space and time. A measurement, for example of precipitation by a rain gauge, may significantly represent only a small area a few metres across.

Consider as an example the specific case of measuring snow water equivalent, an important parameter in stream flow forecasting in northern and mountainous regions. Streamflow forecasting is essential for such operations as hydropower generation, flood forecasting or irrigation. Current operational procedures involve the use of snow courses by which point samples are taken at spatial intervals. A water equivalent measurement is then used as an index in a regression equation to predict watershed runoff. In some river basins in mountainous regions, the number of sample points may be one or even zero with a measurement made in an adjacent region. Moreover measurements are usually made in low lying accessible regions. The result is that any particular point measurement may be of variable value depending on how representative it is of the basin runoff characteristics. It may even have negative information value if it gives misleading information which affects a decision process. Moreover it takes about ten years to find out if the point is representative of the basin. Having once established the representativeness and therefore utility of a particular point, if any significant change occurs in its environment, one is back to where one started. For example, if forest fire or disease or the chainsaw removes trees in the area, the meaning of the measurement in the short term is lost. Alternatively, a point may be lost with the death of the little old lady in the backwoods cabin who was making that measurement.

Examples similar to that above could be given for such environmental parameters as soil moisture, precipitation and others. Satellite data is already showing promise for measurement of some distributed phenomena such as snow cover and precipitation. Remote sensing potential benefit studies indicate that development of remote sensing devices for snow water equivalent and soil moisture would have large economic value.

Remote Sensing Data in Operational Systems

The emergence of a new form of data input for use in an operational system requires tailoring the data to suit the existing system, adjusting the system to accept the new data, or a combination of the two. A wide variety of space and time scales are associated with the environment. The time constants associated with weather phenomena are much smaller than for natural forest changes

which are in turn smaller than for soil changes. In general data
must be treated quite differently for each application, although
there are some common requirements such as correct <u>geographic</u>
referencing.

Considerable success has been achieved with satellite remote
sensing imagery showing rather gross spatial changes in environmental features. With meteorological satellites, cloud features
and cloud motion have been readily interpreted and applied to
weather forecasting problems. With LANDSAT, ice conditions, snow
boundary contours, and fire burn areas have been identified and
used in operational applications. The information contained in
these examples is spatial change with time; the human visual system
is well adapted to analyze such information. Quantitative analysis
of remote sensing data such as spectral reflectivity data (e.g.
multispectral scanner data) of complex vegetation patterns, for
example, has proved much more difficult, and operational application has only progressed a little way to date.

Reliance of many traditional operational methods on point
data has resulted in development of models which take point data
such as temperature, precipitation, or soil moisture and (perhaps
merely by some extrapolation technique) produce an area map output
product. Point measurements may be well suited to models where a
grid system is used such as in numerical weather prediction. However, truncation errors result from coarse resolution in a numerical mode. Errors result mathematically from the use of finite
differences on a coarse grid to approximate derivatives at points
and physically from a failure to account for many processes which
could affect the dynamics and energy balance in the atmosphere.
In the case of a remote sensing measurement or any other measure-
ment <u>it</u> <u>is</u> <u>only</u> <u>useful</u> <u>if</u> <u>it</u> <u>is</u> <u>used</u> <u>properly</u>. Data must be tuned
to the resolution and physics of the model. Error may be generated
by trying to make a remote sensing measurement look like a traditional meteorological measurement for example. It is crucial to
look at the required <u>output</u> <u>information</u> <u>product</u> and develop processing and analysis machinery to minimize error.

Depending on the required information products, conforming
"field" data to a grid point system may not be desirable. The
emergence of quantitative remote sensing data requires extensive
research in data assimilation, model development and model tuning
for many applications.

While theoreticians in meteorology, hydrology, and oceanography seek to develop deterministic models for environmental
prediction, the potential value of <u>real</u> <u>time</u> remote sensing observations of dynamic conditions such as sea state, snow melt or ice
movement to operators affected by these conditions can be very
high. <u>Reliability</u> of such data from microwave sensors on

satellites is expected to provide a large advantage over other measurement techniques.

In making any sound forecast about future conditions, it is essential to have an understanding about the current situation. Many of the benefits of the proposed SEASAT program are expected to come from real time observations permitting knowledge of existing conditions, short term prediction of ocean surface conditions, and initial conditions for various forecast models. The term "nowcasting" may be more appropriate than "forecasting" in many applications.

Learning Experience

The advent of satellite earth observation systems has introduced new spatial scales and new types of information about environment and resources. This new "dimension" of information will only slowly be accepted and incorporated in operational systems. Bertrand Russell wrote in 1925 in his ABC of Relativity 2/:

"Many of the new ideas can be expressed in non-mathematical language, but they are nonetheless difficult on that account. What is demanded is a change in our imaginative picture of the world....Einstein's ideas, similarly, will seem easier to generations which grow up with them...."

A similar learning process is required for data from satellite earth observation systems. It may require development of whole new systems. For example, the potentially powerful techniques of multistage sampling must be developed and applied. These have already shown promise for natural resource inventories.

Satellite earth observation systems provide an overview of environmental conditions. They help to alert us to the existence and to a lesser extent nature of environmental interrelationships and to monitor changes in those relationships brought about by nature or human action. Earth observation systems in the 1980's will present severe data processing, analysis, and storage problems. Appropriate data filtering techniques must be developed.

In terms of data analysis, the human brain is programmed to be alerted to certain things, to see certain things and process that data. What a human "sees" depends on his learning experience. For example, for a child looking at the world around him, many things are new and he questions them - their appearance, activities and relationships. Perhaps Wordsworth 3/ had this in mind when he wrote, "The Child is Father of the Man". As the child grows into adulthood, he becomes more selective in what he observes and questions. He may observe flowers by the roadside but not "see" them. Filters have been developed to reduce the processing of input data. These filters

change with age, custom and culture and form a vital part of the
evolution of man.

Many interesting examples of cultural differences may be given
- for example, the response of a group of African non-literates to
a film 4/. A moving picture of about 5 minutes duration was shown
in slow time to a group. It featured a labourer carrying out some
chores. When the audience were asked what they saw, all gave a
first response, "We saw a chicken". Careful inspection of all
frames showed that in one frame a chicken flew across the bottom
segment. The duration of this event was one second. The non-
literates were conditioned to look for such sudden movements.

"Literacy gives people the power to focus a little way
in front of an image so that we take in the whole image
or picture at a glance. Non-literate people have no such
acquired habit and do not look at objects in our way.
Rather, they scan objects and images as we do the printed
page, segment by segment. Koestler writes, "It is as
difficult to alter our way of seeing the world as it is
to alter our signature or accent of speech".

Parallels may be drawn between the diversity and evolutionary
nature of the human learning process in these examples and the
development of processing, analysis, and utilization stages of
remote sensing systems. Adoption of any new technology takes time.
Considerable selective research must be done before much of the
data can be used properly. It is a process governments must support
with patience. Commitments of managers to "traditional" methods
will inevitably delay acceptance of radically new forms of data.

As the data volumes from satellites escalate in the future,
automatic filtering techniques must be developed to reduce data
at various stages of the processing. Data archival will require
a high degree of selectivity and, like the human memory, methods of
storing cross referenced information are required.

The Essence of Earth Observation Systems:

The Extension of Human Consciousness

Sensors on earth observation systems provide an extension of
man's senses and the systems provide an extension of man's
consciousness of his environment. Consciousness in human experience
implies more than isolated sensory inputs of sight, taste, sound,
etc. It implies an interplay of information inputs - some from
these senses, some from information stored from past experiences.
The information is filtered and processed according to methods
which have evolved with the experience of the individual and the

development of man throughout history. Koestler 1/ writes:

"Memory is not based on a single abstractive hierarchy but on a variety of interlocking hierarchies - such as those of vision, taste and hearing. It is like a forest of separate trees but with entwined branches".

Also,

"When we listen to speech or music, the nervous system extracts patterns in time by bracketing together the present with the reverberations of the immediate past, and with memories of the distant past, into one complex process occurring in the specious present in the three dimensional brain. It constantly transposes temporal into spatial patterns, and spatial events into temporal sequences".

And

"The data are collected on the lowest, local levels. They are then stripped of irrelevant detail, condensed, filtered and combined with data from other sources at each higher echelon, as the stream of information flows upward along converging branches of the hierarchy. Here we have a very simplified model of the working of the sensory-motor nervous system".

Clearly, we can learn a great deal about structuring earth observation systems from nature's greatest achievement - man himself. It is obvious in the first instance that we must talk about total information systems. Sensor data used in isolation from other information has very limited value. The synergistic use of data from several sensors can greatly amplify the information value of each used individually. Can one enhance the value of an earth surface thermal emission measurement from satellite, for example, by simultaneously monitoring the conditions of the intervening atmosphere? A study of the human system indicates the requirement for a hierarchy of processing agencies from sensory inputs to "meaning". It also indicates a requirement for interlocking hierarchies.

Examples from many different fields can be given which illustrate the necessity for integration of information inputs from many sources. For example, in a crop information system, it is unwise to build a remote sensing based system and compare with a "conventional" system. Rather efforts must be made to integrate earth resource and meteorological satellite data with other more conventional forms of input each of which. have their place in an overall optimum system. In ocean environmental monitoring, with

the development of automated data collection systems (particularly
satellites and data buoys) ships, aircraft and fixed land and ocean
stations are expected to perform the same types of functions as at
present, but on a more selective and focussed basis.

Manned vehicles will go where an unknown has been detected,
where more detailed evidence or observations are required, where
regulations require enforcement - in short, in all those situations
where the human presence is genuinely needed and irreplaceable.
Each platform has its strong points and makes a contribution to
the optimum mix. Moreover, most platforms can perform several
functions. For example, a data buoy can provide continuous environmental data at a point and a "control" point for satellite or
aircraft observations. The effective integration of multi-level,
multi-temporal information in such operation and management systems
will evolve in time at a rate dependent on the realization of managers that such an integration is inevitable from economic, energy,
and environmental considerations.

Transformation to an Information Society

Industrialized societies are currently undergoing a radical
transformation because of rapidly evolving technology for the
acquisition, processing and trasmission of information. The
global information revolution and the sealing of the entire human
family into a single global tribe, due to the electronic age - the
"global village" discussed by Marshall McLuhan 5/ - are becoming
fact. Advances in communication technology have much to do with
this transformation. Communications can replace much of the
transportation of people, goods and information producing a timely
integrating effect, together with considerable saving of energy
and other limited resources. Computer data networks provide the
same integrating function that canals, railroads and highways did
in an earlier phase of development. In future it is anticipated
that fully switched data networks will operate worldwide.

"Electric circuitry has overthrown the regime of "time"
and "space" and pours upon us instantly dialogue on a
global scale. Its message is, Total Change, ending
psychic, social, economic, and political parochialism.
The old civic, state, and national groupings have become
unworkable. Nothing can be further from the spirit of
the new technology than "a place for everything and everything in its place". You can't go home again". 6/

The concept of the global village is being further developed
by earth observation systems. Areas which were remote, relatively
inaccessible by land, or characteristically cloud covered can now
be monitored from satellite or with microwave imaging devices on

aircraft. Future land, ocean and atmospheric monitoring satellites with higher resolution sensors and all-weather reconnaissance capability, together with real-time communication will further effectively reduce "time" and "space" dimensions. Technologies tend to create a new human social environment; satellite earth observation systems will do so as well.

Society's demands are changing. One cannot extrapolate society's demands of the 1950's to the 1980's. It used to be that one could build a road or an electric power plant without giving much heed to the environment. Now one must consider both the effect on the environment and on other human activities in the area. Optimization of a route location, for example, requires intimate knowledge of both of these. This means firstly, a requirement for baseline data to understand the environment and environmental interrelationships in the area; it means secondly, monitoring the effect of the development on the environment to see whether and in what manner it has been disturbed.

Environmental indicators such as the health of vegetation, or the quality of water, parameters amenable to satellite monitoring, are increasingly being recognized as indicators of the stress man is placing on the environment. Because of interrelationships, the environment will eventually react back on man to put a stress, perhaps transformed in nature, on him.

Man must learn to live in harmony with nature. Clearly that means, in the first instance, an understanding of its complexities. By providing environmental information on sun, winds, tides, biomass, etc., earth observation systems will help find production methods and patterns of living which do not depend on fossil fuels. Through monitoring and improving our knowledge of soils, vegetation and man-induced changes, they will help in the development of production methods in agriculture which are biologically sound and build up rather than emaciate soil fertility. The real nature of things must be understood.

The world today demands that the method suit the application. For example, in medicine a doctor may indiscriminantly prescribe antibiotics for minor illness. While this works and in some cases is essential, if not used wisely the effectiveness of these drugs in time will diminish because the virus or bacteria adapt and new resistent strains evolve.

The antibiotic attacks the virus. A more reasonable solution is either preventative medicine or a method which will give nature a helping hand and strengthen the body rather than attacking the virus. This principle can be extended to all aspects of nature including those aspects of the environment which can be monitored by earth observation systems.

Perhaps of all aspects of man's environmental abuse, that of chemical spraying of pesticides and herbicides should be singled out. Warnings of the consequences of this practice have been given in Rachel Carson's, "Silent Spring" 1/, one of the great books of this century. In Eastern Canada, millions of acres of forest are ravaged annually by the spruce budworm. Extensive spray programs have had some short term success. However, there is worrying evidence of a possible link between the spray and a new disorder affecting some children in the vicinity. Moreover, with most spray programs two major effects result: (1) Natural predators such as birds suffer heavy loss of life. When a more resistent strain of insect appears, nature is left even less able to cope with it. (2) Nature is defenceless against many of the foreign chemical substances used today. They are cumulative and poisonous.

The solution is clearly one of understanding the ecology and exercising proper management with the long term good of society as the top priority. Earth observation systems play a vital role in these processes.

Earth Observation Systems in a Changing Society

The coming of the global village has increased our understanding of all people and of the limited nature of the earth, its resources and its capacity for abuse. The pattern for the rest of man's history must be to use more common-sense methods - that is our actions must not perturb nature at a faster rate than she can adapt. Life evolves and can adapt but that takes time. A poisoned environment will attack first the most vulnerable part of a species - often the reproductive process. For bacteria or insects the adaptation can be fast in calendar time; for man, it will be much slower. Already fetal deformities and sterilization are on the rise in highly vulnerable areas.

The message for technology is that technology must suit the application. For example, in "traditional" ocean surveillance systems, high level and low level surveillance, long range and short range surveillance are generally not most effectively satisfied by the same vehicle. Attempts to match readily available technology to a requirement may result in use of a fast, high energy consuming military aircraft for a station keeping mission that could be performed with a balloon. Speed costs. In oil pollution monitoring for example, is a fast transit of an area by aircraft while at the same time exhausting more hydrocarbons into the atmosphere the most "effective" way of carrying out that particular mission? Requirements must be assessed more carefully and technologies adapted which satisfy energy, environmental and ethical as well as economic considerations. Intelligent development of earth observation systems is required as one small part of the solution.

Among the social forcing functions of future earth observation programs is that of resource conservation. The philosophy of Limits to Growth 8/ can be extended well beyond the population context. The finiteness of non-renewable resources, particularly petroleum, is now well recognized. This has been referred to 9/ 10/ as irreplaceable "capital" provided by nature, which we are depleting. There are other forms of "capital" which are also limited, however. Clearly "renewable" resources such as fish, wildlife, crops and trees are limited and the term "renewable" must be closely questioned. Perhaps more important is the tolerance nature can endure to man's exploitation. For example, in clear cutting trees, erosion losses are heavy, a slow but definite loss in capital. Use of heavy machinery and herbicides in agriculture result in short term productivity increases but are harmful to the abundant, complex, and necessary life near the earth's surface. Burying productive lands under concrete by urban sprawl is clearly a loss in capital.

The old working rules of justifying development on purely economic grounds are being supplanted by a realization that this generally ignores man's dependence on the natural world. Quality of life is taking on increasing importance in our scheme of values, and understanding and detecting abuses of our environment is part of the role earth observation systems must play. In the new global village one segment of the world's society can no longer go off and use up a disproportionate share of the earth's resources and nature's store of "tolerance to abuse". Ignorance is no longer an excuse. A new responsibility rests on the shoulders of all people of the developed world to lead by example as only we can do.

"Civilized man has despoiled most of the lands on which he has lived for long. This is the main reason why his progressive civilizations have moved from place to place". 11/ 12/

Today with accerlerating abuse and no new lands to move to, the warning is clear. With a better understanding of the interrelationships within nature, of the relationship between man and nature, and with monitoring systems to detect pollutants, effluents and discharges, the life-style of society can be expected to change in the future.

An awareness of the principles of self-limitation is something man must learn. Just as nature is self-balancing and curbs growth, man enveloped by a finite environment with a finite capacity to cope with interference, must impose limits on his technology, his expansionist appetites, his demands on a diminishing "capital". Throughout history man's loyalties have had to expand to include a larger and larger group of people. So they have moved from the family to the village to the city-state to the nation state. Today man's loyalties must be to the global community if he is to survive.

Planning the Future

The big problems of the world today are problems facing all mankind and it is appropriate that men of many nations should work together in solving them. The problems of energy supply, food supply and environmental preservation are in the forefront. While the United Nations has had limited success in facing these problems, a smaller group of nations such as a North Atlantic Community for a Better World without the vast political and economic disparities of the UN might lead effectively by example. Clearly this applies to earth observation systems - one small part of the solution of these problems.

Co-operation is desirable at all levels; consider, for example, sensor development. The geographical and climatic conditions in Canada vary drastically from the fruit orchards in southern Ontario to Arctic tundra and ice-covered waters in the north. A wide range of sensors are required to effectively monitor such a variety of conditions. Sensors to monitor permafrost conditions are not as likely to be developed by other countries as sensors to monitor wheat fields. Clearly co-operation is desirable to develop and utilize data acquisition systems to monitor effectively such a diversity of conditions.

Hardware oriented programs such as construction and launch of satellites seem to attract government support more easily than research programs to develop proper analysis techniques and integrate the data into total information systems. The satellite program provides a well defined focus while analysis and applications development techniques are more diffuse.

The multiplicity of satellite programs now planned suggests that data acquisition capabilities could race far ahead of man's ability to utilize the data effectively. A proper assessment of the value of a particular spaceborne sensor with particular characteristics takes time. Applications research programs must be set up carefully, realizing that while technological milestones such as putting a man on the moon can be scheduled, research advances slowly, often unpredictably and often by serendipity. In the area of applications research, governments must be convinced not to discount the future and to have patience. With closer oversight of and tighter purse strings on programs in all countries, the need to jointly tackle some of the major problems increases. International workshops in such important areas as multistage sampling, pattern recognition and data processing are desireable.

Finally, just as the world is becoming a "global village" through the media of television and communications, and we are now able to place our lives in the context of a global community, similarly earth observation systems permit detailed observations of earth resources and environmental phenomena to be placed in a broader, more meaningful global context.

Context is vital in understanding environmental phenomena. Relationships may be more meaningful than the phenomena themselves. Science is continually striving to discover and elucidate inter-relationships. The universe is one.

> "To see a World in a Grain of Sand
> and a Heaven in a Wild Flower,
> Hold Infinity in the palm of your hand
> and Eternity in an hour". 13/

The challenges of solving the world's problems of energy, food and environment are great. The challenges of solving the problems of earth observation systems - the high data volumes and rates, the complex pattern recognition requirements, the difficult information integration requirements - are also great. The hope however is greater. Through international teamwork, we can be optimistic that that hope can be realized.

REFERENCES

1/ Koestler, Arthur, The Ghost in the Machine, Hutchinson Publishing Group Limited.

2/ Russell, Bertrand, ABC of Relativity, (revised edition) London: Allen and Unwin, 1956.

3/ Quotation from "My Heart Leaps Up When I Behold", by William Wordsworth.

4/ "Film Literacy in Africa". Canadian Communications, Vol. 1, No. 4, 1961.

5/ McLuhan, Marshall, The Gutenberg Galaxy, University of Toronto Press, 1962.

6/ McLuhan, Marshall and Fiore, Quentin, The Medium is the Message, Bantam Books Inc., 1967.

7/ Carson, Rachel L., Silent Spring, Fawcett Publications Inc., Greenwich, 1962.

8/   Meadows, D.H.,   The Limits to Growth, A report for the Club of
        Rome's Project on the Predicament of Mankind.   The New
        American Library Inc., New York, 1972.

9/   Fuller, R. Buckminster, Uptopia or Oblivion, the Prospects for
        Humanity, Bantam Books Inc., New York, 1969.

10/ Schumacher, E.F.,   Small is Beautiful, Sphere Books Ltd., London
        1973.

11/ Dale, T. and Carter, U.G., Topsoil and Civilization, Univer-
        sity of Oklahoma Press, U.S.A. 1955.

12/ de Bell, Garrett,   The Environmental Handbook, Ballantine Books
        Inc., New York, 1970.

13/ Quotation from Auguries of Innocence , by William Blake.

THE POTENTIAL ECONOMIC BENEFITS OF AN OCEANOGRAPHIC

SATELLITE SYSTEM TO COMMERCIAL OCEAN OPERATIONS

B.P. Miller
Vice-President, ECON, Inc.
Princeton, N.J., U.S.A.

Abstract

In 1978 the U.S. National Aeronautics and Space Administration
will launch SEASAT-A, the first satellite dedicated to the remote
sensing of the ocean environment. Concurrently with the development of the technical concepts for SEASAT-A, a two-year study has
been performed to estimate the potential economic impact of an
operational SEASAT system on ocean-based industries. The results
of this study show that substantial, firm benefits could be obtained
in areas that are extensions of current operations such as marine
transportation and offshore oil and natural gas exploration and
development. Very large potential benefits from the use of SEASAT
data are possible in an area of operations that is now in the
planning or conceptual stage, namely, the transportation of oil,
natural gas and other resources by surface ship in the Arctic
regions. In this case, the benefits are dependent upon the rate
of development of the resources that are believed to be in the
Arctic regions, and also dependent upon the choice of surface
transportation over pipelines as the means of moving these resources to the lower latitudes. Our studies have also identified
that large potential benefits may be possible from the use of
SEASAT data in support of ocean fishing operations. However, in
this case the size of the sustainable yield of the oceans remains
an unanswered question; thus, a conservative viewpoint concerning
the size of the benefit should be adopted until the process of
biological replenishment is more completely understood.

Preliminary estimates have also been made of the costs of an
operational SEASAT program that could be capable of providing the
data needed to obtain these benefits.

Introduction

The successful conduct of commercial operations in the ocean
environment is to a great extent dependent upon the ability to
forecast the state of the ocean environment as a function of time.
A statistical knowledge of the expected waves, winds and temperature
is necessary in order to design ships and structures to safely
operate in the ocean environment. The ability to measure these
parameters, and to forecast their state as a function of time, is
an important factor in determining the economic efficiency with
which commercial operations such as marine transportation, offshore
oil exploration and development, and ocean fishing are conducted.

During the past decade numerical forecasting techniques have
assumed a dominant role in the forecasting of ocean conditions and
weather. It is generally considered that in the next decade improvements in ocean condition and weather forecasting are most likely to
come from upgrading and further development of numerical models of
the atmosphere and the oceans. Insight into the opportunity for
improvement can be gained through examination of the errors which
cause forecasts to fail, even in the short range. These errors can
be classified into three major groups, namely,

- misrepresentation of the physics of the atmosphere and the
  oceans,

- truncation errors associated with the coarse resolution of
  numerical models, and

- initialization errors associated with the insufficiency and
  imbalance of data needed to specify the initial conditions
  for numerical models.

Although errors associated with the physics of the models and
truncation errors can be reduced by improved scientific understanding
and increased computational capacity, reduction of the initialization
errors requires the capability for increased data collection on a
global basis. In addition to reducing initialization errors, increased data collection can also provide the empirical information
to improve physical representations of the oceans and atmosphere,
thus further contributing to the reduction of errors due to the
misrepresentation of the physics. The lack of data at present is
simply due to the lack of meteorological observing stations in
unpopulated areas. In the case of the oceans, surface data is
concentrated in the major shipping lanes, thus leaving large portions of the oceans unsampled, and even where surface data is
provided, upper air data may be missing. Of all observing systems,
satellites are probably the most promising for increased data collection because of their capability for global coverage.

While present weather satellites have made many contributions
to the measurement of global weather phenomena, their ability to
provide initialization data to numerical forecasting models is limited to wind as inferred from time lapse photography of clouds and
vertical temperature profile measurements. In 1978 NASA will
launch an oceanographic satellite, SEASAT-A. In addition to a
conventional visible and infrared imaging sensor, SEASAT-A will
also carry four active and passive microwave sensors. These sensors
will provide the first nearly global capability for the measurement
of waves, winds and temperatures over the oceans as well as the
capability for the high resolution, all-weather imagery of selected
ice and coastal phenomena. These SEASAT data can be used to supplement data from other sources for the initialization of forecasting models, for the production of nowcasts (i.e. the present conditions) and to verify model output.

In order to identify the potential users of SEASAT data as
well as the relative economic importance of system characteristics
to commercial users, and to evaluate the desirability of using
public funds for the development of SEASAT-A, an economic assessment of an operational system was performed. The economic assessment examined the impact of the data produced by a hypothetical operational oceanographic satellite system on many types of commercial
ocean-based activities. Substantial potential economic benefits
from improved ocean condition and weather forecasts and improved
ice information were identified in the areas of Arctic operations,
marine transportation, ocean fishing and the exploration and development of offshore oil and naural gas. The costs of the hypothetical systems were examined along with the benefits, and it was
concluded that the potential benefits of improved ocean condition
and weather forecasts greatly exceed the estimated costs of the
satellite system needed to produce this data.

The following sections of this paper briefly describe SEASAT-A,
the hypothetical operational system that served as the basis for
estimating the system benefits and costs, and the significant results of the economic studies of the uses of SEASAT data.

The SEASAT System

SEASAT-A

SEASAT-A is scheduled for launch in April 1978. The objectives
of the SEASAT-A program are:

(1) To demonstrate the capability for:

- Global monitoring of wave height and directional spectra,
  surface winds, ocean temperature and current patterns

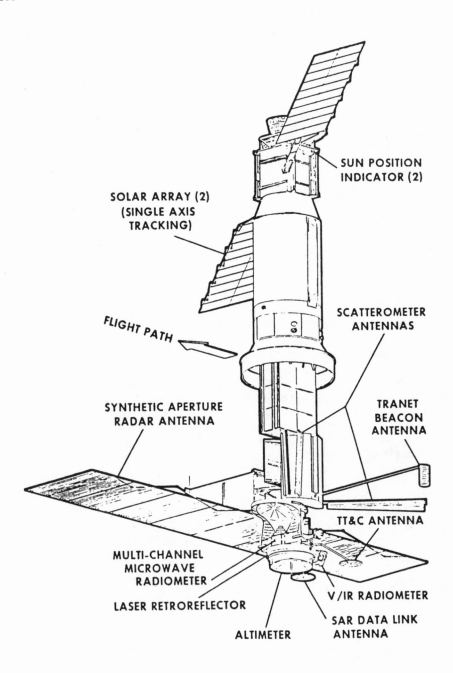

Figure 1.   SEASAT-A

  - Measuring precise sea-surface topography

  - Detecting currents, tides and storm surges

  - Charting ice fields and navigable leads through ice

  - Mapping the global ocean geoid

(2) To provide for user applications such data as:

  - Predictions of wave height, directional spectra
    and wind fields for ship routing, ship design,
    storm-damage avoidance, coastal disaster warning,
    coastal protection and development, and deep water
    port development

  - Maps of current patterns and temperatures for ship
    routing, fishing, pollution dispersion and iceberg
    hazard avoidance

  - Charts of icefields and leads for navigation and weather
    prediction

  - Charts of ocean geoid fine structure

(3) To determine the key features desired in future operational
     systems for:

  - Global sampling

  - Near realtime data processing and dissemination

  - User feedback for operational programming

(4) To demonstrate the economic and social benefits of user agency
     products.

The SEASAT spacecraft bus, the Agena, first flown on military
space missions in 1959 and subsequently on over 300 missions, has
been configured to support the oceanographic mission requirements.
The overall configuration of SEASAT-A is shown in Figure 1. The
SEASAT spacecraft comprises a standard satellite bus and a custo-
mized sensor module which supports and accomodates SEASAT sensors
and their antennas.

The bus is an Agena spacecraft. The Agena main engine propels
SEASAT-A from separation of the Atlas booster to insertion in circular near-polar orbit at 800 km. On orbit, Agena provides 626
watts average, 1180 watts peak electrical power, stabilization and
attitude control including orbit trimming, command reception,
storage and execution, and data storage and transmission. Tracking
aids including S-bands, doppler and laser reflectors assist the
ground system in determining where the satellite is and where the
sensors are scanning.

Table 1.   SEASAT-A sensor characteristics.

SENSOR	COMPRESSED PULSE ALTIMETER	MICROWAVE SCATTEROMETER	SYNTHETIC APERTURE IMAGING RADAR	MICROWAVE RADIOMETER	VISIBLE & INFRARED RADIOMETER
SENSING OBJECTIVE	GLOBAL OCEAN TOPOGRAPHY GLOBAL WAVE HEIGHT	GLOBAL WIND SPEED AND DIRECTION	WAVELENGTH SPECTRA LOCAL HIGH RESOLUTION IMAGES	GLOBAL ALL-WEATHER TEMPERATURE GLOBAL WIND AMPLITUDE GLOBAL AT-MOSPHERIC PATH COR-RECTIONS	GLOBAL CLEAR-WEATHER TEMPERATURE GLOBAL FEA-TURE IDENTI-FICATION CLOUD COVERAGE
FREQ/ WAVE LENGTH	13.9 GHz	14.6 GHz		6.6, 10.69, 18, 22.235, 37 GHz 1.275 GH3	0.52—0.73 μm 10.5—12.5 μm
ANTENNA/ OPTICS	1 m PARABOLA	5—2.7 m STICK ARRAYS	11 x 2.2 m PHASED ARRAY	0.8-m OFFSET PARABOLA	12.7 cm OPTICS
POWER	125 W AVE	165 W AVE	200—250 W AVE	50 W	10 W
DATA RATE	8 kb/s	2 kb/s	15—24 Mb/s	4 kb/s	12 kb/s
HERITAGE	SKYLAB/GEOS-C	SKYLAB	APOLLO 17	NIMBUS G	ITOS

Table 2.   Postulated SEASAT inflight schedule.

CALENDAR YEAR |75| | | | |80| | | | |85| | | | |90| | | |

DEVELOPMENTAL SEASAT

   SEASAT-A

INTERIM OPERATIONAL SEASAT

   UNCHANGED DEVELOPMENTAL VERSION

   NO R&D EXTENSIONS

   SIMPLE R&D EXTENSIONS

FULL CAPABILITY OPERATIONAL SEASAT

   FULL CAPABILITY R&D

   NEW GENERATION (TECHNOLOGY OBSOLESCENCE ~6 YEARS)

As specific requirements evolved within the using community, candidate remote sensing instruments were evaluated jointly by the users and NASA for SEASAT-A application. A set of three active radars and two passive radiometers were ultimately selected each with direct heritage from space and aircraft programs. The characteristics of these sensors are summarized in Table 1. The sensors include a pulse-compressed radar altimeter, a microwave radar scatterometer, a synthetic aperture imaging radar, a scanning multifrequency microwave radiometer and a visible/infrared scanning radiometer. The SEASAT radar altimeter serves two functions: it monitors average wave height to within 0.5 to 1 meter along its narrow (2 to 12 km) swath by measuring the broadening of its returned echo caused by increased surface wave actions; it also measures to a precision of 10 cm the changes in the ocean geoid due to gravity variations and ocean tides, surges and currents. As surface winds increase, so does surface roughness or chop. The radar scatterometer measures the signal strength of its returned echoes. Signal strength increases with the increase in wind-driven waves, which can be converted directly into wind speed and direction. The scatterometer measures wind speeds from 3 to 25 m/sec within 2 m/sec and direction within 20 deg over two 500 - km swaths on either side of the spacecraft ground track. The five-frequency microwave radiometer serves four functions: (1) it measures surface temperature by measuring the microwave brightness of the surface to within 1 deg C, (2) it measures foam brightness which can in turn be converted into a measurement of high (up to 50 m/sec) wind speed, (3) it maps ice coverage and, (4) it provides atmospheric correction data to the active radars by measuring liquid and gaseous water content in the upper atmosphere. The surface swath of the microwave radiometer is 920 km. The visible and infrared radiometer will provide clear weather surface temperature data, cloud coverage patterns, and corroborative images of ocean and coastal features with a resolution of 5 km over a swath of 1500 km. These four sensors, known as the global sensors, will monitor the oceans and adjacent coastal waters globally. Their data will be recorded on magnetic tape recorders on board SEASAT and played back while the satellite is over one of the tracking stations supporting SEASAT. Nearly global coverage is achieved every 36 hours.

The fifth sensor, the synthetic aperture radar, will provide all-weather imagery of ocean waves, ice fields, icebergs, ice leads, and coastal conditions and dynamic processes to a resolution of 25 m over a 100 - km swath. Because of the very high data bandwidth of the radar imagery, this sensor with its own separate data system will be operated only in realtime while within line of sight of specific tracking stations equipped to receive and record its data.

The current plan for SEASAT-A data processing includes verification of the sensor performance by the NASA Jet Propulsion Laboratory. Global sensor data will be received at the NASA station at Fairbanks, Alaska, and relayed via communications satellite to the U.S. Navy Fleet Numerical Weather Centre (FNWC) at Monterey, California. FNWC will process the data from the altimeter, the scatterometer and the radiometers with less than eight hours turnaround time for distribution to the operational meteorological and oceanographic communities.

Operational SEASAT System Concepts

In order to provide a basis for the analysis of the economic potential of the SEASAT Program, it is necessary to hypothesize both schedule and capability for the operational systems that will follow SEASAT-A. Planning of this nature for the proposed operational system is necessary as the level of benefits is dependent upon the technical capabilities of the operational system, and the timing of the benefits is dependent upon the dates of inception of the operational system. Thus, the benefits ascribed to SEASAT are those associated with an operational system that will provide the continuity of service required to obtain full utilization of the data products by potential government and private users. The general SEASAT schedule shown in Table 2 delineates a set of developmental and operational systems that could fulfill presently understood user requirements. Only the first element of the program, SEASAT-A, is approved at the present time. The remaining elements of the program, namely the interim operational and full capability operational SEASATs, have been hypothesized to provide a basis for evaluation of the economic benefits and are not approved programs. The specific configurations of these systems will evolve from an improved understanding of user requirements gained with SEASAT-A, its follow-ons, and the supporting aircraft and sea-truth program.

The first developmental SEASAT (SEASAT-A) is to be launched in mid-1978, and is a single satellite with a one-to-three year life. In the 1980-1983 period, an interim operational SEASAT system is possible, with three satellites providing twice-a-day global coverage of accurate sea-surface winds, waves, and temperatures, plus several sitings a week of specific ocean features at 20-m resolution. As indicated, several interim capability possibilities with differing investment implications are also available. Some combination of these alternative, interim three-satellite systems is probably needed to provide the user community with the continuity of information necessary to realize their potential economic and social benefits. The full-capability (six-satellite), operational SEASAT system could become viable in 1985. A new SEASAT generation could then come into being about every six years which represents both a reasonable life expectancy for this time period and a typical technology-obsolescence period where part and component availability will force creation of new design even without the pressures of remote-sensing improvements.

Specific measurement capabilities for each of the SEASAT developmental and operational stages are somewhat speculative at this time, but some assessment of these potentials is probably appropriate. Table 3 provides an indication of the kind of sensor capabilities that might be expected in an operational system.

It is anticipated that major changes could occur in the nature and usage of the data products produced by an operational SEASAT system as compared to SEASAT-A. Table 4 identifies some of these data products. Other changes in the ground data processing system could provide improved user accessibility by providing direct readout of certain data that could be processed into user products on board the satellite.

Economic Analysis

Methodology

The economic benefits of an operational SEASAT system were estimated by the use of interrelated micro and macroeconomic studies. The studies were performed on a by-industry or by-sector basis. The micro studies consisted of case studies, each case study being an in-depth examination of the potential benefits that might be obtained by the use of SEASAT data in a specific application. The results were then generalized on a by-industry or a by-sector basis using appropriate econometric or economic models. Figure 2 is an overview of the methodology used in the economic assessment. The case studies, or application areas to be studied, were selected after a survey of the potential uses and users of the SEASAT data. The survey was performed by reviewing pertinent literature and by interviewing personnel from the prospective user organizations. The survey led to the identification of the prospective users of both the SEASAT data and the resulting improved weather and ocean condition forecasts. Users were identified within governments, institutions and industries. In each case an effort was made to understand how the user would apply the SEASAT data (or the improved weather and ocean condition forecasts) and what the expected areas of economic benefit would be from the application of these data. Although it was recognized at the onset that improved weather and ocean condition forecasts could affect both land based and ocean-based operations, *it was decided to restrict the attention of this study to only ocean-based operations*. The results of the survey led to the identification of specific industries or sectors whose economies could be affected by improved weather and ocean condition forecasts as candidates for case studies.

Each case study then became an in-depth examination of the operating parameters, constraints and structure of a selected operation. The parameters of the selected application areas were

Table 3. Geophysical oceanographic-measurement capabilities for a full capability operational SEASAT (6 satellites in orbit).

MEASUREMENT			RANGE	PRECISION/ACCURACY	RESOLUTION, km	SPACIAL GRID, km	TEMPORAL GRID, km LESS THAN 2 MONTHS
TOPOGRAPHY	GEOID		5 cm – 200 m	< ±10 cm	1.6 – 12	~10	
	CURRENTS, SURGES, ETC	ALTIMETER	10 cm – 10 m				4/d
SURFACE WINDS	AMPLITUDE	MICROWAVE RADIOMETER	7 – 50 m/s	± 1 m/s OR ±10%	25	25	4/d
		SCATTEROMETER		± 1 m/s OR ±10%	25	25	4/d
	DIRECTION		0 – 360°	±15°			
GRAVITY WAVES	HEIGHT	ALTIMETER	0.5 – 25 m	±0.5 m OR ±10%	1.6 TO 12	~900	1/d
	LENGTH	IMAGING RADAR	50 – 1000 m	±10%	50 m	50 m	2/d
	DIRECTION		0 – 360°	± 10°			4/d
SURFACE TEMPERATURE	RELATIVE	V&IR RADIOMETER	-2 – 35°C CLEAR WEATHER	1°			
	ABSOLUTE			1.5°	5	5	4/d
	RELATIVE	MICROWAVE RADIOMETER	-2 – 35°C ALL WEATHER	1.5°			
	ABSOLUTE			2°	50	50	4/d
SEA ICE, CLOUD LOCATIONS, AND OCEAN FEATURES	EXTENT	MICROWAVE RADIOMETER		10 km	10	10	4/d
		V&IR RADIOMETER		1 km	1	?	4/d
	LEADS	IMAGING RADAR	> 50 m	25 m	25 m		2/d
	ICEBERGS		> 25 m	25 m	25 m		2/d
ATMOSPHERIC CORRECTIONS	WATER VAPOR AND LIQUID	MICROWAVE RADIOMETER		20%	25	25	2/d
ATMOSPHERIC TEMPERATURES	VERTICAL PROFILE	IR SOUNDER		2°, 5 km	10	10	4/d
ATMOSPHERIC HUMIDITY	VERTICAL PROFILE	IR SOUNDER	0-6 g/cc	20% OF COLUMN	10	10	4/d
SURFACE PRESSURE	g OF PERCIPITANT			2-4 mb	5	5	4/d

Table 4.   Potential data products.

	A	SEASAT OPTIONS	
	A	IN-TERIM	FULL CAPA-BILITY
SURFACE WIND FIELD MAPS	36h	2/d	4/d
SURFACE WIND FORECASTS - ONE-DAY AND TWO-DAY FORECASTS	36h	2/d	4/d
WAVE FIELD MAPS	2w	1/2-5d	2/d
WAVE FORECASTS - ONE-DAY AND TWO-DAY FORECASTS OF WIND-DRIVEN WAVES; LONGER FORECASTS FOR ESTABLISHED SWELLS	–	1/d	2/d
SURFACE TEMPERATURE FIELD MAPS	36h	2/d	4/d
SURFACE TEMPERATURE FORECASTS - ONE-DAY AND TWO-DAY FORE-CASTS OF CURRENT AND UPWELLING BOUNDARIES	36h	2/d	4/d
WEATHER MAPS (WIND, WAVE, AND TEMPERATURE, PLUS CLOUD MOVE-MENT, RAIN, ETC.)	–	2/d	4/d
WEATHER FORECASTS - ONE-DAY AND TWO-DAY FORECASTS	–	2/d	4/d
ICE MAPS - EXTENT, LEAD LOCATIONS	2w	1/2-5d	2/d
COASTAL MAPS - EROSION PROCESSES, SHOAL MOTION, WAVE REFRACTION PROCESSES	2w	1/2-5d	2/d
OCEAN DYNAMICS MAPS - CURRENTS, UPWELLINGS, TIDES, SURGES, ETC.	–	1/2-5d	2/d
GEODETIC MAPS	YRLY	6/y	MONTHLY

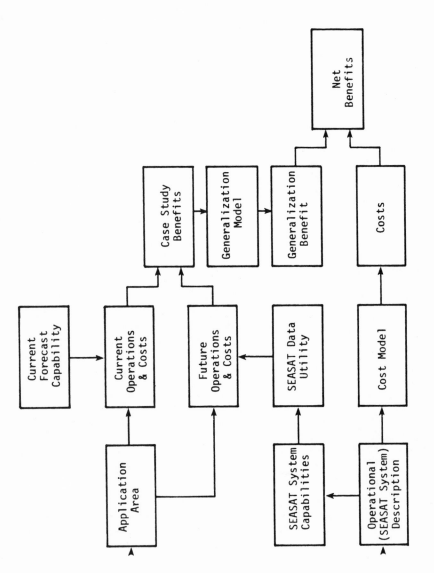

Figure 2. Study methodology.

then evaluated using current knowledge and predictive capability for weather and ocean conditions, and again using the expected improved capability for measurement and prediction of ocean conditions and weather as indicated by the SEASAT data utility. The incremental parameter changes attributable to the use of SEASAT data, or to the improved weather and ocean conditions derived from SEASAT data, were then estimated for each case study. These operating parameter increments then became the basis of the estimates of benefits to the operation under consideration.

The process of generalization was based upon the fact that each selected case study was one of a set of operations with generally similar technical and operational characteristics. With this condition, a generalization can be performed using appropriate econometric models and economic projections to bound and control the process.

Generalization of the case studies necessitated a careful formulation of each case study structure and its parameters that transform SEASAT derived information into economic benefits. Generalization also required that the class of operations represented by the case study be examined for each member of the class to determine the relationship between information and economic benefits. The process of generalization also required the extension of the results of the case studies to the dimension of scale (i.e. the relationship of samples to a population), time (the establishment of a valid planning horizon and forecasting quality variation) and geographical location (for example, extension of results obtained in the North Sea to other geographical sites for offshore oil exploration and development). This required the construction of appropriate physical models of weather effects and econometric models, and the collection and processing of data for use in these models. The planning horizon for these generalizations extended to the year 2000.

It is recognized that SEASAT data could be one of many important contributors to the improvement of ocean and weather condition forecasts in the period of 1985 to 2000. Thus, in each of the major expected benefit areas, two levels of benefits were estimated. An upper level of benefits was established on the basis of the expected improvements in ocean and weather condition forecasts from all sources, and the SEASAT specific benefits were then determined by estimating the level of improvement believed to be attributable exclusively to SEASAT data. In many of the areas examined, an additional element of uncertainty was introduced by the uncertainty in the rate of development of the industry or sector under consideration. In these cases, upper, lower and most likely ranges of benefits were estimated.

Table 5.   Summary of total SEASAT exclusive benefits.

Integrated Benefits

- Planning Horizon to Year 2000
- 1975 $
- 10% Discount Rate

Areas of Maritime Applications	Integrated Benefits
Offshore Oil & Natural Gas	214 - 344
Ocean Mining	not estimated
Coastal zones	3 - 81
Arctic Operations	96 - 288
Marine Transportation	215 - 525
Ocean Fishing	274 - 1432
Ports and Harbors	.5
Sea Leg of Trans-Alaska Pipeline	14
Unclassified Military	31
Iceberg Reconnaissance	12 - 39
TOTAL	859 - 2709

Table 6.   Generalization results for offshore oil and gas industry.

Discount Rate %	0	5	10	15
	Integrated Benefit 1985-2000 ($1975 Millions)			
Seasat Exclusive	464-2656	195-1113	87-501	43-244
Seasat Exlcusive Most Likely	1136-1824	476-765	214-344	104-168
All Sources	1547-8853	650-3710	290-1670	143-813
All Sources Most Likely	3787-6080	1587-2550	713-1147	347-560

Economic Benefits

The cumulative gross benefits attributable exclusively to the
use of SEASAT data and to the improved ocean and weather condition
forecasts derived from the SEASAT data are estimated to be in the
range of $859 to $2709 million ($1975 at 10 percent discount rate).
This is considered to be the most likely range of the SEASAT exclusive benefits for civilian applications. Table 5 disaggregates the
benefits by industry. As may be seen, the main contributors are
offshore oil and natural gas exploration and development, Arctic
operations, marine transportation and ocean fishing. These four
areas account for more than 90 percent of the estimated benefits.
The results of the benefit studies in each of these four areas are
briefly discussed in the following paragraphs.

Offshore Oil and Natural Gas Exploration and Development

The use of improved weather and ocean condition forecasts by
the offshore oil and natural gas industry accounts for approximately
13 to 25% of the total most likely SEASAT exclusive benefits. These
benefits were derived from case studies in which operational data was
obtained for production platform installations in the North Sea in
1971, pipelaying and trenching in the North Sea and Gulf of Mexico
in 1971 and 1972, and exploratory drilling by a drillship in the
Celtic Sea during 1970. The generalization was performed by estimating the size of the offshore exploratory drilling rig population
and the amount of pipe to be laid and trenched in the development
of the offshore fields in the time period extending from 1985 to
2000. Geographically varying weather and ocean conditions were
accounted for by the expected geographical distribution of the
offshore exploration and development activities. The results of both
the case study and its generalization were substantiated by an
independent study of Canadian Arctic oil and gas exploration and
development. Table 6 summarizes the generalization results for
the offshore oil and gas industry as a function of discount rate.
The most likely range of the cumulative SEASAT exclusive benefits
is $214 to $344 million ($1975 at a 10% discount rate). Those
benefits identified as attributable to "All Sources" and "All
Sources Most Likely" represent the estimated range of benefits to
the offshore oil and gas industry as a result of improved weather
and ocean condition forecasts from all potential sources (including
SEASAT) in the time frame under consideration.

Arctic Operations

The use of ice coverage information and improved ocean and
weather condition forecasts in the transportation of resources from
both the eastern and western Arctic regions accounts for approximately 11% of the total most likely SEASAT exclusive benefits.
The case study considered the development of Arctic oil and gas

Table 7.   North American Arctic Transit Case Study Results

Case Study Results

● Western Arctic 1992-2000

  + Transport 1.5 - 6.3 Billion Barrels of Oil
  + Number of Voyages 7900
  + Quantity of Oil Per Tanker 1.6 Million Barrels
  + Annual Benefit $26 Million to $494 Million[*]

● Eastern Arctic 1990-2000

  + Transport Oil, LNG, Hydrocarbons
  + Production Consistent With Alberta Commission Findings
  + Number of Voyages About 1300
  + Average Annual Benefit $152 Million[*]
  + Beaufort Sea Benefit $70 Million Annually Because of Ecology[*]

[*]Annual benefit ($1975) from all sources of improved weather and ocean condition forecasts and ice reconnaissance.

resources and the transportation of these resources from the Arctic regions to the U.S. and Canada by a fleet of icebreaking tankers. As opposed to the offshore oil and gas industry case study that represents an existing industry, this case study was concerned with resource development that may take place in the future and also with an icebreaking tanker fleet that is yet to be designed and built. The results of this case study are summarized in Table 7. The case study indicates an annual benefit attributable to SEASAT data applications of $64 to $234 million. The cumulative benefits, shown in Table 8 as a function of discount rate, are in the range of $96 to $288 million ($1975 at a 10% discount rate).

Ocean Fishing

The use of SEASAT data by the ocean fishing industry accounts for 32 to 53% of the most likely cumulative SEASAT benefits. In this application the case study was based upon the use of improved weather and ocean condition forecasts to improve the safety of ocean fishing operations, as well as the use of SEASAT-provided data on ocean currents and temperatures to improve the forecasting of the fisheries population. It was further assumed that international cooperation would be achieved in the management of ocean fisheries to obtain the maximum sustainable yield. Because the conditions needed to achieve the maximum sustainable yield are not yet understood and the needed international cooperation has not yet been obtained, the benefits in this application are considered to be somewhat speculative. If these conditions can be fulfilled, the possibility of a 1 to 4% improvement in the maximum sustainable yield was indicated by the case study results. Based upon these results, the cumulative benefit to the U.S. was estimated to be in the range of $30 to $157 million ($1975 at a 10% discount rate). Considering the fact that the U.S. fishing catch (1974) is less than 5% of the world total, the potential world-wide benefits in this area are believed to be very substantial if the above stated conditions can be met.

Marine Transportation

The benefits to all segments of the marine transportation industry from the use of improved weather and ocean condition forecasts account for approximately 20 to 25% of the most likely cumulative SEASAT exclusive benefits. Three separate studies were conducted in the marine transportation industry. The first dealt with the benefits to the operation of the tanker fleet now under construction to transport oil from Valdez, Alaska to the U.S. West Coast. The second considered container ship crossings on the North Atlantic between the U.S. and the United Kingdom on Trade Route Number 5. The third study considered world-wide tanker operations. In each of the studies two conditions were considered; the first was the reduction in transit time (and its associated cost saving)

Table 8. Generalization of Arctic operations results.

DISCOUNT RATE %	0	5	10	15
REGION	INTEGRATED BENEFITS TO SEASAT			
	$1975 MILLIONS			
WESTERN ARCTIC	88 – 1600	29 – 551	11 – 203	4 – 79
EASTERN ARCTIC	547	219	85	35
NORTH AMERICAN ARCTIC	635 – 2147	248 – 770	96 – 288	39 – 114

Table 9. Marine transportation summary.

Operation	Cumulative Benefit($1975 at 10% discount rate)
U.S. Trade Routes - Dry Cargo	$113 million
Canadian - Trade Routes	$50-$208 million
Worldwide Tanker Operations	$52-$204 million
TOTAL	$215-$525 million

that could be achieved by the use of improved weather and ocean condition information in optimum track ship routing; the second, the attendant reduction in damage or casualty losses associated with the reduced exposure to severe weather and ocean conditions. The generalization of the container ship case study employed an econometric model to forecast the demand for shipping on all U.S. trade routes to the year 2000. With consideration for the differences between trade routes, the benefits to shipping on all U.S. trade routes were estimated to be in the range of $27 to $49 million per year ($1975).

The tanker study was performed by constructing profiles of each major world tanker route with respect to weather and weather-dependent sailing alternatives. A comparison of routed and unrouted tankers for each of these major routes led to an estimate of time saving by route due to ship routing. Qualitative estimates of the potential incremental improvement possible with SEASAT data in this system were made using the route profiles and the routed/unrouted ship analysis. An analysis of tanker damage and losses by major route, according to standard publications, was performed to yield estimates of benefits from loss avoidance and damage avoidance. Finally, forecasts of tanker traffic on the major routes were made and the results were generalized. Table 9 summarizes the results of the marine transport case studies and generalization. Approximately $19 to $94 million of the benefit of $52 to $204 million to world-wide tanker operations' cumulative discount (10% discount rate) benefit, (1985-2000) is attributable to time savings and $33 to $110 million to prevention of catastrophic losses. The final figure does not include consideration of costs of cargo loss, environmental damage and clean-up, loss of life, or vessel losses that are less than total losses.

Estimated Systems Costs and Benefit Comparison

Preliminary estimates have been made of the costs of an operational SEASAT program that would be capable of producing the data needed to obtain these benefits. The hypothetical operational program used to model the costs of an operational SEASAT system includes SEASAT-A, followed by a number of developmental and operational demonstration flights, with full operational capability commencing in 1985. The cost of the operational SEASAT system through the year 2000 is estimated to be about $753 million ($1975 at a 0 percent discount rate) which is the equivalent of $272 million ($1975 at a 10 percent discount rate). It should be noted that this cost does not include the costs of the program's unique ground data handling equipment needed to process, disseminate or utilize the information produced from SEASAT data. Figure 3 illustrates the net cumulative SEASAT exclusive benefit stream (benefits less costs) as a function of the discount rate.

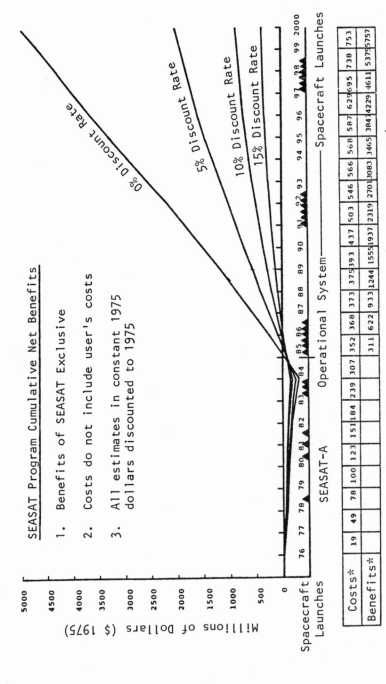

Figure 3. SEASAT program net benefits, 1976-2000

Current Status

The SEASAT-A program is currently on schedule for the launch of SEASAT-A in April 1978. Advanced studies and planning are now in process for systems that could follow SEASAT-A; however, at the present time no committment has been made by the U.S. government for these potential follow-on systems.

The economic studies in support of the SEASAT program are currently investigating the possibility of performing economic verification experiments using SEASAT-A. The primary objective of these experiments would be to provide information that could be used to help design follow-on SEASAT systems to meet user economic needs. Other objectives of these experiments are to begin the process of technology transfer from NASA to the expected users of an operational SEASAT system, and to obtain experimental information that could be used to validate previous benefit estimates. At the present time several potential experiments have been identified in each of the major benefit areas projected for the operational SEASAT. Current studies are aimed at defining the level of user interest or participation in these experiments, the manner in which the experiments would be performed and the feasibility of performing these experiments within the SEASAT-A program. Current planning calls for a decision by NASA in the Spring of 1977 on the implementation of these proposed economic verification experiments.

RESEARCH, APPLICATION AND REQUIREMENTS

OF A MAJOR USER OF LANDSAT DATA

R. Mühlfeld
Head of Photo and Astro - Geology Section
Bundesanstalt für Geowissenschaften und Rohstoffe
Hannover, West Germany

Introduction

The Bundesanstalt für Geowissenschaften und Rohstoffe (BGR)
is a government institute in the Federal Republic of Germany which
roughly corresponds to a federal geological survey in other countries. Important parts of its task consist of support for mineral,
fuel and groundwater exploration as well as investigations into
the quality and distribution of soils and improvement of their use
for sustainable production in agriculture and forestry. Most
activities in these fields are connected with technical aid projects
in developing countries.

Based on experiences with interpretation of aerial photographs
it was expected right from the beginning of manned space flight that
images of the earth's surface taken from space would provide information important to the activities of the Bundesanstalt. We soon
learned that they were indeed a valuable tool. Thanks to the very
helpful and cooperative attitude of NASA and the U.S. Geological
Survey, we could undertake some early studies with Gemini and Apollo
photographs, mainly in India and different parts of Africa. (See,
e.g., Bannert 1/.) Since 1972 the Bundesanstalt acted as principal
and coinvestigator for NASA in connection with Landsat 1 and Skylab
projects.

Research

The BGR is affiliated to the Federal Ministry of Economy. As
a consequence of this affiliation, basic research within the
Bundesanstalt is limited to areas promising practical application

within a relatively short time. In view of this policy, research
in remote sensing was soon directed to investigation of Landsat
imagery.

Landsat imagery appeared promising for practical application
due to the combination of information content with the availability
of almost the total land surface of the earth. However after a
short time it became clear that evaluation of Landsat imagery was
in many respects different from normal photo/interpretation.
Compared to an aerial photograph a Landsat image contains less
information due to the lack of stereoscopic viewing capability
and the poorer resolution. It contains more information with
respect to the multispectral recording including infrared, the
greater dynamic range and the synoptic view over a large area,
all three conditions resulting in a very large amount of data.

Only through systematic research is it possible to develop
the methods necessary to derive geologic information from this
new type of imagery. Several of the Bundesanstalt projects have
contributed to this development.

In the Pampa of Argentina, a relationship was established
between differences in vegetation cover and land use, as observed
on Landsat imagery, and the salinity of the near surface groundwater, as determined from chemical analysis of the groundwater and
from resistivity measurements. (Kruck 3/.)

In Central Europe the relationship between bedrock, soil and
vegetation cover was studied. In humid areas the earth's surface
usually has an almost complete cover of vegetation. Therefore all
radiation received by Landsat is reflected from the vegetation
cover. A proper interpretation of Landsat images in terms of
geology is only possible with a good understanding of the relationship between bedrock and vegetation. With the study in Central
Europe it was possible to determine several rocktypes and soil
units of different quality from Landsat imagery. In addition
structural observations could also be made. (Mühlfeld 4/.)

A study in the eastern part of the Canadian shield showed
the possibilities and limitations in the use of Landsat imagery
for geological investigations. With the adverse conditions of
a heavy cover of forest and glacial sediments, it was almost impossible to discriminate between different rock types. However
structural information derived from Landsat images can be used for
mineral exploration.

Overriding the results described above, these studies provided
a distinct common lesson: Information on geology, soils and groundwater can be derived from Landsat imagery only through indirect
reasoning. The quality of the results depends largely on the

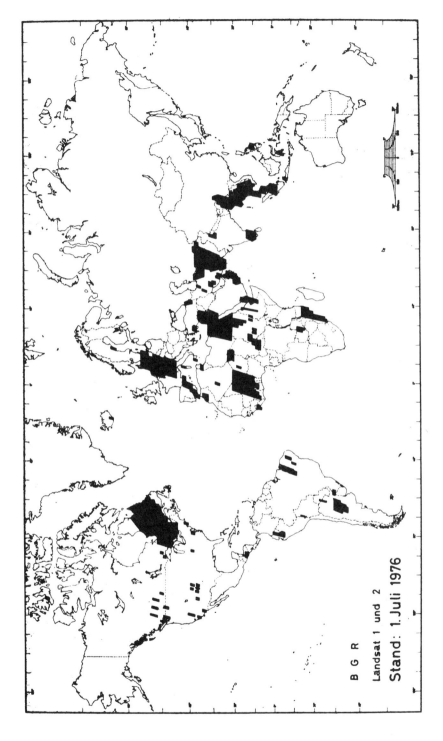

B G R

Landsat 1 und 2

Stand: 1. Juli 1976

Figure 1

interpreter's understanding of the complete ecology of the area under investigation. Especially important is a knowledge of the composition and seasonal behavior of the vegetation cover, natural vegetation as well as field crops and cultivated forests. Since ecologic conditions can vary considerably over relatively short distances, it is necessary for Landsat evaluation to compile a description of the ecology for all areas of interest, which is the practice in many institutions worldwide.

Within our most recent research projects we are studying a semiarid area on the southern fringe of the Sahara Desert. From Landsat images we try to obtain information on the interrelation between soil, groundwater, vegetation or land use and changing meteorologic conditions. The vegetation, regarded as "noise" in numerous Landsat studies in semiarid regions, appears as an indicator of soil quality as well as of the availability and quality of groundwater. However it seems to be rather difficult to discriminate between different plant societies and different stages of seasonal development with simple visual interpretation. We are faced with the necessity to progress to computer processing of Landsat data. The Bundesanstalt is now making a major effort in the study and development of computer processing in connection with different fields of possible application.

Despite the progress made in using Landsat data, we are still far from the point of having developed all the methods necessary for a complete evaluation of Landsat data.

Application

Quick application of Landsat evaluation for practical projects is easiest in arid regions. In connection with a geological survey of a 600,000 km^2 area in the Bayuda Desert, Sudan, a Landsat mosaic served as the best available base map, thus facilitating the planning of the project in its entirety. As its base a drainage network map was prepared for geochemical prospecting. In addition a first geologic interpretation of the Landsat imagery gave valuable information for the most effective continuation of the work with aerial photographs, and for field checking.

In the tropics, useful application of Landsat studies is also possible. In Northern Burma information on the structure and on the facies of sediments in a sedimentary basin was obtained, which was of great value for petroleum prospecting.

Even in a geologically well known area like Germany, structural information derived from Landsat can be used for mineral exploration. (See, e.g., Hoppe 2/.)

The results of the hydrogeological study in the Pampa of
Argentina, described above, are readily available for a planned
groundwater survey. Their application will result in a considerable reduction of the most expensive aspects of groundwork (chemical
analysis of wells, resistivity measurements).

In practice Landsat images are used for each regional project
of the BGR, at least in the initial phase, for purposes of planning
and providing a first general view of the area under investigation.
The BGR index map of Landsat imagery available in inventory for
such planning purposes shows a coverage of large areas of North and
South America, Europe, Africa and Asia.

Requirements

Despite the fact that Landsat has been conceived as an experimental system, Landsat data are already widely used in a routine
way. The main reason for this is the information content and the
availability of Landsat data for almost the entire land surface of
the world. However utility of Landsat data can be further increased
by a number of appropriate developments and improvements.

An important step forward will be a considerable improvement
of image quality, now in hand at the EROS Data Center in Sioux
Falls. In connection with actual projects available, Landsat coverage is sometimes insufficient with respect to distinct seasons
or actual situations such as conditions of vegetation or the extent
of surface water. The value of Landsat data could be increased
by providing a service on a short term basis for such exceptional
cases. On the other hand, the data collected since 1972 have already proven to have an important historical value for the investigation of changes. The above mentioned improvement of image quality
should therefore also be available for earlier Landsat data.

Further, an improvement of classification through computer
processing seems to be feasible, especially if the indicator properties of vegetation cover are used.

A systematic compilation of ecologic data relevant to Landsat
evaluation throughout the world will also increase the practical
value of Landsat data.

A prediction of requirements for a post-Landsat system is
beyond the scope of this paper. Experiences gained with Landsat
point in the direction of spectral band selection improvement,
especially in the infrared, and provision of all-weather capacity
through radar. Since the tasks of the Bundesanstalt are mainly
connected with survey type investigations, most of the needs can
be fulfilled with satellite data gathered almost at random over

a longer period of time. In the field of hydrogeology only, a
monitoring type of observation may be useful in connection with
actual availability of groundwater. This would be a potential
field of application for an operational satellite system with short
time intervals of registration.

The requirements described above are strictly seen from the
point of view of the BGR. However, it seems necessary to increase
the participation of investigations and institutions from developing
countries in the use of satellite data. In this connection a whole
field of additional requirements for optimum use of satellite data
has to be taken into account.

Summary

The use of Landsat data has proved to be most valuable for
major tasks of the Bundesanstalt for Geowissenschaften und Rohstoffe
in the fields of mineral, fuel and groundwater exploration, as well
as in pedologic investigations and improvement of land use. Landsat
investigations have become routine for most survey type projects.
Research in improving evaluation methods is still going on in
close connection with the international scientific community. A
number of requirements for better use of satellite data are
described.

References

1/ Bannert, D., "Afar Tectonics Analyzed from Space Photographs".
     American Association Petr. Geol. Bulletin, 56, 1972.

2/ Hoppe, P., "Relationship between the tectonic Lineaments and
     Ore districts in the Eastern part of the Rhenish Massif
     compiled from interpretation of ERTS-imagery". NASA SR
     No. 328, Type III Report, 1974.

3/ Kruck, W., "Hydrogeological Investigations in the Argentine
     Pampa Using Satellite Imagery". Natural Resources and
     Development 3, 1976.

4/ Mühlfeld, R., "Relationship between vegetation, soil, bedrock,
     and other geologic features in moderate humid climate
     (Central Europe) as seen on Landsat-1 imagery". Geol.
     Jahrbuch, A 33, 1976.

A CROP PRODUCTION ALARM SYSTEM

Archibald B. Park

System Overview

Agricultural reports are often phrased in terms of "normal" years, but the real world experience is that "normal" is really quite abnormal. Climate and technology may produce long-term trends in yield; but in any one year drought, floods, winds, hail, frost, insects and disease may greatly affect production. The effects of these events can vary widely over large areas. Many events can be recovered from by replanting or good weather conditions later in the year. Present agricultural reporting procedures which rely on sampling techniques frequently over or under-estimate the extent of these events. The repetitive and synoptic coverage of earth resources provided by Landsat and later operational satellites can potentially provide a supplementary means of monitoring and assessing these events more accurately and timely.

An Alarm System has been designed and a study performed to demonstrate the utility of satellite observations for crop monitoring. The system utilizes Landsat imagery in conjunction with weather data and field reports to monitor crop development, identify anomalies, assess their impact, and adjust crop production forecasts.

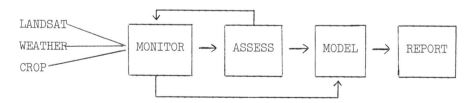

Figure 1. A top-level functional overview of the system

The system has three major functional elements: monitor, assess, and model. Of these, the monitor functions operate continuously throughout the growing season. The assess functions are initiated when an event is identified by the monitoring process. After each event is assessed, the model is used to adjust forecasts of crop production.

The monitoring process is the heart of the system. It is this part of the system which "alarms" events or conditions which may affect crop production. Landsat images are received and analyzed in near real-time. These images are in the form of black and white photo products and transparencies. Images of the current growing season are analyzed for changes occurring between each satellite pass, and are compared to images from previous seasons in the Alarm System Data Base (ASDB). The ASDB contains keys and models for interpreting crop development patterns from the imagery, as well as current and historical information on available satellite imagery, weather conditions, crop calendars, yields, and production. In addition to the Landsat images and ASDB, current weather and crop reports are monitored. Alarms are generated from either changes observed in the Landsat imagery, or conditions extracted from the weather and crop data. When an alarm has been generated, the assess function of the system is initiated.

The assess function takes the alarm related information output by the monitor function, and initiates a series of higher order interpretation and analysis procedures. If a change in the Landsat imagery is observed, an attempt is made to ascertain its probable cause. If an alarm is generated from the weather or crop reports, the Landsat imagery is examined for an observable effect. In either case, more detailed collateral data is acquired. This may mean getting local weather station reports, contacting county agricultural representatives, or a site inspection. Landsat computer compatible tapes (CCTs) may be acquired for more detailed digital analysis using the IMAGE 100 system. The outputs of the assess portion of the Alarm System are estimates of the extent of crop damage due to the alarmed event.

The model uses the assess function outputs to adjust forecasts of crop production. In the present study, estimates of the yield or crop area reduction due to an assessed alarm are used to modify Department of Agriculture forecasts. If an alarm has no assessed impact on production, no adjustment is made.

The Alarm System has one additional characteristic. After the initial assessment of an alarm, the system utilizes the Landsat repetitive coverage to monitor replanting and/or recovery. From this follow-up monitoring, the event is reassessed and production estimates updated periodically.

Alarm System Data Base

A key element in the Alarm System is the Alarm System Data Base (ASDB). The ASDB contains the keys, models, and collateral data used to interpret and monitor the Landsat imagery. In addition, the ASDB contains current and historical information on crop calendars; yields, and production; and on climatic conditions. These data are used to assess and model the impact of crop "alarms". The ASDB has four major components: crop data, Ag/Met data, Landsat.data, and collateral data.

Crop Data

In order to monitor a crop effectively and assess the impacts of events on production, it is important to understand the growth stages the crop goes through and the factors which affect production. Rates of development and yield vary for different varieties, locations, and weather conditions, but all plants progress through the same series of physiological stages during the growing season. A numbering system has been developed to identify these different stages of plant development. The stage at which the plant tip emerges from the soil is stage 0 and the stage when the plant is mature is stage 10. Intermediate stages are assigned numbers between 0 and 10. Table 1 is presented to exemplify a typical input to the ASDB.

In Iowa, the average number of days required for physiological development of corn from stage 0 through 10 is 126 days. The table lists average dates and days from emergence required to reach each stage. This table indicates that on the average a corn hybrid adapted to conditions in central Iowa planted on May 15 will emerge by May 24 and be physiologically mature on September 28. This hybrid should be ready for harvest (for grain) by mid-October.

Although 126 days is considered to be the average length of time required for development of an Iowa full season hybrid, the length of time required for a corn plant to complete its life cycle is more closely associated with the accumulation of growing degree units (GDU):

$$GDU = \frac{Tmax + Tmin}{2} - 50^\circ F$$

Tmax equals the maximum daily temperature and Tmin equals the minimum daily temperature. If Tmax is greater than $86^\circ F$, Tmax is set equal to $86^\circ F$. If Tmin is less than $50^\circ F$, Tmin is set equal to $50^\circ F$. These daily readings are accumulated from planting data until growth ceases in the fall. Table 2 shows expected number

Table 1.   Average Dates and Days from Emergence for the
           Different Stages of Growth of Corn.*

Growth Stage	Date	Days**	Identifying Characteristics for Field Use
0	May 24	0	Plant emergence. Tip of coleoptile of plant visible at soil surface.
1	Jun 8	14	Collar of 4th leaf visible.
2	Jun 22	28	Collar of 8th leaf visible. Leaves 1 and 2 may be dead.
3	Jul 6	42	Collar of 12th leaf visible. Leaves 3 and 4 may be dead.
4	Jul 20	56	Collar of 15th leaf visible. Tips of many tassels visible. Leaves 5 and 6 may be dead.
5	Jul 30	66	75% of plants have silks visible. Pollen shedding.
6	Aug 11	78	12 days after 75% silked. Kernels in "blister" stage.
7	Aug 23	90	24 days after 75% silked. Very late "roasting ear" stage.
8	Sep 4	102	36 days after 75% silked. Early dent stage.
9	Sep 16	114	48 days after 75% silked. Full "dent" stage.
10	Sep 28	126	60 days after 75% silked. Grain physiologically mature.

*   Average for adapted hybrids in central Iowa. Appropriate modifications should be employed for other hybrids and other areas. Planting date assumed to be May 15.

** From emergence.

Table 2. Expected Number of Growing Degree Units That Would
          Accumulate (50 Percent Probability) in a Given Season
          For Various Planting Dates*.

	APRIL 19	MAY 3	MAY 17	MAY 31	JUNE 14
Mason City	2647	2545	2412	2236	2008
Ames	2969	2849	2695	2500	2259
Clarinda	3199	3057	2882	2663	2397

* Based on 1941 to 1970 data.

Table 3. Iowa Corn Grain Yields and Acres of Corn Harvested for
          All Purposes.

YEAR	YIELD (BU/A)	ACRES (x 1000)	YEAR	YIELD (BU/A)	ACRES (x 1000)
1939-48	51.6	10,226	1962	77.0	10,121
1949	48.0	11,471	1963	81.5	11,144
1950	47.0	9,865	1964	79.0	10,252
1951	43.5	10,190	1965	82.0	10,457
1952	62.5	10,750	1966	89.0	10,666
1953	53.0	11,180	1967	88.5	11,733
1954	54.0	10,453	1968	93.0	10,325
1955	48.5	10,767	1969	98.0	10,119
1956	53.0	10,067	1970	86.0	10,625
1957	62.0	10,218	1971	102.0	12,196
1958	66.0	10,065	1972	116.0	11,225
1959	65.0	12,481	1973	107.0	11,940
1960	63.5	12,607	1974	80.0	12,740
1961	75.5	10,338	1975	92.0	12,950

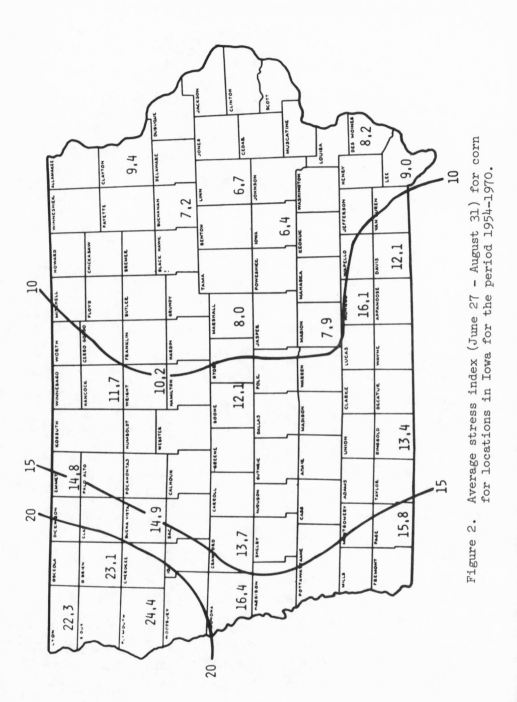

Figure 2.   Average stress index (June 27 - August 31) for corn
            for locations in Iowa for the period 1954-1970.

of growing degree units that would accumulate for various planting dates for three sites in central Iowa. The data shows that earlier planted corn can expect to accumulate more growing degree units over a season; and that corn in the southern part of the state (Clarinda) receives more sun than northern parts of the state (Mason City).

Long term studies in Iowa indicate a yield advantage for planting by May 10. Presumably the major reason for this is the reduction in GDUs as a result of the later planting. On the average, yields have been found to decline nearly 1 bushel per day from May 10 to May 30. In fact, when planting is delayed past May 20, hybrids classed somewhat earlier than "full season" (approximately 126 days from emergence to physiological maturity) are recommended by the Iowa State University, especially for northern Iowa.

Table 3 shows trends in Iowa corn grain yields for the last 35 years. Yields have increased dramatically since the early 1950's. More fertilizer, better hybrids, earlier planting, higher plant populations, better weed, insect and disease control, improved harvesting methods and improved timeliness, have all contributed. During the 1960's and early 1970's weather was more favorable than for most periods. Depressed yields in 1974 and 1975 are evidence that weather still has a major influence on corn production.

In addition to year to year variations in the state-level average, yield varies widely from one section of the state to another during a single year. These variations are due mainly to soil differences and intra-state·variation in climatic conditions and moisture levels. Table 4 gives some estimates of average yields over a five-year period for some good corn soil in different parts of Iowa. Although these yield levels are not absolute, the relative ranking of yield from the soils listed tends to remain constant. Top producers growing corn will generally do better than these averages; farmers with below these values should look critically at their production practices.

Over the years, moisture stress has been found to be the most significant environmental factor decreasing corn yield in Iowa. A 50 to 75 percent reduction in yield can be expected from severe moisture stress during the week preceding or following silking. In general, two to three week moisture stress periods experienced during the vegetation stages only, stages 0 to 4, influence final yield least, usually a maximum of 10%. Subjection of corn to brief periods of moisture stress during stages 6 through 10 causes an intermediate reduction in yield. Figure 2, taken from Shaw and Felch (1972), shows that average stress index is highest in the western and northwestern portions of the state.

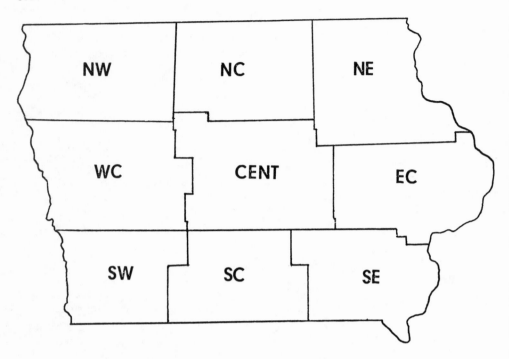

Figure 3. Nine Crop Reporting Districts in Iowa

Table 4. Estimated Five-Year Average Corn Yields for Selected
          Soils.

PART OF STATE	SOIL TYPE	SLOPE (%)	YIELD (BU/A)
Northwest	Primghar silty clay loam	1-3	100-110
North Central	Nicollet loam	1-3	115-125
Northeast	Floyd loam	1-3	105-115
West Central	Monona silt loam	2-5	95-105
Central	Muscatine silty clay loam	1-3	125-135
East Central	Tama silty clay loam	2-5	120-130
Southwest	Marshall silty clay loam	2-5	105-115
South Central	Grundy silty clay loam	2-5	105-115
Southeast	Mahaska silty clay loam	1-3	120-130

Agrometeorological Data

In addition to crop calendar and yield data, the ASDB also includes crop and weather reports published by the U.S. Department of Agriculture weekly during growing seasons. Reports in the ASDB include the 1974, 1975 and 1976 crop years. Data in the weekly USDA reports is summarized by the nine Crop Reporting Districts in Iowa (Figure 3). Format varies but district level data reported generally includes crop development stage, average weekly temperature, precipitation, growing degree units, and soil moisture. These data are recorded on wall charts and are used to generate potential crop alarms.

Collateral Data

Collateral data in the ASDB include soil maps, topographic maps, highway maps, and indexes of available aerial photography. These data are used primarily during the assessment process after an alarm has been generated by the Landsat imagery or weather and crop reports.

Landsat Data

Landsat data in the ASDB consists of bands 5 and 7 photographic prints (1:1,000,000 scale). Imagery was collected over the test areas for the April through November growing seasons for 1974 and 1975. Landsat images of the 1976 season are compared against those in the ASDB in the monitoring process.

Alarm System Interpretative Techniques

In addition to the establishment of the Alarm System Data Base, it was necessary to develop certain interpretative techniques for analyzing the Landsat images in near-real-time. Qualitative techniques were developed for photo-interpreting the Landsat photographic products; and quantitative techniques were developed for analyzing Landsat computer compatible products on the IMAGE 100 system. The quantitative techniques resulted in a temporal relative reflectance profile model for corn, but the qualitative analysis was unable to develop reliable photointerpretative keys.

Qualitative: Montaging

As described in the discussion of the Alarm System Data Base, montages were made of all the 1974 and 1975 images for which there were sufficient useable images. In addition to providing a convenient means of storing the images, the montages permit inspection of the images for synoptic features or changes.

Qualitative: Stereographic Viewing

Temporal changes in the Landsat photos were also detected by
registering the two dates under a stereographic viewer.   Alternately
closing the right and then the left eye with the opposite one open
causes changes in the scene to appear as sudden movement to the
interpreter, thus facilitating the detection of subtle changes.

Qualitative: Negative/Positive Sandwich

The stereographic viewer technique is slow, however, and tends
to result in interpreter fatigue.   Thus, when the 1976 data began to
be received, a second technique was developed.   By registering
negative and positive transparencies, areas of no change cancel-
out and appear gray; areas of change appear lighter or darker.

The negative/positive sandwich technique is quicker and less
straining than the stereo viewer procedure, but to date has only
been applied to selected images because the request for 9 by 9
inch negative transparencies could not be fulfilled for the 1976
image standing order.

LANDSAT DIGITAL IMAGE ANALYSIS

Temporal Spectral Reflectance Profile

To construct a temporal spectral reflectance profile for
vegetation, it is necessary to develop procedures for processing
multi-date images.   Problems in processing multi-date images are
caused by geometric and radiometric differences among the images.
Geometric differences between Landsat scenes are caused by changes
in satellite attitude and orbit and by the location of synthetic
pixels added during ground processing.   Since the processing required to create temporal spectral profiles requires accurately
locating fields in multiple images, considerable processing time
was saved by geometrically correcting the Landsat images before
analysis.   Radiometric differences between Landsat scenes are
caused by changes in sun angle, dynamic effects of the atmosphere,
sensor changes, and changes in data calibration.   Since the object
of a temporal profile is to measure the temporal dynamics of a
crop, a method was developed for converting Landsat gray levels
(MSS radiance) to normalized relative reflectance.

Geometric rectification was accomplished using GSFC "Digital
Image Rectification System" (Van Wie, 1976).   The DIRS system
offers extensive capabilities for geometric mapping and image
resampling.   The geometric correction algorithms are driven by
scene data in the form of landmarks or Ground Control Points
(GCP's).   The regular road grid and prominant section lines of

Iowa proved ideal for selecting ground control points. Initially
20 control points were located on county highway maps and in each
image. Sub-pixel accuracy was achieved through the use of "shade
prints". The number of points was reduced to 15 by rejection of
the points with the greatest position error. Resampling was done
with the TRW "cubic convolution" algorithm at 50 meter steps.
Resampling of the picture elements was performed so that they
would be consistently positioned in a ground based coordinate
system.

The final result of the DIRS rectification and resampling
was four registered images of the 30 by 30 kilometer Madison County
test site. Two of the Landsat images were not resampled. The
July 2 image (2161-16214) was found to contain extensive haze not
detected in the 70mm film clips; and the Landsat 1 image (5101-
16083) could not be properly registered with the other scenes.
Time restrictions and other problems required that only one of the
orginal test sites could be resampled. Registration of the four
resampled images proved to be $\pm$ 1 pixel over the test area.

A simple and effective method for converting Landsat radiance
values to reflectance can be developed from the following model,

$$E\lambda = K(L_p T_a + L_s) \tag{1}$$

which states that sensor irradiance ($E\lambda$) is the sum of the spectral
radiance of the ground feature as seen through the intervening
atmosphere ($L_p T_a$) and the upwelling atmospheric path radiance
(L_s) multiplied by K a factor associated with sensor gain and
illumination angle. If we now assume that

(a) atmospheric effects, path radiance, and attenuation
    are constant over the scene for each band and each
    date,

(b) Landsat CCT counts have a linear relationship to
    sensor irradiance, and

(c) all objects studied exhibit similar illuminative
    angles vs reflectance characteristics over the
    period analyzed,

then all of the parameters in equation (1) except reflectance (L_p)
can be approximated by a constant. The model becomes:

$$E\lambda = M\lambda\ R\lambda + A\lambda \tag{2}$$

Landsat data from different dates can now be normalized by transform radiance ($E\lambda$) to reflectance ($R\lambda$). This is done by solving

Table 5.   Estimates of the Path Radiance.

IMAGE	DATE	MSS 4	MSS 5	MSS 6	MSS 7
2125-16213	May 27	14	12	7	0
2179-16213	July 20	14	12	11	0
2197-16210	August 7	14	11	8	0
2233-16203	September 12	9	7	2	0

Table 6.   Landsat Relative Reflectance Functions for Madison
County Site 1975.

IMAGE	DATE	MSS 4	MSS 5	MSS 6	MSS 7
2125-16213	May 27	.80E-12	.84E-10	.79E-3	.82E
2179-16213	July 20	1.03E-14	1.18E-14	1.05E-12	.93E
2197-16210	August 7	E-14	E-11	E-8	E
2233-16203	September 12	1.63E-16	1.59E-11	1.27E-3	1.29E

Relative Reflectance Function: $R_{ij} = mE_{ij} + a$

R = Relative Reflectance for the i image
and j band

E = Landsat Radiance for the i image and
j band

for Rλ in equation (2); that is:

$$R\lambda = \frac{1}{M\lambda} E\lambda - \frac{A\lambda}{M\lambda} \qquad (3)$$

In order to use the model in equation (3) to normalize a series
of images, it is necessary to locate reference areas which have
maintained a constant reflectance. Non-vegetated features
like bare soil, roads, buildings, and water provide useful targets.
These targets can be easily recognized in Landsat images by
slicing or thresholding the high end of a MSS 5/MSS 7 band ratio.
Using this technique "barren" areas were located in the four
resampled images of the Madison County site. The features mapped
included ponds, roads, inner city, and construction sites. The
total area mapped was 3% of the test area or about 7,500 picture
elements.

The non-image forming radiance or atmospheric component
was estimated from the Landsat signals of the barren targets in
each band. By assuming that the minimum signal represents an
object of low reflectance, the additive component (A) was estimated at just below the minimum signal. For example, standing water
has a low reflectance. Any radiance associated with such a feature will be primarily due to the intervening atmosphere, and
would be a good estimate of path radiance. In this way, the
additive factor A in equation (2) was calculated for each band
and image. Values are given in Table 5.

Since in-situ measurements of field reflectance were not
available for determining the multiplicative component M, a relative reflectance normalization was calculated. The August 7 image
(2197-16210) was selected as the standard image by which to normalize the other three images. For the August image, the multiplicative component was arbitrarily assumed to be unity (Figure
4). From this graph the relative reflectance for 15 features
in the August image was determined. The multiplicative component
for the other images was found by a linear regression of the
radiance values of the 15 features with the relative reflectances
established for the August reference image (Figure 4). Accuracy
of the resulting reflectance functions was improved by weighing
the ordinate intercept to the previously determined additive
component value. Figure 4 shows the reflectance functions for
the Landsat images in band 4; Table 6 lists the functions for
all four images in all bands.

Figure 5 shows an IMAGE 100 display of a portion of the resampled Madison County site for all four dates before normalization. Clockwise from the upper left, the images displayed are:
May, July, September, and August. Slight gain differences can

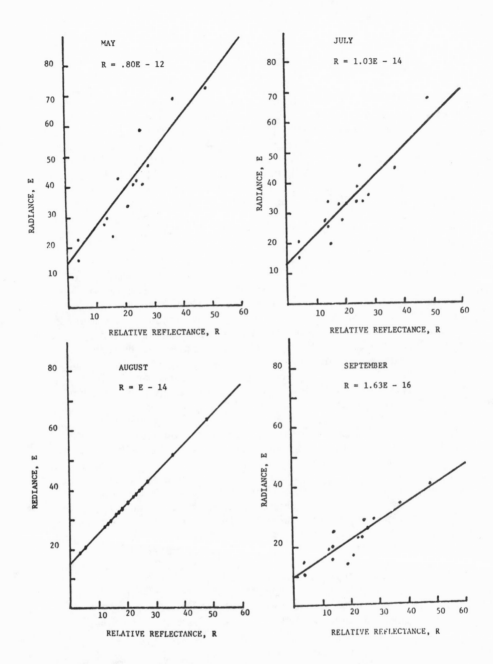

Figure 4   Relative Reflectance Functions Normalized
to August Image for Band 4

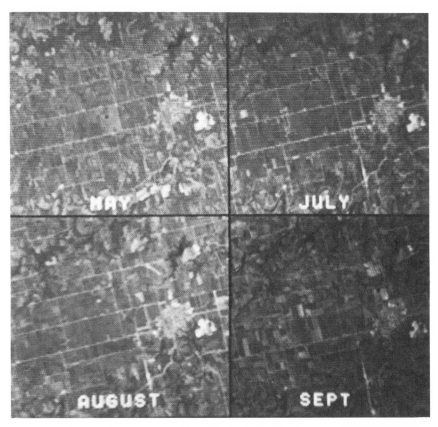

Figure 5   IMAGE 100 Display of Portion of Resampled Madison County
            Site. (The contrast of the original color photograph is
            lessened in this reproduction.)

be noticed between the May, July, and August images; and a dramatic
intensity reduction in the September image. The relative reflectance functions were applied to each band of each date via an IMAGE
100 radiometric correction program. (Ed. Note: Figure 5, requiring color reproduction, is not included in this publication).

Since the images had been resampled it was possible to simultaneously train on corn and soybean fields in all four quadrants
(dates) by using the IMAGE 100 16-channel classifier algorithm.
The outputs for corn and soybeans are shown in Tables 7 and 8.
From the training sample, mean radiance and reflectance profiles
for corn and soybeans were constructed (Figure 6 and 7).   Comparison of the radiance and relative reflectance profiles shows that
normalization had the greatest effect on bands 4 and 5, the bands
most effected by atmospheric noise. The normalization also had a
large effect on the September image. Looking at the radiance
profile soybeans appear to peak in August, but the relative réflec-
tance profile shows that soybeans continue to increase reflectance
in all bands through September when leaves are turning yellow and
beginning to shed.

From the relative reflectance profile, it is possible to construct a three-dimensional temporal model of reflectance for corn
and soybeans (Figure 8 and 9). When correlated with phenological
changes these models, or profiles, explain the Landsat observable
manifestations of crop development. Corn, for example, peaks in
infrared reflectance in July at the time of silking, and gradually
decreases infrared reflectance as the ears mature. Soybeans, on
the other hand, continues to increase in infrared reflectance
through September. In September when soybean leaves begin to shed
a slight increase in visible reflectance is noticed.

A detailed model could not be hoped for from just four observations of one test site. Nevertheless, the validity of temporal reflectance profiles was established; and techniques for
digitally monitoring crops with Landsat data were developed.

Several tentative conclusions can be made about the operation
of the 1976 Alarm System for Iowa corn to date. Alarms triggered
by weather and crop reports are generally more assessable than
the image generated alarms. This is in part because the weather
and crop reports are received more timely than the Landsat images.
The average delay in receiving Landsat images has been 20 days
after satellite acquisition. Frequently, this is too late to
interpret and assess dynamic local events. Also, it has been
necessary to develop photointerpretative keys during the near
real-time monitoring process, since keys could not be developed
from the retrospective data.

Table 7. Image 100 signature extraction for corn in four resampled Landsat images of Madison County, Iowa, 1975.

16-Channel 1-D Training

CH	QUAD	BAND	LB	UB	DEL	RES	PEAK	MEAN	VARIANCE
1	1	M4	8	15	8	64	490	10.81	1.78
2	1	M5	9	20	12	64	431	13.41	3.77
3	1	M6	11	43	33	64	271	17.66	14.29
4	1	M7	11	49	39	64	264	17.52	18.12
1	2	J4	7	13	7	64	877	8.50	0.79
2	2	J5	6	16	11	64	995	7.65	1.70
3	2	J6	21	40	20	64	303	29.99	5.97
4	2	J7	20	47	28	64	223	24.17	11.29
1	3	A4	7	14	8	64	889	9.55	0.72
2	3	A5	6	18	13	64	839	9.41	1.76
3	3	A6	24	40	17	64	354	29.48	4.03
4	3	A7	25	46	22	64	277	32.56	7.46
1	4	S4	5	11	7	64	1012	7.01	0.51
2	4	S5	6	12	7	64	794	7.81	0.70
3	4	S6	8	34	27	64	292	14.46	11.12
4	4	S7	7	37	31	64	269	14.19	16.14

Training Area (Pixels)   1683.

Corn Before Normalization

CH	QUAD	BAND	LB	UB	DEL	RES	PEAK	MEAN	VARIANCE
1	1	M4	0	6	7	64	696	2.47	1.15
2	1	M5	2	11	10	64	514	6.12	2.57
3	1	M6	7	32	26	64	392	12.32	9.02
4	1	M7	9	40	32	64	265	13.98	12.55
1	2	J4	0	7	8	64	876	1.51	0.82
2	2	J5	0	12	13	64	995	1.80	2.46
3	2	J6	16	36	21	64	306	25.41	6.70
4	2	J7	18	42	25	64	226	31.29	9.93
1	3	A4	0	7	8	64	383	2.55	0.73
2	3	A5	1	13	13	64	706	3.89	1.87
3	3	A6	20	36	17	64	357	25.48	4.02
4	3	A7	25	46	22	64	271	32.56	7.48
1	4	S4	1	10	10	64	544	3.53	1.36
2	4	S5	4	13	10	64	574	6.96	1.62
3	4	S6	8	42	35	64	294	16.87	18.13
4	4	S7	9	47	39	64	269	17.94	26.87

Training Area (Pixels)   1683.

Corn After Normalization

Table 8.   Image 100 signature extraction for soybeans in four
           resampled Landsat images of Madison County, Iowa, 1975.

16-Channel 1-D Training

CH	QUAD	BAND	LB	UB	DEL	RES	PEAK	MEAN	VARIANCE
1	1	M4	8	16	9	64	491	9.98	1.43
2	1	M5	6	19	14	64	351	11.96	3.50
3	1	M6	4	49	46	64	240	18.37	46.70
4	1	M7	8	58	51	64	214	18.86	76.97
1	2	J4	7	18	12	64	490	9.02	3.04
2	2	J5	5	26	22	64	323	9.06	11.61
3	2	J6	17	45	29	64	161	28.53	15.78
4	2	J7	16	51	36	64	120	30.29	25.97
1	3	A4	7	15	9	64	529	8.90	1.59
2	3	A5	2	19	18	64	611	8.31	4.84
3	3	A6	22	41	20	64	132	32.42	12.64
4	3	A7	22	47	26	64	91	36.55	28.94
1	4	S4	4	12	9	64	514	8.02	1.13
2	4	S5	5	13	9	64	437	7.65	2.45
3	4	S6	11	39	29	64	167	30.85	17.91
4	4	S7	9	44	36	64	167	34.53	25.91

Training Area (Pixels)   1305.

<u>Soybeans Before Normalization</u>

CH	QUAD	BAND	LB	UB	DEL	RES	PEAK	MEAN	VARIANCE
1	1	M4	0	6	7	64	507	1.78	0.95
2	1	M5	0	11	12	64	349	4.85	2.56
3	1	M6	6	37	32	64	261	12.91	28.88
4	1	M7	7	47	41	64	218	15.13	52.56
1	2	J4	0	11	12	64	494	2.03	3.29
2	2	J5	0	24	25	64	326	3.52	16.11
3	2	J6	12	41	30	64	179	23.88	17.44
4	2	J7	15	47	33	64	220	27.87	21.43
1	3	A4	0	8	9	64	536	1.90	1.57
2	3	A5	0	13	14	64	475	2.86	4.66
3	3	A6	18	37	20	64	134	28.37	12.66
4	3	A7	22	47	26	64	94	36.46	28.98
1	4	S4	0	11	12	64	521	5.00	2.76
2	4	S5	0	15	16	64	274	6.61	6.15
3	4	S6	12	48	37	64	163	37.57	29.41
4	4	S7	11	56	46	64	169	44.01	44.18

Training Area (Pixels)   1335.

<u>Soybeans After Normalization</u>

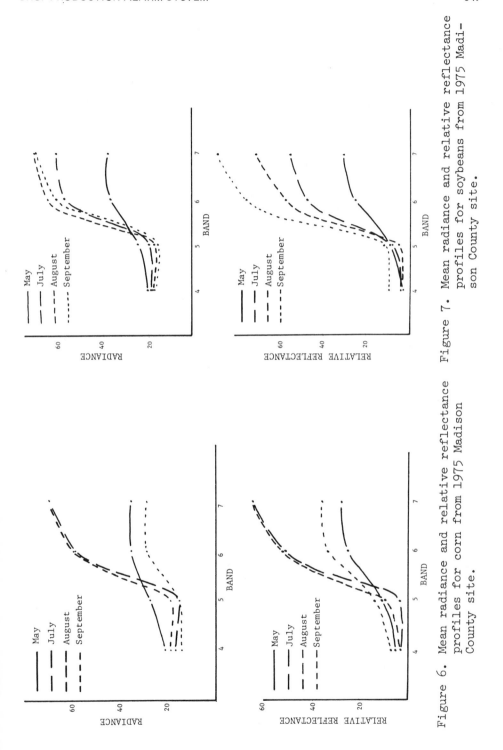

Figure 6. Mean radiance and relative reflectance profiles for corn from 1975 Madison County site.

Figure 7. Mean radiance and relative reflectance profiles for soybeans from 1975 Madison County site.

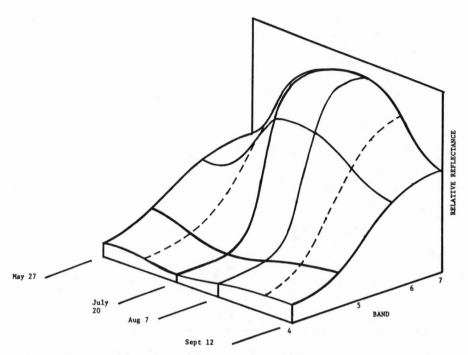

Figure 8.   Three-dimensional temporal model of reflectance for
            corn, Madison County test site, 1975.

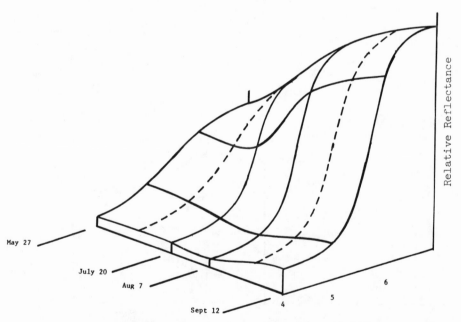

Figure 9.   Three-dimensional temporal model of reflectance for
            soybeans, Madison County test site, 1975.

Most of the events alarmed to date have been local and have little effect on state-level production. Many events have occurred during the early stages of crop developed before vegetative cover can be observed in the Landsat photo products. Many events early in the season, however, are recovered from later in the year if good weather is received.

Cloud cover and the 18-day coverage (9-day if both Landsats combined) are also limiting factors for any satellite crop monitoring program. Several alarms to date have not been assessable because of cloud cover in the subsequent images. There are several limitations with interpreting the Landsat single-band photo products. The digital data and analysis techniques appear most promising, however, for quantitatively assessing the impact of an alarm on crop production.

APPLICATIONS OF REMOTE SENSING TECHNOLOGY AND INFORMATION SYSTEMS

TO COASTAL ZONE MANAGEMENT

Eric S. Posmentier
Institute of Marine and Atmospheric Sciences
City University of New York
Wave Hill, New York, U.S.A.

Abstract

Much of the responsibility for natural resource management, particularly for coastal zone management, lies with local, state, and regional agencies in the United States. Remote sensing data have the potential of making a major contribution to these management efforts. However, few of these agencies have the technical manpower and facilities necessary to make use of existing remote sensing data in formulating management plans and decisions. A data base management system capable of producing remote sensing information products useable within existing agencies must take into account the agencies' hierarchy of information format requirements and programming sophistication. Specific components of such a data base management system should include consistent geo-referencing, a relational data base manager integrated with a "query by example" system, a natural language interface, and a "context script".

Introduction

Numerous local, state, and regional agencies in the United States have information needs which can be best satisfied by the use of remote sensing among other data sources. In many ways, coastal zone management (CZM) agencies, with which federal legislation has placed the responsibility for planning and undertaking CZM, is a representative subset of such agencies. CZM can be divided into a number of tasks. Ten of these are listed in Table 1, together with their respective information requirements. All of these information requirements can be satisfied in part by remote

Table 1:   Remote Sensing Information Needs Related to Coastal Zone Management Tasks

Data Needs Which Can Be Addressed By Remote Sensing

Data Need	A	B	C	D	E	F	G	H	I	J
1. Temperature	X	X	X			X	X	X	X		X
2. Salinity	X	X	X			X	X	X		X	X
3. D.O.	X	X	X			X	X	X	X		
4. Turbidity	X	X	X	X		X	X	X		X	X
5. Sediment Load	X	X	X		X	X	X	X	X		
6. Organic Load	X	X	X	X		X	X	X			
7. Nutrient Concentration	X	X	X	X		X	X	X		X	X
8. Chlorophyll	X	X	X	X		X	X	X		X	
9. Water Chemistry (Trace Met & pH)	X	X	X	X		X	X	X			
10. Surface Oil	X	X	X	X		X	X	X		X	
11. Currents		X	X		X	X	X	X	X	X	X
12. Water Depth	X	X			X	X	X	X	X	X	X
13. Water Volume	X	X	X		X	X	X	X	X		
14. Precipitation	X	X			X		X	X		X	
15. Snow Cover	X	X			X		X	X			
16. Ice Cover	X	X			X		X	X	X	X	X
17. Surface Runoff	X	X	X		X	X	X	X			
18. Soil Moisture	X	X			X		X	X			
19. Groundwater Supply	X	X	X		X		X	X			
20. Point Source Location	X	X	X	X	X		X	X		X	X
21. Outfall Monitoring	X	X	X	X	X		X	X	X		X
22. Boundaries	X	X	X		X		X	X	X	X	X
23. Topography	X	X	X		X		X	X	X	X	
24. Land Cover	X	X	X			X	X	X	X	X	X
25. Land Use	X	X	X	X	X	X	X	X	X	X	X
26. Demography	X	X	X	X	X	X	X	X	X	X	

Agency Tasks
A. Water Supply
B. Pollution Control
C. Waste Disposal
D. Industrial Cooling
E. Flood Plain Management
F. Shoreline Management
G. Wetlands Management
H. Fisheries
I. Navigation
J. Recreation

sensing data. The present and potential utility of remote sensing
data in satisfying the information needs of CZM is surveyed in the
following section. Although there is the potential for remote sensing data to make a major contribution to the information needs of
local, state, and regional agencies in general, and of CZM agencies
in particular, these agencies are not prepared to use the research
methods and techniques which are necessary to utilize remote sensing
data with existing information management systems (Shinnick et al,
1974; National Research Council, 1974).

A number of existing information systems have had various
degrees of success in facilitating the access to and interpretation
of Landsat MSS data. For example, the Bendix M-DAS system has been
used by at least one regional planning program (Anonymous, 1975).
Other operational systems include Image 100 (General Electric Corp.),
LARSYS (Purdue University), and MIDAS (Environ. Res. Inst. of
Michigan). However, future systems apparently will have to go still
further in the direction of satisfying the needs of non-programmer,
non-technical users (Eastwood et al, 1976). All the basic components
of such an information system have been developed. These are surveyed below. Their integration into an operational system remains
to be accomplished.

Applications of Remote Sensing Data in Coast Zone Management

As discussed in the preceding section and summarized in Table 1,
at least 26 different information needs arise from tasks related to
coastal zone management. In this section, we shall review the utility of remote sensing data in satisfying these information needs.
Remote sensing data which can be applied to these information needs
are tabulated in Table 2. These remote sensing data include five
types which are available from the Landsat Multi Spectral Scanner
(MSS), ten from Skylab or other platforms, and one which can be
relayed from a variety of in-situ sensors by a data collection
platform (DCP).

The Landsat frequency of coverage is 18 days unless cloud cover
interferes, which is adequate for most CZM tasks. However, the
detection of transient phenomena for enforcement tasks or other
tasks requiring fast responses would necessitate a higher frequency
of coverage. The spatial resolution of the present Landsat MSS is
80m, and is expected to improve to 40m with Landsat C. Resolution
in this range is adequate for most CZM tasks, but not for the detection of small-scale changes or for the accurate location of boundaries of reflectance zones. Photographic data from manned aircraft
can be obtained with spatial resolution adequate for virtually any
CZM application, but is prohibitively expensive if it is required
frequently over extensive areas. The use of in-situ sensors in
combination with DCPs is valuable because they are capable of

Table 2: Remote Sensing Technology Related to Coastal Zone Management Information Needs

Agency Data Needs

1. Temperature
2. Salinity
3. D.O.
4. Turbidity
5. Sediment Load
6. Organic Load
7. Nutrient Concent.
8. Chlorophyll
9. Water Chemistry
10. Surface Oil
11. Currents
12. Water Depth
13. Water Volume
14. Precipitation
15. Snow Cover
16. Ice Cover
17. Surface Runoff
18. Soil Moisture
19. Groundwater Supply
20. Source Location
21. Outfall Monitoring
22. Boundaries
23. Topography
24. Land Cover
25. Land Use
26. Demography

Remote Sensing Device

Landsat

A. MSS Undifferentiated
B. MSS Band 4 0.5–0.6μm (Green)
C. MSS Band 5 0.6–0.7μm (Red)
D. MSS Band 6 0.7–0.8μm (IR)
E. MSS Band 7 0.8–1.1μm (IR)

Aerial (Incl. Skylab)

F. MSS Visible
G. MSS Near IR
H. MSS Thermal IR
J. Passive Microwave
K. Active Microwave
L. B & W Photo
M. B & W IR Photo
N. Color Photo
P. Color IR Photo
Q. Other

In Situ Data Coll. Platform

R. Various Types

Device	1	2	3	4	5	6	7	8	9	10	11	12	13	14	15	16	17	18	19	20	21	22	23	24	25	26
A				X	X	X		X			X	X	X		X	X	X	X	X		X	X		X	X	
B				X	X	X		X			X	X	X		X	X	X	X	X	X	X	X		X	X	
C	X			X	X			X		X	X	X				X	X			X	X	X		X	X	
D				X	X			X								X	X			X	X	X		X	X	
E				X	X			X							X	X	X				X	X		X	X	
F	X			X				X		X	X	X			X	X	X	X	X		X			X	X	
G								X			X	X				X	X	X		X				X		
H	X				X					X	X					X	X	X	X			X			X	
J	X	X								X	X								X						X	
K										X	X	X				X	X		X	X	X				X	
L				X							X	X				X	X		X		X	X	X	X	X	
M	X			X	X	X		X			X	X				X	X		X		X	X	X	X	X	
N					X			X			X	X				X	X		X		X	X	X	X	X	X
P	X			X	X			X			X	X				X	X	X	X		X	X	X	X	X	X
Q																							X	X	X	
R	X	X	X	X					X				X	X					X			X	X	X	X	

providing frequent ground truth with less ambiguity and more precision than remote sensors. However, they are not well-suited for the acquisition of geographically extensive and spatially dense information.

Temperature information is readily obtainable from thermal infrared data (Anding, 1975), although atmospheric interference is not easily removable. This has been demonstrated with Nimbus and Skylab data, and thermal infrared is an anticipated addition to Landsat C. Also, temperature is often correlated with parts of the spectrum other than thermal infrared. Water surface temperature is important for detecting thermal pollution, coastal upwelling, and current patterns. Surface temperature on land is a useful component of the information used in land-use identification for CZM.

Salinity, which is of use in surveying and monitoring changes in tidal wetlands and estuaries, can be sensed by passive microwave sensors (Thomann, 1975) if other water properties are known.

The turbidity of water is observable by the Landsat MSS and by any of the other remote spectometric or photographic techniques. Klemas et al (1974) reported that patterns of sediment concentration in the upper meter of the Delaware Bay, determined by using a previously derived relationship to find the sediment concentration (mg/ℓ) from the ERTS-1 MSS band 5 (red) data, agreed qualitatively with tidal current predictions. By using more than one band, it is possible not only to infer suspensoid source locations and current vectors from observed dispersion patterns, but to infer the nature of the suspensoid. For example, chlorophyll concentrations, which are indicative of phytoplankton density, and thus of nutrient loads, upwelling, and eutrophication processes, have been studied by Landsat MSS band 4 (green), (Szekielda, 1973). There are also reports of correlations between MSS data and the suspensoid mineral content, nutrient concentration and trace metal concentration. Where the visible light attenuation between the water surface and the bottom is not very great, the MSS data can be used as indicators of water depth (Ross, 1973).

Ocean current monitoring based on sun glint variations associated with surface roughness changes caused by wave-current interactions has been suggested by Strong and DeRycke (1973). They demonstrated feasibility of this technique by using the 0.6 to 0.7 µm detector aboard the NOAA-2 satellite to observe the Gulf Stream.

Oil on the water surface affects the optical properties of water, and also the surface capillary wave spectrum via surface tension effects. As a result, oil slicks are observable in the visible spectrum (Kondratyev et al, 1975) and by passive and active mocrowave sensors (Kuitenburg, 1975).

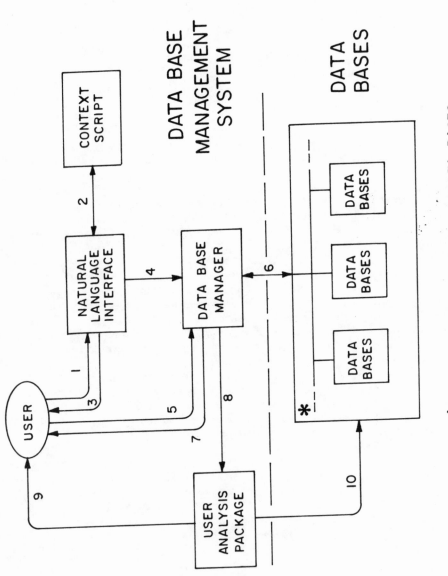

Figure 1

A number of land surface properties and land uses can be studied by means of MSS and other remote sensing data. Land surface properties and land uses identified by remote sensing are valuable in CZM because they are basic to the estimation of permeability and runoff (Ragan and Jackson, 1975). This same information is necessary to the land use planning and management task of CZM.

A major limitation of most of the inferences discussed above is the need for ground truth information. Geographic patterns, boundaries, and temporal changes are direct results of remote sensing data, but the identification and quantification of the water (or land) properties require concurrent ground truth data in the same geographic area. Other limitations on the use of remote sensing data arise because of variations in sea state, whitecaps, atmospheric properties, and sun angle. To use remote sensing data in combination with ground truth, and to apply the resulting information to CZM or any other management area, require that all data be referenced to a common georeferencing system. Unless this situation prevails for the raw data, the problem of common georeferencing of data collected by different means or at different times becomes one component of the DBMS.

In summary, information cautiously inferred from remote sensing data can be used qualitatively, and can now be used quantitatively only in combination with concurrent local ground truth. However, there is considerable potential for improving the reliability of information inferred from remote sensing data, by further studying the correlations between the end-use information and the remote sensing data.

Information System Modules for Coastal Zone Management

User requirements for information and information systems, as discussed above, may be viewed as having a hierarchy analagous to the hierarchical structure of the user group, which is assumed here to be a CZM agency. The configuration of a model Data Base Management System (DBMS) designed to satisfy these user requirements is shown in Figure 1. This model DBMS contains the following modules:

Natural Language Interface - handles discourse with the user and the Data Base Manager in ordinary conversational "Keywords" whose meanings are self-evident to the non-programmer user. Can be bypassed by the sophisticated user.

Context script - contains vocabulary and information format descriptors keyed to different users within a user group's managerial hierarchy.

User Analysis Package - software for the analysis
of retrieved information keyed to different users.

Data Base Manager - software for storage, retrieval,
and manipulation of the data base(s) using a "Query
by Example" system and based on the concepts of
"relational" data bases.

The Natural Language Interface is provided to meet the needs
of those users not able to address the DBM (path-way 5 in Figure 1)
directly in a programming language. Natural language processors
have been thoroughly studied (see, for example, Weizenbaum, 1966).
The language is actually a data sublanguage (the subset of the
language concerned with information retrieval and storage) embedded
in a host language. The vocabulary by which each user communicates
with the DBM through the Natural Language Interface is one part of
the Context Script. Together with the Context Script, the Natural
Language Interface is able to select among different kinds of
prepared responses to the user. For example, the user may request
"a list of all places where turbidity increased". A stroking system
response might be "We have that kind of information, but please be
more specific." A contributive system response might be "We have
that kind of information for 476 pixels." The most sophisticated
kind of response is a clarification response, such as "Do you want
a list of all such pixels, counties, or states?"

Sophisticated user analysis packages have been developed by
a number of groups for a variety of purposes. These include
software for the classification of spectral signatures either with
or without ground truth "training areas", for change detection, for
boundary detection, for pattern recognition, and for all the more
specific kinds of data interpretation discussed in the preceding
section. The state-of-the-art in data analysis software is
sufficiently advanced for the needs of CZM agencies. As commented
on in Section 1, however, most of these agencies require that such
software be available within an integrated information management
system comprised of additional software such as that in Figure 1.

The DBM which can interact most effectively with the user
through the Natural Language Interface, and which can make access
to the data base efficient, independent of the source of the query,
and independent of the information storage system, is a "Relational"
DBM which is a combination of a "Query By Example" system and the
"Relational Data Model". A few Relational DBM's are in use, but
are not as widely used as they should be, and have not been used
in combination with a Natural Language Interface and a Context
Script.

The "Query by Example" system (Zloof, 1974), allows the user
to use simple instructions to define his perception of the tabular

arrangement of the data (which need not coincide with the arrangement of data within the storage medium), and to identify those
data which he wants retrieved, rearranged, processed, or outputted.

The "Query by Example" system combines naturally with "Relational Data Model" (Codd, 1970) to comprise the DBM. The relational
data model is the mathematical theory of "relations" upon which
is based the DBM software for defining, storing, retrieving, and
rearranging multi-dimensional intersecting matrices of information.
The relational data model allows the user to perform these operations through the simple "Query by Example" system, and to perform
them without being at all concerned about the actual formats,
order, indexing, or other particulars of the data stored within
the data bases.

The mathematical theory of "relations" is an application
of set theory. A "relation" is a set of "components". For an
example, the relation "zone" might have 100 components, each one a
different zone. Each relation has a number of corresponding
"domains", which are attributes of the components. For example,
the relation "zone" might have six domains, such as identification
number, latitude, longitude, use, oxygen, and spectrum; so each
of the 100 different zones would be associated with its six respective values of these six attributes. Note that the sixth domain,
spectrum might be a relation, itself comprised of its own four
domains.

Mathematical operations can be performed on relations to
produce new relations. For example, operations referred to as
"Permutation", "Projection", "Join", "Composition", and "Restric-
tion" are defined operations. These operations are performed by
the DBM. Singly or combined, they readily accomplish tasks such
as listing the identification numbers of all the zones which have
values of "spectrum" which also appear in the corresponding domain
of another specified relation, such as "training area". Another
example of a task which would be simple for any relational DBM
to perform would be listing the latitude and longitude of any zone
in which "spectrum" changed between two dates, and in which no
construction permit was issued. The complexity of performing this
task with ordinary programming language -- if all of the different
types of information used in the task were stored in many different
formats, different files, different groupings, and different indexing systems -- is not easily imagined!

In summary, a variety of extremely important DBMS components
has been developed to the level of operational feasibility, and
several combinations of these components have been integrated into
DBMS's which are partial implementations of the system in Figure 1.
However, the full system in Figure 1, which is necessary for the
non-programmer to use remote sensing data (or any large quantities of many types of data) in CZM, remains to be implemented.

Conclusions

Local, state, and regional agencies in general, and CZM agencies in particular, have needs for many different kinds of information in a variety of formats. Remote sensing data have a potentially major role in satisfying these information needs. The data analysis software required for integrating remote sensing data with ground truth and other data has been developed adequately for many operational uses, but further development would increase the potential uses for remote sensing data.

In order for the role of remote sensing data to realize its potential in CZM, information systems will require some additional development. All the components of an information system which could be used by non-programmer, non-technical CZM staff, and which could produce information products appropriate for use in CZM, have been developed. However, no operational information system for CZM integrates all of the components necessary to encourage the use of remote sensing data by CZM agencies.

Acknowledgements

Many of the ideas and concepts discussed in this paper were developed or brought to the author's attention by his colleaugues on Project ASTRO (Applications of Satellite Technology to Regional Organizations), which is sponsored by NASA under contract NSG 7144. Some of these colleagues, without whom this paper would not have been possible, are Professors S. Wecker, D. Weiss, H. Rubin, and C. Hamlin.

References

Anding, D.C., "A procedure for estimation of sea surface temperature from remote measurement in the 10-13μm spectral region". Abstracts of the NASA Earth Resources Survey Symposium, June 8-12, 1975.

Anonymous, "208 Area scanned from space". Waterwise, No. 3, 1975.

Codd, E.F., "The relational model of data for large shared data banks". Communications of the ACM, Vol. 13, No. 6., 1970.

Eastwood, L.F., Jr., et al, "Preliminary needs analysis report". Summary Report of Program on Earth Observation Data Management Systems, Center for Development Technology, Washington University, St. Louis, Missouri, 1976.

Klemas, V., Bartlett, D., Philpot, W., Roger, R., and Reed, L., "Coastal and estuarine studies with ERTS-1 and Skylab". Remote Sensing Environment, 3, 1974.

Kondratyev, K. Ya., Rabinovitch, Yu. I., Melentyev, V.V., and Shulgina, E.M., "Remote sensing of oil on the sea surface". 10th International Symposium on Remote Sensing Environment, October 6-10, 1975.

Kuitenberg, J. van, "Radar backscatter measurements of the water surface and oil slicks in the wave tank". 10th International Symposium on Remote Sensing Environment, October 6-10, 1975.

National Research Council, "Remote Sensing for Resource and Environmental Surveys". Report No. CORSPERS-74-1, NTIS number PB-237 410, 1974.

Ragan, R.M., and Jackson, T.J., "Use of satellite data in urban hydrologic models". Journal of Hydraul.Div. ASCE, Vol. 101, No. HY 12., 1975.

Ross, D.S., "Water depth estimation with ERTS-1 imagery". Symposium on significant results obtained from the ERTS-1, Vol. 1, Section B., 1973.

Shinnick, W., Morain, S., Grogan, N., and Inglis, M., "A study of the needs and problems of state agencies in the area of natural resources and the environment to which remote sensing could contribute". Report to Director of User Affairs, NASA Office of Applications, June, 1974.

Strong, A.E., and DeRycke, R.J., "Ocean current monitoring employing a new satellite sensing technique". Science, 182, 1975.

Szekielda, K.H., "Distribution pattern of temperature and biomass on the upwelling area along the N.W. coast of Africa". American Society of Photogrammetry, Pt. II, Fall Convention, October 25, 1973.

Thomann, G.C., "Remote sensing and sea surface salinity". Abstracts of the NASA Earth Resources Survey Symposium, June 8-12, 1975.

Weizenbaum, J., "ELIZA" Communications of the ACM, Vol. 9, 1966

Zloof, M., "Query by Example IBM Res. Center". Report RC 4917.

AN OPERATIONAL EARTH RESOURCES MONITORING AND MANAGEMENT SYSTEM:

SOME PROBLEMS AND SOLUTIONS RELATING TO THE LAND AREAS

R.A.G. Savigear
Department of Geography
University of Reading, U.K.

If a monitoring and management system using remote sensing is to be developed and is to be successful, it must be at least as efficient as the existing systems. Nevertheless, there are no absolute measures of efficiency and the criteria by which efficiency is judged will vary in time, and from place to place, as will the requirements that will be expected of the system. Indeed, these requirements will differ according to the environments (*sensu lato*) of the people who will carry out the monitoring and management programmes. In addition to their differences in objectives, there will be a variation in the nature and extent of the type of restraints that will limit the use, or applicability of, some or most, parts of any earth resources monitoring system established. The development of a world-wide operational system can only be achieved, therefore, if we are able to identify both the requirements of the users of the system and are able also to overcome whatever restraints that may make it a less efficient system than that, or those, already in operation. There is no single requirement and there is no single restraint; nor are we able to identify all the requirements or all the restraints. Can we, however, identify enough requirements, and are we able to push the restraints to limits, to enable us to achieve a system that will be more efficient for most of the land surfaces than those currently operating?

The Concept of the 'Restraint Scale'

Initially we have to identify what technical requirements a potentially global operating system may necessitate and whether, during the period in which it will be operative, it will provide information that is both required and will be used adequately enough to make it an efficient application of world energy.

Constraints to the development of an operational system will
vary from environment to environment; a term now used to encompass,
for example, political, social, economic, climatic, and physiographic environments. Clearly these categories may be further
subdivided and other categories may be added. It is important,
however, to recognise that such environments, or environmental
complexes, vary over the land surfaces and provide complex
interrelated and interacting restraints to the efficient operation
of any system proposed or developed. Their characteristics also
vary temporally; some diurnally, some seasonally, some annually,
some quinquennially, decennially, etc.

To determine what is necessary to establish an efficient
operational monitoring and management system, the land area
environments should be located within such a multi-dimensional
restraint matrix. By this approach we shall be able to identify
the problems that must be solved in each environment if its requirements are to be met, and if the data provided by the system are
to be used. The point may be made simply in case it is not adequately emphasised. It is that, for example, a system that may work
for Arizona may, or may not, work, or be the most appropriate for
either, say, Maine, or Chad, or the Guinea Coast. We have to
determine what needs to be done to make it workable in each one
of these areas and ultimately for every part, or most parts, of
the land surfaces.

Conceptually, at least, we can envisage a sensor package in
a satellite system that could meet the several requirements of
such diverse environmental complexes. In practice, however, we
need to know what will be technically and economically possible
in 1990 and 2000, since the identification of such system restraints
will make it possible to define the associated sub-systems that
will be required during each ten-year period so that ultimately
we may achieve a truly global, and not merely a national or regional, monitoring and management system.

To develop the environmental restraint matrix requires a
major effort, but the necessity for such an analysis appears when
an attempt is made to select subjectively the constraints to the
development of a monitoring system, since the interconnection of
the components of each environmental complex makes the identification of logically reached conclusions very difficult to achieve.
The comments which follow, therefore, represent only an attempt
to illustrate the application of the method by identifying some
major restraints to the development of a world-wide system. The
restraints recognised indicate only an analysis framework within
which many other categories and sub-categories are potentially
recognisable. Their recognition leads to the identification
of problems and so to analyses leading, hopefully, to solutions.
An extended and rigorous analysis, which is what is required,

would provide the information from which we may determine whether
a global land monitoring remote sensing system is feasible and, if
so, what needs to be done to make its potential a reality.

Operational Limits Within the Restraint Matrix

The operational limits within the restraint matrix relate to
those imposed by the monitoring system selected (*system restraints*),
by the characteristics of each environmental complex within which
the system will be required to operate (*environmental restraints*)
and by the objectives of the users of the system (*requirement
restraints*).

System Restraints

If we assume that the first operational satellite system will
be comparable to the Landsat experimental system, we see that the
major restraints will be the eighty to one hundred metre resolution
(which cannot be improved beyond thirty to forty metres in the
foreseeable future), and the long periods between sensor scans of
the same area, currently eighteen days. Aircraft may be used to
support the system by providing both spatially and temporally
stratified sampling surveys, by underflying sample areas within the
scanned zone during the approximate time of the satellite passage,
and at selected time intervals between the satellite passes for the
same sample areas.

Aircraft alone constitute very flexible platforms that may be
used in any area at any time. The major restraint to the aircraft
system relates to its inability to provide for more than a severely
limited cover per unit area by comparison with that available from
satellite platforms. Additional serious restraints relate to the
availability and characterisitics of the sensors, each of which is
limited by different restraints. For example, some scanners provide
data which need special ground-based sub-systems if they are to be
reconstituted as imagery: presently available imaging radars on
aircraft are limited in resolution and provide low oblique imagery;
aircraft cameras provide hundreds of photographs with varying tonal
or colour properties for any large area flow, etc., etc.

Environmental Restraints

Political restraints are likely to isolate certain land areas
from a global monitoring system. Will, say, Russia or China or Argentina find it politically acceptable to collaborate on the development of a land monitoring satellite system? It seems hardly
likely that all states will collaborate. Although smaller aircraft
and cameras may be available in most, if not all, states, only the
larger countries will possess the more sophisticated aircraft and

sensor systems and may not make these available to any other country. We can conclude also that (for the foreseeable future) some countries will wish to obtain and analyse their own data without external or international commitments or assistance. Political factors may be a potent inhibitor to the establishment of a global monitoring and management system.

Social Restraints cannot be ignored. The concept of monitoring land areas and the development of data banks for management purposes is abhorrent to some sections of many communities. Farmers, large commercial and industrial national and international companies, and others, may well oppose monitoring systems which make all data universally available.

Economic Restraints are likely to be severe in most developing countries. In developed countries also the cost effectiveness of remote sensing will not be easily demonstrated, if indeed it can be proved to be as cheap or as efficient as existing monitoring and survey systems, particularly in small countries such as the United Kingdom. The availability of money will limit both the provision of satellite and aircraft systems and also, and perhaps more importantly, the provision of the more sophisticated analysis and interpretation equipment.

Technical and Educational Restraints limit both the development and use of remote sensing equipment and the analysis and interpretation, and the application, of the data produced by the sensors. The inability of politicians and of 'senior' scientists and of 'resource managers' to understand the potential and limitations of the new methods and techniques provides an inertia that may well be more restrictive in developed than in the developing countries. The former have long established and efficient monitoring and management systems, whereas many of the latter are only now taking responsibility for the running of systems without expatriate assistance, and may be more receptive to the application of remote sensing if methods can be developed that may be used where there is a small educated work force.Nevertheless, there are severe limitations to the use and application of a monitoring system in developing countries because of the relatively small number of scientists and technicians available to carry out the several functions that must be performed if a working monitoring system is to be established. Apart from the physicists, engineers, mathematicians required to obtain, receive and manipulate the data (and perhaps these need be only a relatively small proportion of the whole number involved in a working system), there is currently, and there is likely to be in the foreseeable future, a very limited number of field scientists interested in and able to develop and implement the field traversing and sampling methods necessary to make a monitoring system workable.

Climatic Restraints are complex, but the most significant is provided by cloud cover, preventing the use of most sensors, apart from radar, for a large proportion of the land areas for long periods. Such restraints may make it impossible to use a Landsat type satellite system in an operational situation in any but the arid and sub-arid tropical land areas and only at certain seasons (usually the periods when vegetation is dormant). Aircraft can take advantage of cloud free periods but are able to survey only relatively small areas in the time available. Climatic variability throughout the land areas, and daily, and through the seasons, and from one year to another, results in changes in landscape properties that affect both emitted and reflected radiation. In particular, in the humid areas it is the occurrence and movements of water at the land surfaces, in the soils and rocks affecting the growth and characteristics of vegetation that make the prediction of sensor signals for any but limited areas, and for a limited time, very difficult, or generally impossible; so that the identification and classification of any but the most general categories of landscape properties cannot be undertaken by automated, by semi-automated, and even by subjective interpretation methods.

Physiographic Restraints currently limit the ability of the interpreter to relate a specific radiation signal to specific land properties because of the spatial variability of rock, soil, form, plant and water properties and relationships, and because their radiation properties change through time as the occurrence and movement of water, sun angle, etc., affect them differently. The fact that the non-geophysical sensors are unable to penetrate the land surfaces, and to provide radiation measures that may be related unambiguously to the presence of, say, specific rock, or soil types, or other properties, such as the presence of shallow aquifers, means that the identification of such properties and their distributions depends on the ability of a scientist to interpret them and this based on his experience of the relations of landforms and/or vegetation to rock or soil or water, etc., in the area being examined. The inability of interpreters accurately to relate such properties to physiographic conditions without ground traversing and sampling programmes constitues also a major restraint to the establishment of automated, semi-automated or even subjective interpretation procedures and monitoring programmes. An associated limitation to the establishment of a global monitoring system arises because there is currently an insufficient number of interpreters available with an adequate knowledge of the characteristics of landscape patterns and of their relations both to photographic and image patterns and to the properties of landscape materials and radiation signals.

Requirement Restraints

Clearly not all the objectives of all potential users of a

monitoring system can be satisfied, and their requirements will
vary from one environment complex to another. Probably the more
general the requirement, the more the likelihood that it can be
satisfied.

  Dimensional and Temporal Restraints can be recognised. The
Landsat type of system is appropriate for large area analyses
that do not require information to be available at a specific time.
When, however, information is required at a specific time, or
for a small area and with high resolution, the system is unable
to meet such a requirement. In areas where, for example, the
geology or the soils are unknown, or inadequately mapped, and
any imagery at any time will be useful, the Landsat system meets
the needs of the surveyor. If, however, crop surveys, and certain
types of land use surveys are required, that need to be carried
out at specific times and in areas where the fields are small,
the Landsat system cannot provide the necessary information because
it is uncertain whether cloud cover will blanket the area and,
in any case, the spatial resolution will be too poor.

  A *Mobility Restraint* may also be recognised. Its identification draws attention to the problem of obtaining information about
crops, or to provide solutions to urgent problems, when high resolution imagery and/or data may be required for both small and for
large areas. Such requirements may necessitate the provision of
narrow spectral resolution which can only be derived through sensors mounted on platforms provided especially for such purposes.

                            Some Solutions

  To determine the feasibility of establishing a global land
monitoring system that is capable of being integrated with a system
for the management of natural resources, and to identify the areas
where it will be applicable and/or used, the characterisitics of
the land areas must be examined and classified geographically,
country by country and region by region. It is necessary to
identify the monitoring and management requirements of each country,
and within it of each region and its associated physiographic
sub-divisions, so that all the major, and most of the minor,
restraints to its development and establishment may, as far as
possible, be identified. Although political, social and economic
requirements and restraints are likely to relate more commonly
to countries or to politically significant states within them,
regional requirements and restraints are also likely to be identifiable. The boundaries between political units, however, hardly
ever coincide with the boundaries of physiographic units, yet it
is by an examination of the restraints within these that the
scientific problems will be identified which must be solved if
a monitoring system is to be established. It is nevertheless

likely that the existing management systems will relate to political
rather than natural units, and these differences between political,
administrative and natural units may result in difficult analysis
and correlation problems.

Limitations of Satellite and Air Platforms

For each political unit and each physiographic unit requirements and restraints should be estimated for each decade for the
twenty years between 1990 and 2010. The potential and limitations
of each of the following systems and sub-systems should be evaluated
for each of these subdivisions of the land surfaces. They are
satellite systems with ground sampling sub-systems: satellite
systems with aircraft and ground sampling sub-systems: aircraft
sampling systems with ground sampling sub-systems. Such an ordered
examination should make it possible to determine the most efficient
combinations of platforms and sensors for each physiographic sub-
division, for each region and for each country, so that each local,
regional and national sub-system may be integrated into a global
monitoring system. In the course of such an analysis it is believed
that the inability of satellite and aircraft to provide large area
high spatial resolution imagery, within specified and short time
periods, will be defined and the value of developing rocket systems
to meet this requirement will be recognised.

A Physiographic 'Geo-referencing' Framework

One of the major restraints currently operating to limit the
potential for establishing an effective monitoring, as distinct
from a survey, system is due to the limited knowledge we have of
the distributions and characteristics of the physiographic subdivisions of the land surfaces. For certain regions this is because
it has never been collected, although the essential information is
available; for others the basic survey has still to be carried out.
The occurrence and movement of water, which relates to the geomorphological properties of the land surfaces and which influence
the spatial and temporal growth of plants, condition the distribution patterns of, for example, healthy and stressed vegetation.
The understanding and recognition of the relations and distributions
of landforms and plants (the physiography of the land surfaces)
provide, therefore, the spatial framework for the identification
of areas within which radiation measures may be expected to be
susceptible to prediction. Such land surface and landscape patterns
and their associated properties relate to the evolution of landforms and landform components, which thus condition the characteristics and the distributions of site properties that provide the
environments for plant growth. Physiographic properties constitute,
therefore, landscape indicators that may be used to infer earth
material properties, or plant species, or areas liable to flooding,
etc. If monitoring is to become semi-automated then the framework

for identification and prediction should be the physiographic sub-
division, since this is likely to contribute the basic unit for
predicting the spatial limits of identifiable radiation measures.
This framework can be achieved only by the application of integrated survey methods based on geomorphological and geo-botanical
analyses. Given such spatially defined physiographic subdivisions,
and with data derived from climatic and meteorological monitoring
systems, it becomes possible also to see how, for example, crop
yield, or floods, etc., may be predicted by semi-automated methods,
and such units therefore constitute the most appropriate 'geo-referencing' framework for a monitoring and management programme.

Educational and Training Programmes

     There is an inadequate appreciation of the integrated relation-
ships of properties and processes within and at the land surfaces.
This relates to the relatively recent development of educational
and training programmes in geomorphology and geo-botany. It is
a matter of considerable importance, therefore, that seminars
and educational courses be established, to disseminate knowledge
of these fields of study to those who will be developing the use
of remote sensing data for the solution of their monitoring problems. Associated with such programmes is a requirement to demonstrate the value of applying ordered methods and techniques for
traversing and sampling to establish the meaning, significance, and
relationships of photographic and image patterns. It is also
important to improve the knowledge field scientists have of the
characteristics and value of using physiographic subdivisions to
evaluate the relations of radiation properties to, for example,
the occurrence of soil moisture, crop vigour, shallow aquifers,
crop yield, etc.

A Fundamental Research Problem

     Research programmes should be established to examine the
problems and possibilities of achieving a landscape spatial pattern
classification related to rock, and/or soil, and/or water, and/or
form, and/or plant conditions and distributions. It is believed
that an investigation of these problems is an essential prerequi-
site for the development of semi-automated, and ultimately automated, monitoring programmes, since it is the landscape patterns
which determine the radiation measures and so the tones, colours,
textures, and patterns on photographs and images. Such work will
be best allied to the pattern recognition interests of the physicists and mathematicians and requires the establishment of integrated research teams composed of mathematical, physical, biological, earth and applied scientists.

OPERATIONAL REMOTE SENSING SYSTEMS FOR ENVIRONMENTAL

MONITORING AND REGIONAL PLANNING

Professor Dr. Sigfrid Schneider
Bundesforschungsanstalt für Landeskunde und Raumordnung
Bonn-Bad Godesberg, West Germany

Environmental monitoring and regional planning are the instruments for creating living conditions that respond to up-to-date social, economic and cultural requirements.

Remote sensing including aerial photography offers the data needed for information on:

- inventory of land use,
- monitoring of land use changes,
- detection of land use conflicts,
- detection of air and water pollution as well as of land-scape damages,
- decision process for planning measures.

There can be no doubt about the fact that the great interest of regional planners in the systematic application of remote sensing processes - including data from satellites - is based on a compilation of data gained by various remote sensors and by conventional statistical methods.

Nevertheless, the utility of remote sensing systems for regional planning seems not be fully acknowledged by all planners. Four important advantages of satellite remote sensing need to be reiterated:

1. the repeated coverage of the earth surface by instantaneous imagery facilitates monitoring of all actual features and dynamic processes,

2. the synchronous overview of large areas permits a
better perception of the landscape situation and the
development of rural and urban areas and their inter-
dependencies,

3. the surveying of features that up-to-now have been
unknown or only partially known is now feasible,

4. multispectral surveying in relatively narrow bands
facilitates most exact differentiations of emission,
reflection and thermal state of objects, thus permitting emphasis of special features such as haze and
smoke in the blue band, land-use and water permeation
in the green band, vegetation and floods in the infrared
band.

Operational satellite remote sensing systems, augmented by airborne
systems, seem to promise the most cost-effective means to meet the
growing needs of environmental monitoring. The following four
examples are cited to support this point of view.

Case 1. Several regional planning authorities of the Ruhr-
district, the Frankfurt area and the Rhein-Neckar district have
made inventories of their territories by remote sensing methods,
mainly by multispectral photography and infrared thermography.
In comparison with the conventional official land-use statistics,
which are limited to certain traditional land-use classes, the
remote sensing data have enlarged the information to encompass
ecological factors as well as to socio-economic factors in urban
areas. The quality of remote sensing data for the measurement
of certain residential areas, the different sorts of fallow-land,
the green areas in cities, and the recultivated areas of open-pit
mining, for example, exceeds the quality of conventional statistical data. The topicality and reliability of such remote sensing
data has justified their application in regional planning. Total
intepretation accuracy for these projects was about 95% which,
considering current land use planning criteria, can be considered
as fulfilling most of the requirements of an up-to-date land-use
inventory for regional planning purposes.

In the Rhein-Neckar region, areal differences of several land-
use units of more than 25 percent have been detected between the
old cadastral classification and the actual remote sensing data.
Moreover the land-use data of this region have been digitized in
a very short time. A statistical tabulation as well as a thematic
map in the scale 1:50,000 have been printed. On this map, complete
information on the land-use structure of non built-up areas has
been represented for the first time.

The progress in evaluation and interpretation of Landsat-imagery provided the basis to start a systematic study on land-use. We proposed that the German planning authorities order a land-use map of the scale 1:200,000 as a first sample sheet of a series covering the Federal Republic of Germany. This map will be the first sheet of a thematic map series interpreted from satellite data.

Case 2. Another problem is the monitoring of land-use conflicts such as the detection by Landsat imagery of gravel pits within the water protection zones of the Rhein-valley, where several gravel pits have been transformed to waste dumps; or the detection of new housing projects in the neighbourhood of big chemical plants. According to the law, the planning authorities have to establish an up-to-date regional development program. This requires a regional plan which should indicate the basic development of future structures for the area, and proper use of the land, particularly in respect to requirements of settlement, agriculture, forestry, industry, energy, use of natural resources, traffic, preservation of the historic landscape, and recreation.

We suggested that some federal institutes monitor the upper Rhein valley where land-use conflicts between gravel-extraction and recreation have been observed. Recreation, whether active or passive, uses an area in such a manner as to subject a land-scape to greater or lesser stress, depending upon the activities preferred by the user.

The regional planner has the responsibility of providing a balanced level for various types of land use, to answer the needs of recreationists and at the same time to solve the attendant environmental problems. In order to fulfill these demands, the requirement is for concrete facts which can only be obtained from reliable data. The studies on the Upper-Rhein have shown that aerial infrared color photography and thermography provide invaluable sources of such data. They provide up-to-date and quantitative data on the actual state of a recreation area for actual use of the entire region as well as for the use of special bathing, camping, fishing and boating areas.

At the same time, an inventory of landscape quality (nature conservation areas, landscape or water protection areas), in relation to future locations of industrial or power plants, has been compiled by infrared-colour aerial photography. This seems to be a way to avoid future land-use conflicts. More and more, we observe that in our country these means of a documentation of .the actual landscape situation are used by planning boards as well as by industries.

Case 3. The third example of the application of remote sensing methods in regional and environmental planning is the ther-
mographic study of urban areas. The three planning boards mentioned above were encouraged to study this problem. The most
interesting results were presented in the polycentric Ruhr district
which includes several large cities and between them dividing
zones which hitherto have been protected from excessive development. These dividing zones represent predominantly farm land
with small wood areas, public parks, cemeteries, allotments,
playing and sports grounds in between. Since the founding of the
Ruhr planning board in 1920, this board has been responsible for
the configuration and limitation of green areas and zones, which
are important to all housing estates.

It has been proven that continuous green areas filter and
renew the air. The green areas filter the dust particles out
and cause air currents, widely distributing air pollutants with
a low sedimentation trend. Even the noise can be reduced by a
corresponding configuration of the green areas. Due to its close
proximity to the residential areas, the regional green zone system for the central part of the Ruhr district is to a large extent
suitable as a local recreation area.

The thermographies made by line-scanner, and presented as
false-color images, clearly delineate the regional green zones.
The thermography of the central Ruhr district is an impressive
proof of the fact that the continuous open areas effect a cooling
of the city. Generally speaking, the infrared thermography corresponds with the ground truth measurements of the city climate,
namely that the temperature of the surroundings also rises with
an increasing density of buildings. By differentiating the
radiation temperature in stages of 0.5^{0} C, not only the exact
position but also the geometrical pattern of buildinga and other
objects can be accurately determined, and heat effects calculated.
The urban thermographies allowed a differentiation between areas
covered with 3-floor closed building blocks and those covered
with separate 2-floor buildings. At the same time the images
delineated the isolation of the building elements (after Ruhrs-
iedlungsverband).

A similar observation as in the Ruhr district was made in
the Main basin near Frankfurt and in the Rhein-Neckar region near
Mannheim. In both cases, regional green corridors have been established after gaining a complete picture of local climatic conditions by short-interval flights with infrared linescanners.

It will no longer be permissible to make land-use changes
affecting large areas without simultaneously making an environmental impact statement. It has now been shown conclusively that
remote sensing is an invaluable aid to environmental assessment.

Case 4. The last example concerns the monitoring of polluted rivers. The Department of Interior asked our Federal Institute of Regional Planning to propose models of an operational remote sensing system for monitoring and estimating the water quality. Two test areas were chosen: the small Saar river with an extreme width of 30 to 40 m, and the Upper Rhein valley with a width of about 240 to 280 m. We have used different types of line-scanners with and without blackbodies, and have preferred a combination of infrared radiometer-thermometer and infrared linescanner in order to avoid atmospheric influences as far as possible and to get a distinct differentiation of surface temperature across the river. A helicopter, equipped with an infrared radiometer for measuring the temperature above the water surface, has flown in a longitudinal axis over the Rhein. Over small rivers one single temperature curve in the middle may be sufficient to represent the temperature above the water surface. Over big rivers, streams and estuaries it is necessary to obtain water surface temperature data at least along the middle of the river and along both sides. This situation has been shown by a temperature diagram of the Rhein near Mannheim-Ludwigshafen.

Conventional methods of measuring the water surface temperature are helpful, but are limited in comparison with the aerial overview given by remote sensing methods. All essential discharges into the Saar river as well as into the Rhein river have been detected, by combining the results of multispectral photography and infrared scanning. In one case of a big cellulose plant, the heavily polluted and heated discharge plume was clearly imaged by both sensors: color camera and infrared scanner.

The form and spread of pollution plumes depend on water quantity and flow velocity. Therefore the discharge plumes of the Saar and the Rhein have usually been quite different. After a short distance, the plumes of the Saar cover the whole river surface. But the plumes of the Rhein are pressing along the banks in long, thin bands reaching in some cases more than 30 km in length. Only seldom has the surface temperature of the mid-river of the Rhein been influenced by the heated discharge plumes near the banks. That means that the water surface temperature of big rivers in industrial regions is usually quite differentiated from bank to bank.

An overheating of river water generally has the following consequences. When the water temperature rises, the oxygen demand of fish and other organisms is increased. Overheating of the Rhein for example, would probably result not only in the death of the last fish in the river, but also in the danger that the biological balance in the water would be severely disturbed, causing a loss of the remaining self-cleaning power of the river. Ships would

be in trouble by the heavier development of mist. The recreation
value of the river landscape would be lowered by increasing turbidity and visible algae development. In extreme cases, mass
formation of algae would lead to intensive decomposition processes
with decay and odour problems, and with dangerous effects on the
production of drinking water.

The water quality of the small Saar river may be characterized
by interpreting the existence or non-existence of aquatic plants
and plant communities as an indicator of pollution. In the case
of the canalized Upper Rhein waterway, the observation of such
aquatic plants would have to be restricted to the branches and
tributaries where interpretation is based on certain plant communities as an indicator of eutrophication.

The wetland zone of branches and tributaries, most impressive
by its nature, is now more endangered by human activities than
the main waterway itself. The conflicts in using this zone involve
industrial location, power plants, refuse pits, pipe lines and
gravel pits *versus* recreation and sporting grounds, forestry,
agriculture as well as nature conservation areas.

The radiometric temperature curves as well as the linescanner
imagery of these rivers have now been used by local administration
and planning authorities in solving difficult location problems
such as, e.g., the location of power plants.

Data and enhanced images obtained by remote sensors have now
been proven as useful indicators of the type of water, characteristics of flow, source and type of discharge plumes, distribution
of surface temperature, and kind and degree of eutrophication,
for example. *Remote sensing methods are operational for regional
planning and environmental monitoring.*

Editorial Note: Typical river temperature profiles, false-color
imagery of thermal and pollutant plumes in rivers, and tables of
land-use mapping codes were presented by Dr. Schneider to the ARI
participants, for working group consideration. However, because
of space limitations, this illustrative material is not included
here.

DIGITAL ENHANCEMENT AND ANALYSIS TECHNIQUES IN

PROCESSING OF EARTH RESOURCES DATA

Charles Sheffield
Vice-President
Earth Satellite Corporation
Washington D.C., U.S.A.

Abstract

The progress of digital processing of satellite earth resources data is charted over the past four years. From the general trends, probable patterns for future processing are inferred. These lead to suggestions for a new approach to the provision of earth resources services involving satellite data. Mechanisms are proposed and discussed for implementing this approach, and basic questions that must be answered are set forth.

Introduction

When LANDSAT-1 (then known as ERTS-1) was launched in July, 1972, experience in the use of remotely sensed information for earth resources applications was limited to several very diverse data sources: high-resolution images obtained from aircraft using cameras, multispectral scanners, and radar equipment; low-resolution images obtained from meteorological satellites, using a single-spectral-band scanner or television camera; and a heterogeneous group of black-and-white and color photographs taken using hand-held cameras in the manned Gemini and Apollo programs, on a non-systematic basis in terms of targets and repeat coverage.

LANDSAT was planned with two principal observing instruments, a three-camera, three-spectral-band Return Beam Vidicon (RBV) and a single aperture, four-spectral-band Multispectral Scanner (MSS). Although the use of these instruments was intended to be complementary, both instruments had their ardent proponents and detractors in the period prior to the launch of LANDSAT-1. The very early

379

failure of the RBV, for reasons unrelated to the performance of
the television cameras, meant that the battle was over before it
was joined. Almost all analysis of LANDSAT data has been done
using MSS data.

In terms of digital processing, it was clear even before launch
that the MSS had some significant advantages. First, it is a single
aperture device, which makes registration of the several spectral
bands unnecessary; they are registered already. By contrast, it
is necessary to perform band-to-band registration on all RBV data.
Accurate registration of different spectral bands is particularly
important in multispectral classification work. Analysis of
aircraft data at Purdue and other centers, before the launch of
LANDSAT, had shown that small mis-registration in multispectral
images can produce significant errors in classification results.

Second, the radiometric fidelity of the MSS is much better
than that of the RBV. This is important in all work where quantitative measures of relative spectral reflectance are involved,
and is especially necessary in multispectral classification.

Third, scanners can be designed to take measurements over a
much wider spectral range than television or film cameras.

This paper discusses the progress in digital manipulation of
earth resources data, especially LANDSAT data, in the past four
years. Then it looks at the processing trends emerging for the
coming generation of operational earth resources satellites. It
is fair to say that such a paper could not have been written had
the MSS on LANDSAT-1, rather than the RBV, been inoperative. In
terms of processing for earth resources, we can justifiably para-
phrase Leibniz's assessment of Newton's work: "LANDSAT has done
more than had been done in all history before it." It is also
increasingly evident that the real uses of earth resources satellites, and of digital processing methods, are just beginning.

The discussion that follows therefore builds very largely on
the processing of LANDSAT data. Processing of meteorological
satellite data has many similar features, since they too use scanning devices. However, the particular problems of handling several
spectral bands of data predominate in LANDSAT data analysis.

Basic Digital Processing Considerations

It is assumed that the reader is already familiar with the
basic characteristics of the LANDSAT systems from the viewpoint of
hardware, orbital coverage, and available products. If not,
Reference 1 offers a good starting point. Only those elements of

the system that are particularly relevant to digital processing
of LANDSAT digital tapes are described here.

We assume that the reader has had relatively little exposure
to the mechanics of digital processing, and we will go into it here
only in enough detail to establish the main concepts.

The first point to make is that scanning devices and cameras
are *fundamentally different*, even though the results of both are
conventionally displayed in an image format. The basic source of
the difference is the instantaneous nature of the image obtained
with a film or television camera, versus the element-by-element
scan of a ground scene performed by a scanner.

As a simple example of the effect of this basic difference,
consider the result of a change in satellite altitude, produced by
long period variations in the orbit. If the satellite is at greater
altitude, the film or television camera image is simply at a smaller
scale (neglecting second-order effects such as the curvature of the
earth). Without further thought, one might assume that the same
result is true of the multispectral scanner image. In fact, the
change in satellite altitude will result in a differential scale
change in the MSS image, with a smaller scale in the across-track
direction and a larger scale in the along-track direction.

To see why this is so, note that an increase in altitude of the
satellite is accompanied by an increase in its orbital period. This
means that there are more scans of the scanner mirror per orbit,
since the scan period of the mirror is a fixed quantity. As the
ground distance covered in one orbital period is fixed, and equal
to the circumference of the earth, the distance covered in a single
scan of the mirror is decreased, i.e. the along-track scale is
increased.

On the other hand, the mirror sweeps through a fixed angle
across-track, thus as the satellite moves higher the across-track
scale decreases. In practice, this differential change of scale
may amount to 2.5% between two images taken a year apart. Since
it is a differential change, it is very difficult to account for
using conventional photo-processing techniques. Such effects are
best removed by <u>digital</u> transformations of the recorded MSS data,
prior to moving to an image format.

Other digital pre-processing needs arise in a similar way. For
example, it is necessary to "deskew" the MSS image because the earth
rotates beneath the spacecraft in the time it takes to image a
single scan line of a scene; it is necessary to square the image
because the picture elements after line data are sampled are rectan-
gular, not square; and it is necessary to compensate for distortions

produced by the roll, pitch, and yaw of the spacecraft.

These processes are so basic that they are all usually lumped together as "pre-processing" and considered of no interest to the data user. All image products ordered from the Sioux Falls Data Center, for example, should have these corrections made to them. The resulting images should be close to orthographic and able to be overlaid on conventional maps with little distortion.

Users who order digital information directly, in the form of computer-compatible tapes, are on their own. Since no geometric corrections have been applied to these tapes, it is a case of *caveat emptor*. If the user is not knowledgeable enough to make all necessary corrections, he will generate image products that still contain image distortions. The information necessary to make such corrections is not to be found in the Data User's Handbook or other standard reference works, and must be obtained by direct questioning of Goddard Space Flight Center. Corrections may call for a knowledge of the LANDSAT orbit.

It has been a central part of U.S. Earth Resources development philosophy that *information extraction* should be performed by data users, and that data provided to these users should therefore be as little processed as possible. The logic is simple: one man's signal is another man's noise, and attempts to process an image in an optimal way for one user may make it useless to another. Therefore, data service centers must either give users the basic data that has had a minimal amount of processing applied to it, or be willing to offer custom-made services, perhaps different for each user. The latter approach would clearly increase the cost.

The geometric corrections referred to above are an exception to the rule. They are regarded as being desired by *all* users, therefore the question of information loss during these corrections has been little considered in the context of LANDSAT-1 and 2 data. Even here, however, there has been a fall-back position: the user who wants minimally processed data can order digital tapes. Users who have done so are well aware of the fact that these tapes contain a greater dynamic range of information than can be represented on a piece of film, and therefore digital tapes are a good starting point for a wide variety of information extraction techniques.

Unfortunately, there is good evidence that the popularity of digital tapes was a surprise to the LANDSAT project manager. There is little information readily available on geometric correction needed for the correct use of these tapes, and what there is must, as mentioned earlier, be sought out by the user.

These problems have not arisen for meteorological satellite data, where most users rarely if ever handle tape data themselves,

and the requirement for accurate geometry is much less because
of poorer resolution and a lesser need to register accurately to
ground references.

We will now review, very briefly, the main digital image
enhancement and information extraction methods that have evolved
over the past few years for use with LANDSAT tape data. Clearly,
some of these were in use long before they were applied to LANDSAT.
This is not an attempt either to assign credit for development, or
to prove that the techniques exist at a higher level of sophistication in LANDSAT data analysis than other application areas.

The techniques fall into two main categories:

(1) processing algorithms designed to enhance the usefulness
of image-format data to a human interpreter or disciplinary scientist, and

(2) algorithms designed for direct extraction of scene
information without human image interpretation, or with
a minimum of the latter intervention.

Before doing this, it is desirable to point the direction
we are heading. In Section IV, we will analyze these techniques
in the context of the next generation of earth resources satellites.
This will lead to certain design criteria that this operational
generation should satisfy, and that is the main objective.

A. Digital Image Processing of Earth Resources Data for Image
Enhancement

Within each spectral band, a LANDSAT image is represented on
digital tape as an array of 3,240 x 2,340 picture elements or
pixels. Each of these is a number between 0 and 255, termed the
grey level or grey value, and representing the reflectance of a
ground area – in the case of LANDSAT-1 and 2, approximately 57 x 79
meters. Thus, a single LANDSAT scene, with four spectral bands,
is represented by about 30 million pixels.

All digital processing operations for image enhancement consist
of either *geometric adjustments* to this array, basically by moving
pixels with respect to each other; *radiometric adjustments,* or *grey
level modifications,* in which changes are made to the stored grey
values; *spectral band combinations,* in which pixels from different
spectral bands are compared or combined; or some combination of
these three types of operation.

1. Grey level modification (radiometric adjustment)

In these operations, the grey level associated with each pixel
of the original image is changed to a new value according to a pre-

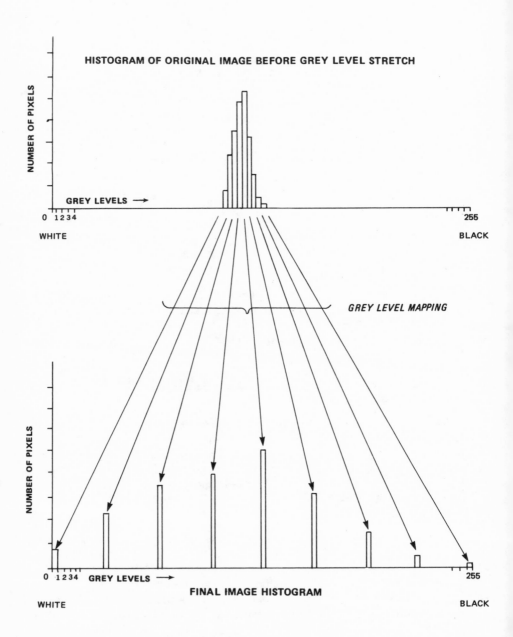

Figure 1.   Grey level stretch operations - the linear stretch.

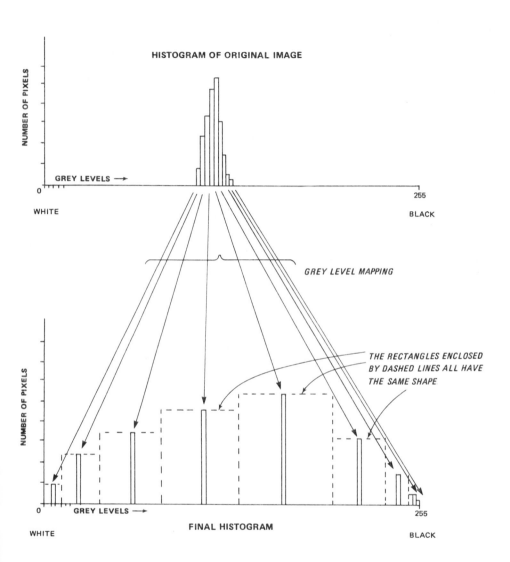

Figure 2. Histogram equalization of an image.

specified scheme. For example, the new value might be the old value subtracted from 255. The optical analogy to that simple operation is a switch from optical positive to optical negative. However, most digital grey level modifications have no optical analog. Digital transformations are much more general than any photo-processing can be, since they permit any grey level value to be changed to any other. Different grey level modification can be applied to different parts of the pixel array to remove, for example, latitude dependence in thermal band images that extend over a large range of latitudes. These, however, are rarely used. The most common grey level adjustments in practical use are:

a. Grey level linear stretch - when only a few grey levels of a scene are populated (i.e. a low contrast scene), a linear transformation is applied that re-allocates these grey levels to occupy the full range available (see the example of Figure 1). The result of such an operation is an image with more contrast. In this process, the ratio of grey levels in the scene is unchanged.

b. Grey level histogram equalization - this is also applied when only a fraction of the available grey levels of a scene is populated. A transformation is applied which re-allocates the grey levels to occupy the full available range, and does so in such a way that the most densely populated grey levels are separated furthest from neighboring populated grey levels. The picture of Figure 2 shows this more readily than words can. Note that this operation is a nonlinear one - grey level ratios are not preserved.

c. Grey level chopping - in this process, grey levels within a specified range are left at their original value, grey levels outside the range are all set equal to an extreme value (0 or 255). The resulting image 'masks out' (i.e. sets to black or white) all the image except those pixels with grey levels in the desired range. Clearly, such a procedure can be combined with a grey level stretch process, so that the remaining pixels are displayed at maximum contrasts.

The objective of these and other grey level modifications is always the same - to render all or part of an image with superior contrast or feature definition. The extreme case is binary image - all pixels of interest black, all others white. Such an image is often useful in displaying, for example, legal or political regions when it can be used in a multi-band algorithm to study those particular regions of an image that correspond to legal or political subdivisions. This is discussed further when we consider multi-band operations.

Since the final criterion of the effectiveness of digital processing for image enhancement resides in the opinion of the final user, there are no universal 'best' procedures. However, working experience built up over four years of LANDSAT data suggests that a histogram equalization process (b above) is the most popular of the commonly used grey level adjustments for most disciplinary users.

2. Geometric adjustments

These operations are not strictly speaking enhancements. They do not render features more readily separable from a background, nor do they emphasize objects of particular shape or grey level. What they do is permit an image to be created that is at a desired scale or map projection, so that it can be more readily overlaid or compared with other image format data.

Two fundamentally different processes are in common use for geometric adjustments. They can be thought of as pixel movers and as pixel interpolators. Pixel movers create a new image by adding or deleting rows or columns of pixels, to make an array with the desired geometry. No new grey level values are calculated, but pixels are moved around and duplicated as required. Pixel interpolators, on the other hand, compute the grey level at a location by an appropriate interpolation among the pixel grey level values of the original array.

Pixel mover algorithm generally require less computation and are therefore often preferred in practice - for example, image squaring, referred to in the first part of this section, is usually achieved by a pixel mover process. However, there are cases where a pixel interpolator is much preferable. For example, in a change of scale (e.g. a digital enlargement) adding pixel-by-pixel duplication will lead to a "blocky" appearance of the resulting image, because arrays of pixels with the same grey level are being created. Digital enlargement by pixel interpolation gives a better - looking product since there is smooth interpolation of grey levels in creating new pixels.

If computer time were not a consideration, pixel interpolator methods would undoubtedly predominate for geometric adjustments.

Geometric adjustment can in principle be used to create images at any scale and any map projection. However, most LANDSAT images are shown in a 'space oblique' projection or (less commonly) a variety of Transverse Mercator projections(for comparison with maps). The space oblique projection is one in which squaring and deskewing processes have been applied, but earth curvature and finite satellite altitude effects remain. These lead to differential scale effects at the center versus the east and west

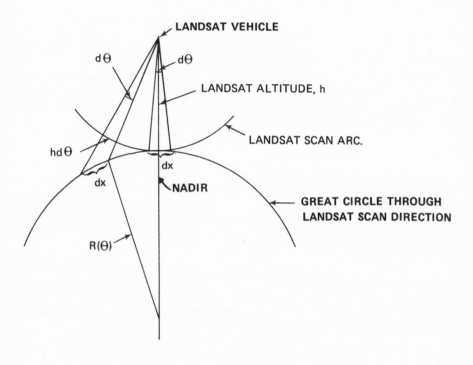

$$\text{Scale ratio at nadir} = \frac{dx_{NADIR}}{h.d\theta_{NADIR}} = \underline{\underline{1}}$$

Scale ratio at scan angle θ from nadir =

$$\frac{1}{h}.\frac{dx}{d\theta} = \underline{1 + (x/h)^2[1 + h/R(\theta)]} ,$$

(approximately 1.012 at the edge of the LANDSAT frame).

Figure 3.   Geometric distortion of MSS imagery due to Earth
            curvature and satellite altitude.

edges of a frame, amounting to about 1.2% scale difference across the image. The *mean* scale error is only 0.4%, in total measured distance from frame center to frame edges. This point is illustrated in Figure 3.

Digital enlargement and digital change to other map projections will become increasingly important as the need to compare and combine LANDSAT with other image data grows.

3. Geometric enhancements

This class of operation uses information derived from the *geometry and radiometry* of a scene to define radiometric adjustments to it. The best-known examples of this are processes for edge and line enhancement.

In edge and line enhancement algorithms, the difference between the grey level values of neighboring pixels is computed. Where this difference is large (which is where we usually say we see an edge in the image) the neighboring grey levels are modified to make the difference larger yet. Thus, edges and lines in the scene are emphasized. (Note: this description is a gross simplification of the real procedure!) Processes can be developed that emphasize lines in any direction equally, or that only emphasize lines in a selected direction. The latter are popular in geological work, where directionality of structure is a key factor.

Note that these images appear to have improved spatial resolution, because there has been a boosting of the high frequency part of the spatial spectrum. The best examples of digitally processed edge-enhanced images appear to give spatial resolution approaching 50 meters. The theoretical best that can be achieved using LANDSAT images, even with a combination of multiple views of the same scene in perfect registration, is about 55 meters - thus, part of the apparent gain in using edge-enhanced images must lie in the linear nature of the features studied. Highly linear, high contrast objects (such as roads and bridges) can often be discerned on LANDSAT even though they are much less than 55 meters wide. Resolution, needless to day, is not an easily defined and invariable thing.

These edge and line enhancement operations have to be used with some care. They tend to increase the amount of 'noise' in the picture. The result often looks grainy, with any imperfections of the original signal (due for example to electronic disturbances in the original recording) emphasized. This takes some getting used to by the interpreter who is working with an edge-enhanced product.

Other geometric enhancement procedures remove lines and edges from a scene, or smooth electronic noise from an image. The best known example of this is the 'de-stripping' procedures, that remove the effects of variable detector sensitivity from LANDSAT images.

4. Spectral band combination for image enhancement

The operations described to this point all work on a single spectral band - they apply basically to black-and-white images. The image enhancement procedures that we describe in this section apply only to multi-band images, and are meaningless if only a single spectral band is available.

The most powerful of the multi-band enhancement processes is probably band-ratioing. Unlike most of the operations referred to already, there is no optical analog of this procedure.

When we have two or more well-registered images of a scene, such as those provided by the four-channel MSS on LANDSAT, corresponding individual pixels in each spectral band can be compared. We can take the grey level values in each band associated with particular pixels, and divide, subtract, multiply, or otherwise combine these numbers. For each operation the result is a new grey level value, which when normalized to lie in the range 0 to 255, can be used as the grey level associated with the pixel and hence can be used to create a new image.

In band ratioing, the grey level in one spectral band is divided by the grey level in a second band (being careful to avoid division by zero), the resulting number is normalized to lie in the appropriate grey level range, and a 'ratioed' image results. From the four spectral bands of a LANDSAT image, 12 images can be made using pairs of bands (we regard the image derived by taking Band N/Band M as different from that found taking Band M/Band N, though they are very closely related). Most commonly, only three are created since the result can then be displayed photographically as a three-color composite.

The great advantage of using band ratios is that *the ratio is independent of the overall illumination level of the scene.* Thus, differences in appearance arising solely from the solar aspect angle of the terrain disappear from the ratioed image, and similar ground covers or lithologies are more readily compared.

Other nonlinear band cominations have been used for particular purposes. Particular band ratios (mainly MSS Band 4/ Band 5) are used for mineral surveys and a more complex band combination and ratio, termed the Transformed Vegetation Index (TVI), is used as a measure of bio-mass.

The flexibility possible via multi-band combinations of this type is enormous. Exploration of these methods is still in its early stages, and will undoubtedly be a fruitful source of research activity for years to come.

A number of other standard image processing techniques, such as Fourier Transform methods, will not be discussed here. They can certainly be applied to resource work, but in our opinion they have not proved to be a useful tool in the main areas of earth resources application.

B.   Digital Image Processing Methods for Direct Information Extraction

In the techniques described in Section A above, the end product was always an image, designed for interpretation by a disciplinary specialist. In this section, we describe processes in which the end produce is not an image for viewing, but basically a *classification* of a scene. It happens that the *results* of the classification are often presented in the form of an image, but this is for convenience.

Classification procedures invariably work with more than one band. Single band classifiers have been developed that utilize the spatial texture of the image, but these are of practical interest only when they are part of a multi-band analysis. The logic is simple: if you are trying to classify a scene automatically, you want to use all the information you can get. Since different spectral bands contain different information, it is desirable to use all available bands. Computer procedures can work as well with ten bands as with one (leaving aside questions of computer time). Humans, on the other hand, are happier with three bands or less - the most that can be effectively displayed with the use of color.

A large number of classification methods have been programmed for analysis of multispectral digital data. Their names are forbidding to the newcomer: maximum likelihood, with and without a prior probability; table look-up; density range selection with and without priorities; discriminant analysis with a variety of decision boundary selectors; and so on. The user can choose from an extensive menu, programmed for a variety of general or special purpose computing devices. The problem is one of choosing rather than finding, and the literature is extensive.

Despite this diversity, all the techniques have a common objective. They seek to associate pixels (which themselves represent small ground areas) into groups. This grouping is performed in terms of the similarity of pixels' spectral reflectances in each of the available bands. Pixel location on the ground is very much a secondary issue, though some algorithms seek to use the information regarding pixel contiguity in deciding the class to which

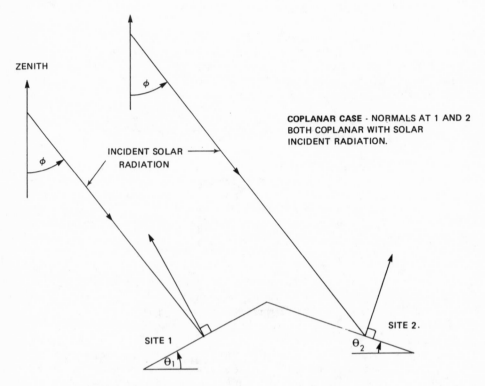

Incident solar radiation at site 1 = $I_0 \cos(\phi - \theta_1)$

Incident solar radiation at site 2 = $I_0 \cos(\phi + \theta_2)$

For an isotropic scattering surface, identical at sites 1 and 2, and neglecting the effects of scattered sky-light, the ratio of reflectances is:

$$R[\text{Site 1/Site 2}] = \frac{\cos(\phi - \theta_1)}{\cos(\phi + \theta_2)}$$

This is about 1.1 for $\phi = 30°$, $\theta_1 = \theta_2 = 5°$.

Figure 4. Grey level variations produced by terrain slopes.

pixels belong, especially pixels that cannot be assigned a group
unambiguously on the basis of spectral reflectances alone. The
pixel groups can be defined in advance ('supervised classification') or determined by the process of grouping ('unsupervised
classification'). When all or nearly all of the pixels in a scene
have been assigned to classes - which may be very difficult to do
in an unambiguous and physically meaningful way - it is easy to
calculate the fractional area of a scene occupied by a particular
class.

Most of the practical experience with classification of LAND-
SAT data has been in application to agriculture. The LACIE project
(Large Area Crop Inventory Experiment) will be discussed elsewhere
in this meeting, but we wish to note here that agricultural applications are in some ways the hardest, and in other ways the easiest,
applications of multispectral classification techniques.

First, why are they the hardest? In agriculture, we are often
trying to separate crops that for much of the growing season are
'look-a-likes' (for example wheat, oats, and barley). Differences
in spectral reflectance among such crops may be so small for most
of the growing season that very minor variations in soils, available moisture, or sun aspect angle obscure and overshadow them. The
multispectral classification technique that works very well in separating water from bare ground, or from forest, must be a much more
refined and delicate tool if·it is to distinguish among very similar
agricultural classes - the crops that we are interested in.

Then why are agricultural applications also described as the
easiest? Because gross variations in slope, soil depth, altitude,
and fertility, that are very common in rangeland situations, are
much rarer in agricultural areas. Agricultural sites, for prefer-
ence, are level, on good soil, with good drainage. That is why
they are prime agricultural sites in the first place.

Workers who have tried to apply multispectral classification
methods to non-agricultural lands have been dismayed by their poor
success. A group in the U.S.A. is currently attempting to classify
a precipitous LANDSAT scene that has 23,000 feet of terrain relief
within it - and they are hoping to develop unique signatures for
particular types of ground cover.

We can assert with confidence that the project is a very difficult one. The reasons: first, a terrain that has five degree
slope changes within it will cause variations in overall reflected
illumination (from varying sun aspect angle) more than enough to ruin
a multispectral classification algorithm (see Figure 4). The variations in overall illumination can be eliminated by use of band ratios
instead of the original spectral bands in the classification program,
but this has not yet been done. Second, reflected solar illumination
that travels through the atmosphere to ground level, and back to the
spacecraft, has a radically different spectrum from radiation that

travels only to 23,000 feet and is then reflected to the spacecraft. Again, this is more than enough to ruin a multispectral classification scheme, unless an explicit allowance for altitude variations in the scene is made at the outset - for example, by carrying a topographic image of the area as an additional "pseudo-spectral band" and incorporating it into the multispectral classification process.

Other information can also be carried as additional bands and used in a multispectral classification scene. For example, a binary image (two grey levels only) can be created to delineate a particular portion of a scene, and used within the multispectral classifier to restrict the classification to that part only. We would not recommend it in such a case - there are easier methods of restricting attention to part of an image, and most multispectral classification schemes are built with assumptions about the way in which pixel groups will occur. However, in principal *any* information that can be presented as a grey level array (soils, temperatures, altitudes, aspects, and so on) and registered with other grey level arrays can be used as one 'spectral band' of a multispectral classification. This discussion seems to have taken us far afield from our objective - the design of the next generation of operational earth resources satellites. It is, however, relevant from two points of view. First, it points out that the important problem of computerized scene classification needs a much more subtle solution than a canned computer program. Second, it has important implications when we consider how and when the processing of acquired data should be done - which is discussed in the next section.

Computational Considerations

It is not realistic to discuss digital processing of earth resources data without explicitly considering data rates, data volumes, and computing capacity. It is also not realistic to indulge in a simple-minded 'bit count' and leap to a conclusion.

An argument is often presented as follows. Assume that the next generation of earth resources satellites has seven spectral bands (somewhere between LANDSAT and the S-192 of Skylab). Assume that it has a ground resolution (i.e. pixel sixe) of 30 x 30 meters, and that it has the same orbital altitude and ground cover swath width as the present LANDSAT - these are reasonable assumptions. Assume that we have a Tracking Data Relay Satellite System (TDRSS) which enables us to get information down to the ground for forty minutes in each orbit (no night-time coverage, and nothing above latitude 70°). Finally, assume a detector sensitivity that permits us to use 1024 grey levels, rather than the present 64 on LANDSAT.

Take out your pocket calculator, and do the arithmetic. You will find that you have 3.2×10^{12} bits per day that need to be

processed, representing about 4.6×10^{10} pixels. Suppose you
intend to perform all digital processing, and you use a computer
that can process 50,000 pixels per second - a good rate, considering
that this includes geometric and radiometric corrections, plus
any enhancement or classification operations that are desired.
Then you readily find you need 11 such computers, running 24
hours a day, to keep up with the processing load.

Such arguments are often presented to justify the acquisition
of equipment - especially computing equipment. What is wrong with
them? Nothing - as arithmetic. However, they have little contact
with reality.

First, will a satellite transmit all the time? Not if it is
sensibly designed. The next generation of satellites should have
a look ahead in the orbital track, that monitors cloud cover and
does not attempt to send back images that are solid clouds. Even
if the satellite does not perform this function, it is no great
feat to filter on cloud cover at the ground receiving station.
Apart from clouds, what about water? Three-fifths of the Earth
is covered by oceans, and 18-day repeat coverage at 30 meter
resolution of all the oceans is hard to justify. If the oceanographers object, let us rephrase: It is perhaps reasonable to
transmit and store this huge mass of data on the oceans - but it
is highly unlikely that we will want to process all of it, especially using enhancement operations or multispectral classification.

Much the same point applies to *all* the data that the next
generation of earth resources satellites will send back. A very
small fraction of it will be processed in detail. However, this
does *not* mean that only a very small fraction should be received
and stored. It is not necessary to read every item in a set of
encyclopedias before they are considered useful or worthwhile,
and in the same way a reference library of earth resources data
has tremendous value - even though only a small part will be
processed, and that part not known in advance.

We can expect that *much less than one-tenth of one percent*
of all the data sent back by our next generation of earth resources
satellites will be examined in detail. A single large computer
could handle this load comfortably, if it were doing straight-
forward 'number-crunching.' In practice, present processing
techniques need a lot of modification to a standard series of
operations, before such a computer could be utilized with its full
potential. There is no way of organizing the input data, or
specifying the required processing to allow a computer to run continuously at its maximum processing rate, *unless* the operation
is programmable completely ahead of time - for example, such pre-
programming is possible for the standard LANDSAT radiometric and
geometric corrections.

We are led to the conclusion that the deciding factor on computational needs is not the data rate of the satellite data acquisition or transmission system; it is the organization of the ground processing, especially the complex decision-making process that decides which scenes will be processed, and how they will be processed. These decisions are unrelated to spacecraft design, or to telemetry systems. They are made in terms of the *applications* of the data. The number of applications and the processing needs of specific applications are the only real factor in sizing and designing the data handling problem.

This conclusion is not new. Unfortunately, it is easy to work with the tangible and measureable idea of bit rates, and hard to work with the unknown and unquantifiable idea of user requirements. Thus, we find lots of discussion of user requirements - but sizing and design are based on data rates and numbers of bits per day.

One additional comment before moving on to discussion of operational satellite systems. A number of special purpose computers and processing devices such as the General Electric Company Image - 100 and the Bendix MDAS take advantage of various forms of parallel processing logic to speed up image classification and analysis. This makes a big difference to the actual calculation time, but leaves untouched the central problem of input organization and specification of required processing. No purely hardware developments can solve that problem; in fact, they tend to make it more apparent as the computational element becomes less of a limiting factor.

Design of Operational Processing Systems

Experience in the digital processing of LANDSAT data provides several clues to the future. First, much more digital processing for image enhancement has been performed in the past four years than most people expected. Second, the sophistication of these processing procedures, described or at least hinted at above, has steadily increased and is still increasing - if anything, more rapidly than ever. Third, we are still on the steep part of the learning curve for classification of LANDSAT images, for incorporation of other image format and non-image format information into the analysis of LANDSAT data; and for combination of multiple source remotely sensed data. Fourth, users are learning that more information is recorded by LANDSAT than can be accommodated on a single piece of film, and this information to be fully used calls for digital processing.

We will by-pass the question of the continuity of the Earth Resources Program - a subject sufficient for several papers in

its own right. We will assume that there will continue to be
satellites - improved versions of LANDSAT-1 and 2, both spatially
and spectrally, and using multispectral scanners in a similar way.
For our discussion, we will consider only the processing of received
data — by-passing the important questions of ground receiving stations, tape recorders versus TDRSS systems, or rights to data on
an international basis.

Scientifically and economically, processing of data from
these satellites represents a tremendous opportunity. Organization-
ally, there will be problems.

Information extraction has been defined to be a user's problem
with LANDSAT data. U.S. Government agencies have carefully steered
clear of that area and have provided only image products, digital
tapes and general training in remote sensing and LANDSAT image
analysis. Now we face a situation in which an increasing number
of users are aware of the advantages of special processing for
particular disciplines. Digital products are in demand, despite
their relatively high price. The standard Sioux Falls image product, highly esteemed four years ago in its own right, is now esteemed by the more sophisticated users as a valuable *starting point* -
to be replaced in detailed work by digital products created for
particular disciplinary work.

*In a sentence, the trend for detailed interpretation work is
away from the standardized product and towards the tailor-made.*

Making these special products is a non-trivial task. It would
be unreasonable to expect every user of LANDSAT data to develop his
own programs and techniques for this, as well as extremely wasteful
of time and money. How then should the creation of these tailor-
made products be handled?

There are several possible answers. First, and perhaps the
path of least effort, each country with the financial and technical
resources can develop a processing facility as part of its government. Such a facility would have the capability to create tailor-
made products, including the combination of LANDSAT data with meteorological satellite data and country-specific data such as political boundary maps, demographic data, etc. This approach is wasteful
in the sense that there will still be a lot of duplication of development effort, but politically attractive, since a country can
still have the feeling of an autonomous program.

A second approach - and in this author's opinion a superior
one - is the creation of an international service center. There
are several ways of doing this. First, under the auspices of some
existing group which already has a worldwide mandate - the one that
comes to mind is the United Nations, in a role similar to the

existing FAO. Second, through creation of a private industry
consortium, in the form of a new corporation. Here, the example
is ComSat. Products would be offered to all countries of the
world, on a pay-for-services basis, in a manner analogous to the
use of communications satellites. There will be difficulties here
at least initially, when users find that they have to pay more for
what is now very inexpensive. However, that problem will inevitably arise at some time, whether services are offered through governments or corporations, because the cost of specially made user
products is high. It is unlikely that operations such as Sioux
Falls can continue long except on a break-even basis for products.
Such a break-even basis becomes more difficult to achieve as the
complexity of the product increases, unless there is also a big
increase and re-structuring of prices.

A new service corporation would be less troubled by this last
point, since comparison with earlier product prices would not be
so easy.

I would now like to analyze the idea of an industrial group
providing services in earth resources in a little more detail,
for two reasons. First, I want to present an industrial viewpoint
of the earth resources program, which may not otherwise be too
strongly represented at this meeting. Second, the timing is appropriate because the direction of the U.S. Earth Resources program
is not firmly fixed, and alternatives can still be seriously enter-
tained on developments after LANDSAT-C.

It is appropriate though perhaps unnecessary to add that the
main purpose of the following comments is to promote discussion
among other workers.

It is unlikely that any consortium of corporations at this
time would accept profit-and-loss responsibility for taking over
the entire operational earth resources program. First, the market
is completely unsized if products are processed to make a profit
(this was never the basis for Sioux Falls pricing). Second, the
considerable financial investment in the space segment would deter
many groups, because of the great impact of a single launch failure.
Third, it is likely that the development of new satellite systems
would render obsolete or non-competitive a series of satellites
based on LANDSAT-1, and research work to explore the next generation
of satellites can only realistically come from government programs.

Pursuing this line of thought, there seems to be a logical
distinction between the space segment and the ground processing
and dissemination segments. This separation has already been recognized with the U.S. program in terms of the roles of NASA and the
Department of the Interior. A system in which NASA's role remains
the same (just as NASA continues to provide launch services for

communications satellites) for the space segment, but product creation, sales, and dissemination becomes the function of an international corporation, appears to be feasible. Development of new technology - better scanners, spaceborne radars, and so on - would continue to be provided by NASA and other national or regional space programs.

All of this appears consistent with and complementary to the recent suggestions of the U.S. Senate's Aeronautical and Space Sciences Committee (Reference 2). Timing, costs and the inter-relationships of governments and industry need to be explored in a great deal more detail, but that process has already begun. (Ed. note: See the paper by J.J. Gehrig in this Proceedings).

Let us raise the natural questions, and then seek to answer them.

Q. If the responsibility for the creation and distribution of earth resources satellite products were in the hands of an industrial corporation, could it get into financial troubles and so no longer maintain a suitable level of service?

A. Yes, it could. The corporation would in no sense have an exclusive right to earth resources satellite data. If service was not up to standard, customers - and governments - would move to alternate sources of supply, perhaps by creating their own centers. Such a danger has always existed with Comsat, and no one seems unduly worried.

Q. What would the role of the existing U.S. service center be in such a new structure? Would Sioux Falls continue in operation, or not?

A. There are two answers here, that both have logic to them. First, however, it should be pointed out that Sioux Falls is already something of an anomaly, in that it is a part of the U.S. Department of the Interior that acts as a service center not only to all other branches of the U.S. Government, but also to the rest of the world. Sioux Falls could continue to serve the United States Government's interior needs, with other U.S. needs and all non-U.S. needs served by the suggested new international corporation. Alternatively, the functions of Sioux Falls could be subsumed by the new organization, including the U.S. Government supply of Earth Resources pro-ducts. Such a decision would depend largely on the actions of the U.S. Government's legislative branch, but should take into account the economics resulting from a single combined operation.

Q.   Would the new service corporation benefit from research work
      done by each country on the development of new remote sensing
      technology, free of charge?

A.   Broadly speaking, yes. However, the corporation would have
      a strong incentive to perform research of its own in the
      processing of new remotely sensed data, since revenues would
      depend on creating as large and diverse as possible a series
      of useful and saleable customer products. Thus there should
      be corporate 'matching funds' applied to ground processing
      and distribution, to parallel NASA and other space agency
      instrument development.

Q.   How much capital would be needed to form such a corporation,
      and how would it be provided?

A.   Forty to sixty million U.S. dollars, over a seven year period
      (the author's personal estimate, unblessed by other industrial
      inputs). Capital could be raised in several ways. Firstly,
      by sale of stock to governments of participating countries.
      At first sight this may appear attractive, since a truly
      multi-national corporation would be the result. In practice,
      it would probably be unworkable because the corporation would
      become a political football. Presumably the Board of Directors
      would be drawn from the participating member countries, and
      division of interests would be a frequent problem, resolved
      by vote. Another method is to invite investments by the
      industrial corporations of the world, through what would
      amount to a public offering. There are clearly risks in
      investing in such a program, but chances are high that many
      groups would be interested in participating - too many, in
      all likelihood. A mechanism for restricting eligible groups
      would be necessary, perhaps by requiring other space and
      earth-resources related activities.

Q.   How long would it take to have an effective group offering
      earth resources services and products?

A.   At least three years, and at most seven years.

Q.   Where would such an earth resources service corporation be
      head-quartered?

A.   Either in North America or in Western Europe, because of the
      access to equipment and personnel. Why not Japan? No real
      reason, except that marketing to and spread through the rest
      of the world would be more difficult - at least to the Western
      eye. The Spacelab/Space Shuttle cooperative effort is already
      making the space program a more multi-national and interna-
      tional effort, and a truly international approach to the use

of earth resources satellite data would take that a step further.

Q. Is it too soon to consider such an international industrial
    consortium, since necessary markets are not well established
    nor the technology completely developed?

A. It is not too soon and if nothing is done, it may well soon
    be too late. Each country is obliged, if no service exists,
    to create its own. Once entrenched, such national programs
    would be impossible to stop and become a real impediment to
    effective international programs. That which is marred at
    birth time will not mend.

Conclusions

Digital processing of earth resources data did not begin with
LANDSAT, but it matured and proliferated rapidly when that program
began. Processing experience over the past four years suggests
that computerized, tailor-made products, for both image enhancement
and information extraction, have great potential and are already
in rapidly increasing demand.

To satisfy that demand, a new type of service will be needed,
one that produces products in terms of specific end-user needs,
rather than in a limited and circumscribed set of standard formats.
It is not clear who is best suited to provide such a new, more
flexible service - government, international government-supported
agencies, or industry.

The idea of a new industrial group has already been proposed;
the questions that must be answered before such a group could come
into being have been asked here, and one set of answers provided.
More detailed answers, giving due weight to practical limitations
of finances, government policies, available existing institutions,
and political and geographic constraints, are the next logical step.

References

Earth Resources Technology Satellite Data Users Handbook, Goddard
    Space Flight Center Document No. 71SD4249; 1972.

"An Analysis of the Future LANDSAT Effort". Staff report prepared
    for the Committee on Aeronautical and Space Sciences, United
    States Senate; C.W. Mathews, U.S. Government Printing Office,
    Washington D.C., 94th Congress, 2nd Session, August, 1976.

REQUIREMENTS FOR REAL-TIME OCEANOGRAPHY

FROM EARTH OBSERVATION SYSTEMS

Robert E. Stevenson
Office of Naval Research/Pasadena Scientific Liaison Office
University of California, San Diego
La Jolla, California U.S.A.

Spacecraft systems have three unique capabilities in collecting data from the ocean: (1) frequent repetitive measurements (daily to weekly), (2) the synoptic view, and (3) real-time data transmission. On the other hand, there are two basic disadvantages; the measurements are only of the sea surface and atmospheric parameters interfere with the quality of the measurements. Consequently, the most suitable ocean conditions to examine from space are gradients; in color, temperature, and surface texture.

Over continental margins, gradients are usually sharp; that is, there are measurable differences over a few tens of meters. They are created by a variety of causes, ranging from river discharge through upwelling to the turbulence resulting from internal waves. Gradients in these waters exhibit daily changes in position and intensity, but also have strong seasonal differences. In some waters, the gradients may occur only in certain seasons.

Because most of the present-day mineral and biologic exploitation of the ocean is in marginal seas, the observation of such ocean areas would have an immediate and obvious impact.

In the open ocean, gradients are less sharp than in marginal seas, with measurable differences being measured over tens of kilometers. The most common scale has dimensions of 100 km and frequencies of tens of days. These turbulent features interact with currents and central ocean gyres to result in the 100-km scales that are more familiar than the meso-scales.

It is clear that ocean currents and gyres lend themselves to classic measurement techniques, assuming an interest in ocean

Figure 1

climatology. If the interest is in forecasting or the transfer
of thermal energy, however, classic techniques do not provide
data at appropriate rates or densities. It is recognized, therefore, that the late twentieth century oceanographer has reached
a state of knowledge whereby any significant further advance in
oceanography can come from data gathered in time and space scales
possible only from orbiting spacecraft.

For some research, the presentation of the space-acquired
data may be delayed substantially from the time of collection.
An impressive advantage of a satellite system, however, over
any other data-acquisition and transmission scheme is the capability of literal, real-time, or quasi-real-time (that is, minutes)
data presentation to the user.

As a result, a whole new direction in oceanographic research
is now possible. Waxing and waning of oceanographic turbulence,
with life histories of hours to days, may now be studied in fine
detail. Furthermore, the sea-going practitioner; the sailor, the
fisherman, and the commercial marine explorer, may have in his
hands, the conditions of the ocean through which he will sail today.

There is, therefore, a wide-ranging suite of satellite data
presentations that can meet a variety of users. The prime system
requirement is to develop data-reduction/presentation and modes
to match the satellite data transmission rates and the desires of
the users. The development of such a system must be matched with
carefully scheduled oceanographic experiments to key the sea-surface
emissions with the related oceanography.

As a typical example, the accompanying Figure shows a photograph and infrared scan of an ocean front area off Spain taken
within one hour of each other. The photograph was taken by Major
General Thomas Stafford, Spacecraft Commander of Apollo during the
Apollo/Soyuz Test Project, on 20 July 1975. The infrared scan is
from the U.S. Defense Meteorology Satellite orbiting over Spain
one hour before the Apollo orbit. In the infrared image, the
computerized depiction of water temperatures covers a 6° C. range
centered around 18° C. (The lighter shades indicate water colder
than in areas of the darker group).

In the photograph, the straight feature in the sun's reflection
in the upper left is a current shear on the edge of the Tarif Eddy;
the north-south trending, thin dark feature is an internal wave
sequence marking the warm-cold boundary of the Huelva Front.

Both are representative of ocean data depiction that can be
made available in near real time.

THE USE OF SATELLITE DATA IN OCEAN ENGINEERING

P.G. Teleki
U.S. Geological Survey
Reston, Virginia U.S.A.

Introduction

Ocean engineering is a composite of several disciplines including physical oceanography, chemistry and metallurgy, marine
geology, structural and mechanical engineering, soil engineering
and system analysis. An integrated approach in designing offshore
structures, selecting safe locations for them, maintaining and
operating them efficiently, and at the same time avoiding unnecessary hazards to the environment or to man, is just emerging. Perhaps
the most significant drawback to such an integrated approach is
the dearth of adequate environmental data; adequate in the sense
of coverage and continuity. Extraordinary costs accompany most
marine operations requiring data collection over long spans of
time; most of these expenses are related to research and development
of reliable instrumentation, to the problems of maintaining sensors
over protracted duration in the ocean and often to the complexity
of reducing and interpreting large data collections.

One of the promising means to reduce costs of acquiring marine
data is to apply remote-sensing techniques. Experience with the
ITOS, NOAA, and Landsat satellites demonstrates the utility of
earth-viewing sensors in observing dynamic changes on the earth's
surface or the variations in the planet's atmosphere. The French
EOLE satellite proved by tracking drifting buoys that ocean currents
can be charted accurately 1/. The dynamics of oceanic phenomena
are comparable with those in the atmosphere; the scales of motion
(length and time) in the oceans are approximately one order of
magnitude less than similar atmospheric scales. Repeated coverage,
at synoptic scales offered by satellite sensors, is well suited
to the study of mesoscale eddy motion, an elementary step toward
the understanding of ocean circulation.

407

Aerospace technology for ocean applications is, however, near the same level of development as the subject of ocean engineering itself. Both technologies are now developing side by side; the connection between them is still tenuous. Conceptually, the utility of spacecraft instruments in viewing the ocean surface has been realized, but the kind of data and the format in which these are useful to the engineering community are as yet poorly defined. The emphasis is still on basic, mainly physical oceanography. The objective at this time is to understand the gross variations in the dynamics of the ocean surface, one parameter at a time, rather than to obtain integrated answers to site-specific problems. The latter should follow naturally. In fact, the progress in instrument development for space oceanography has been quite rapid. Only seventeen years ago the Committee on Oceanography of the National Academy of Sciences issued its first call for applying remote-sensing techniques to the marine sphere, stating: "... it seems likely that aircraft can be used effectively for some research and surveys on the open ocean, particularly for studies involving the joint problems of oceanography and meteorology" 2/. The promise of side-stepping arduous procedures conventional to oceanographic experiments is reflected in the Committee's statement: "Truly synoptic pictures of selected parts of the ocean are needed. For these, the data must be collected quickly, processed rapidly and be ready for interpretation within a few days after collection".

If one queries the oceanographic community, the importance of synopticity is probably exceeded by the importance of repetitivity about oceanic parameters and processes. A similar query of the practicing ocean engineers would result in much introspection, consternation and skepticism because the potential benefits of remote sensing in ocean study have not been demonstrated. The need to bridge the technological communication gap is real because many nations have made claims to larger sectors of their "territorial seas", have begun to expand exploration activities on their continental shelves, and instituted engineering programs in the oceans. A recent study by the National Advisory Committee on Oceans and Atmosphere found that "The civilian effort in ocean engineering, both public and private, appears to be undersupported in view of the rapid expansion of activities in the ocean and little or no reserve of technology to provide the technical alternatives to meet the requirements which thus develop. Use of the oceans is expanding faster than is the knowledge being provided to support it. While the difference in rates of growth may be temporary, it exists now, and creates a gap" 3/. Private industry, particularly in petroleum and shipping, has been meeting practical engineering problems case by case, but the general opinion is that the success criteria for safety of structures is closely related to the tendency to "over-engineer", i.e. design for environmental conditions much worse than are actually occurring in order to reduce

risks. The high cost is partly related to the harsh physical environment, in which risk factors are high, but in part these costs
are directly related to our inadequate understanding of climatological factors and our reaction to "over-design" in face of it.
Naturally, excessive capital investments related to "over-design"
are reflected at the consumer level. Excess costs are not only
associated with design of structures, such as exploration rigs,
production platforms, pipelines, offshore terminals and floating
power plants, but with operations as well. For pipe-laying operations in the Gulf of Mexico, an economic study indicated that 15
percent of daily labor costs could be saved given better environmental data and prediction 4/. In areas that have a high probability of extreme wave and wind conditions (North Sea, Gulf of
Alaska, Georges Bank) or where ice is a factor (Bering and Beaufort
Seas) cost increases because of prolonged standby operations are
considerably higher.

The main issue is generating low-cost oceanographic data.
These data, to be useful to both the oceanographers and to ocean
engineers, must be acquired in a near-continuous manner, that is,
at short-term cyclic rates in order to observe the temporal fluc-
tuations and trends of the dynamic ocean. Point measurements are
useful only if a site has been selected for construction of an
engineering structure; more critical is the need for spatial
information to improve the chances of choosing a safe location for
the site. The data so required are statistical in nature for the
purpose of assembling climatological charts on winds, waves, currents, sea-surface temperature, and sea ice.

The task at present is not to devise new space techniques to
help the ocean engineer. Instead, the real need is for education
that is a viable transfer of existing information into terms useful
to ocean engineering, demonstrations for specific applications, and
the collection of an adequate data base to use in hindcasting and
forecasting. Improvements in maritime communications and in more
frequent relay of data via Data Collection Platforms (DCP's) from
fixed and floating instruments represent increased utility to offshore construction and operators, and much progress has been made
in this area. This is not the case in the transfer of oceanographic knowledge, existing or anticipated, to the engineering community. Remote measurements from space are no less significant than
in the ocean itself, acoustics being a good example. Nonetheless,
the skepticism about the validity and usefulness of space-derived
data has permeated the engineering community. The cause can be
found in the lack of well-conceived demonstrations, tying the
parameters measured by satellite sensors to parameters measured at
the ocean's surface, Rectification of this problem is just beginning, particularly under the aegis of the Seasat-A program.

Data Needs in Ocean Engineering

Three ocean-engineering applications stand out from a wide ranging list: offshore oil and gas exploration and production, shipping, and ocean mining. Experience with designing rigs and ships is extensive and well documented. Ocean mining is emerging as the new frontier in ocean engineering. Design forces and loads affecting safe and economical operations are related to meteorological, oceanographical, and geological variables. Meteorological variable include average and extreme (worse case) wind conditions, storm propagation (probability and frequency of occurrence, tracks and duration), and icing of structures resulting from differences in air-water temperatures. Such factors are critical concerns to ship traffic (routing), operation of offshore structures and to "downtime". Oceanographical variables include forces due to surface-gravity waves (including tides), currents, and the shearing movement of ice fields. These factors are necessary input to the calculation of dynamic response of a structure as well as to fatigue analysis. Climatological data on the range of frequencies and directions of wind waves and swell and on the velocity of currents is important to modelling oil-pollution trajectories. Dynamics of sea-ice motion, thickness and persistence (from first-year ice to multi-year ice several meters thick) are factors in the stability analysis of rigid engineering works and in ship routing. Geological variables include rate, magnitude, and effective depth of sediment transport. Such variables must be determined in order to avoid excessive scour around pile foundations and near buried or surface pipelines. The stability of the seafloor is another consideration. Related factors include the bearing capacity of soils, fault and fracture zones, and seismic risk.

Perhaps more than any one parameter, the interactions of several forces are most difficult to gauge, interpret and extrapolate to other areas and conditions. The boundary flow areas of air-and-sea and sea-and-seafloor are least understood at this time; these are the regions of interaction (energy and momentum transfer) where differential loading of structures affects their stability and longevity most.

The more narrowly defined field of coastal engineering has similar requirements. The activity is concentrated along another boundary, that which is interactive between the sea and the shore. In many ways the problems that engineers face in nearshore waters and on the beach have more immediate impact on man; in addition to safety and economics, social factors also play a dominant role in the coastal zone. Nonetheless, the highest priority is still associated with the requirement for climatological data. Much effort has gone into assembling a coastal wave climate data bank 5/, although it is still far from complete. Other processes like wind- and wave-generated coastal currents are yet to be examined systematically.

Trends in Oceanography

Signs indicate that global wave forecasting, which is about
a decade or two behind global weather forecasting, may be realized
by the 1990's. The aim is to understand the coupling mechanism
between winds and waves in the major oceans and make joint prognoses
operational. Some progress has been made in the northern hemisphere:
00, 24, 36, and 48-hour wind-wave forecasts for the Pacific Ocean
are regularly made, and the evaluation of the North Atlantic model
is underway. Progress in modelling the southern hemisphere has
been less than spectacular. The reason is the relative abundance
of data obtained by ships-of-opportunity along the heavily travelled
northern route and the sparseness of the same in southern oceans.
Existing observational data on waves, winds, visibility and currents
are being assembled for the World Meteorological Organization 6/,
thus assuring the beginning of a global data base.

Several major experiments on ocean dynamics, air-sea interac-
tion, and ice dynamics have either taken place - GARP Atlantic
Tropical Experiment (1974), Joint Air-Sea Interaction (1975), Joint
North Sea-Wave Atmosphere Program (1973-75), Mid-ocean Dynamics
Experiment I (1973), Arctic Ice Dynamics Experiment (1974-76),
Coastal Upwelling Experiment II (1973)-or are scheduled to take
place - Polar Experiment (1977-78), First GARP Global Experiment
(1977-78), Stormfury (1977-78), Monsoon Experiment (1977), and
POLYMODE (1978).

An interesting sidelight is that many of these experiments,
like GATE, AIDJEX, and JONSWAP were not strictly ship- or surface-
based operations. Planning and execution involved aircraft and
remote sensing instruments, both optical and microwave types.
Although aircraft have been used in missions either to test the
performance of sensors or for specific remote sensing applications,
their involvement as complementary data gatherers in major oceanographic experiments for the purpose of correlating airborne to
"surface truth" data is a new development.

To conclude that the apparent acceleration in the study of
ocean dynamics and the use of remote sensing from aircraft are
panacea to the problems of global oceanography is premature.
However, there are trends in spacecraft systems which will supplant
our knowledge with more frequently collected data of broader scope.
The combination promises firm benefits to the ocean-engineering
community, provided the information is transferred in a manner
acceptable to engineers.

Trends in Remote Sensing of Ocean Phenomena

Exclusive of aerial photography, one can consider techniques

to study oceanic processes from airborne platforms to have evolved
during the last 20 years. Many studies used active sensors, i.e.
instruments which emit electromagnetic energy and receive its
return from point or distributive targets. Because energy from
active sensors generally does not penetrate into the water column,
measurements are mainly of surface phenomena and processes, of
which sea-state and sea-ice received most of the attention 7/.
The main effort was in making radar measurements of waves to
determine their growth of height and changes in frequency spectra
as a function of wind speed and direction, incident angle, polarization and wavelength of the radar. Because of the wide variety
of wave conditions and radar types, standardization of the data
began with associating the frequency of the radar with the return
power in units of average radar cross section per unit area (σ_0).
The back-scatter measurements over wind waves gradually expanded
to include swells, but the mechanism for radar return from long
gravity waves is still debated.

Sea-ice studies benefited from both passive microwave
(radiometers) and active microwave measurements (altimeters and
imaging radars). These measurements were used to differentiate
ice types, to determine the age of ice flows, and height variation
in pressure ridges and to chart the movement of ice fields. Microwave measurements at varying frequencies have been used to assess
the water-vapor content (rain rate) in the atmosphere, salinity
and temperature of surface waters and wind speed at the surface.

Currents have been charted successfully by tracking drifting
buoys with a satellite. Indirect measurements using a radar
altimeter to detect changes in elevation because of geostrophic
currents promises a new approach.

Optical instrumentation, such as lasers, multi-spectral
sensors, thermal scanners, and sensors detecting fluorescence,
have been used to observe other oceanic phenomena. Reasonably
good, albeit qualitative, success has been achieved in charting
suspended sediments near the water surface, determining bathymetry
of sediment-free nearshore areas, tracing pollutants such as
thermal effluents and oil and determining the presence of chlorophyll. Up-to-date summaries for applying remote-sensing techniques
can be found in References 7 and 8.

                    Trends in Oceanography From Space

Space oceanography began in 1965 when the U.S. Navy established the Spacecraft Oceanography Project (SPOC) 9/. SPOC was
partly successful in that it defined the needs of the oceanographic
community, evaluated the capabilities of spaceborne instruments
and defined components of a dedicated program in satellite

oceanography. However, it did not establish a satellite program
for the U.S. Navy. Interest in the subject was promulgated by
the early success of meteorological satellites, such as TIROS,
ITOS, and NIMBUS. The oceanographers realized that major improvements in conventional oceanographic measurements was a slow
process and was severely constrained by the high cost of research.
SPOC's recommendations in the 1960's were identical to those today:
to identify, test, and evaluate techniques which can be applied
to earth-orbiting sensors so that they can provide useful oceanographic data, establish the reliability of space-derived data by
comparison with ground-based measurements and to develop and
test new techniques of displaying and using synoptic oceanographic
data on a global basis which can enhance understanding of the
properties and behavior of the oceans. Emphasis was on such
topical interests as sea-state, thermal conditions, currents,
water-mass differentiation, coastal mapping and shoal detection
and biological phenomena 10/.

The first concrete promise to develop a program leading to
satellites dedicated to ocean research, surfaced in the 1969
Williamstown Conference 11/. The catalyst was the National
Geodetic Satellite Program, whose scope was expanded to include
studies on the dynamics of waves and wind 12/.

Skylab was the first space activity in which oceanographers
had the opportunity to participate. Its Earth Resources Experiment Package (EREP) consisting of an X-band radar altimeter,
scatterometer, and microwave radiometer (S-193) and an L-band
radiometer (S-194) provided much useful data and some interesting
findings as well. Winds in the Pacific Hurricane AVA were measured
with a scatterometer 13/. A 15-meter gravimetric change in the
surface of the ocean over the Puerto Rico Trench was determined 14/.
Success with the Skylab altimeter gave impetus to launching GEOS-C,
a geodynamic experimental ocean satellite, in 1976, whose altimeter is collecting data on the ocean topography. These data are
not only useful in determining the marine geoid, but in detecting
currents and tides (by measuring gradients of the surface) and
obtaining rms wave height along the sub-satellite track.

The first satellite fully dedicated to ocean research is
SEASAT-A to be lauched in May, 1978. Housing a complementary
set of active and passive microwave instruments, together with
data processing modelling capabilities and underflight programs,
Seasat-A is designed to evaluate state-of-the-art microwave
technology to both marine and terrestrial applications. Seasat-A
should be viewed as the beginning of a series of satellites flying
microwave sensors and designed to achieve global coverage useful
to climate modelling by the late 1980's. The oceanographic
community has need for an *operational ocean-dynamics satellite
system* having sensors and data density adequate to update fore-

casts every 24 to 36 hours for a week in advance 14/.

Seasat-A instrumentation consists of a 5-frequency scanning microwave radiometer (SMMR), a synthetic aperture imaging radar (SAR), a scatterometer, radar altimeter, and a visible-infrared passive radiometer. Placed into a non-sun-synchronous, circular, 108° inclination retrograde orbit near 800 km in altitude, and orbiting the earth every 100 minutes, the satellite will proceed westward in steps of 18.5° at the equator. Global coverage every 36 hours is achieved, given the wide swaths of the instruments, for all but the SAR. Data from this sensor can only be read out in real time over receiving stations, planned to be located in Alaska, California, North Carolina, and Newfoundland. The express desire of the oceanographers is to obtain SAR data at additional sites especially in areas of intense storm activity (monsoons, typhoons, et cetera). Measurement capabilities of the Seasat-A sensors are shown in Table 3 of the paper by B.P. Miller, in this Proceedings.

In 1978, Nimbus-G will also be launched having an SMMR identical to that of Seasat-A and a multispectral scanner known as the Coastal Zone Color Scanner (CZCS). The latter instrument is specifically designed to detect chlorophyll. The major difference between the satellites is in their orbit; Nimbus-G will be placed into a polar configuration.

The era of the Space Shuttle is also approaching. Throughout the 1980's, several Shuttle missions are expected to be flown for oceanographic applications. There is much interest in using the Shuttle platform to test sensors in various configurations. An example would be a multifrequency, multipolarized SAR in which the look angle in range and azimuth could be varied to maximize the return from different classes of targets on the ground. The TERRSE study 16/, assessing the needs of the marine community, found that Shuttle could be most helpful in obtaining information in times of adversity: storms, inundation of coasts, river floods, pollution. Because the missions are manned, Shuttle is also attractive to oceanographers desiring to conduct real-time experiments, namely, maximizing sensor performance and density of data acquired over "targets-of-opportunity". These missions are not likely to be low in cost or capable of generating an even semicontinuous data set; both criteria are necessary for efficient ocean research and efficient forecasting.

The future of ocean applications of spacecraft data is guided largely by the desire of the scientific community to understand climate. Land-based weather forecasting has matured to an operational status, but global marine weather forecasting and wave forecasting can only be anticipated to this time. Our increased activities in the marine environment cannot now depend on the existing

data and their interpretation, these being too fragmentary, unstand-
arized, and questionable in quality. Safety at sea is an important
requirement met only by better prediction techniques. In a recent
report of NASA 17/, the climatology of oceans and its influence
on agriculture, optimum ship routing in these times of increased
frequency of major ocean crossings, the need to reduce accidental
oil spills, complex engineering problems on continuental shelves,
and the prediction of storm tracks, their magnitude and duration,
were identified as the major tasks to which remote sensing technology could usefully be applied.

Low-cost data acquisition, globally every 24 to 36 hours,
without interference from man or the elements is the first step
in this direction and space systems can help in this regard.

Outlook for Ocean Engineers

The trepidation of engineers in accepting new technology
has not helped with applying remote sensing to ocean engineering.
Until now, much of the data obtained from spacecraft was qualitative,
visually pleasing but not directly introducable to engineering
nomograms and manuals. Introduction of microwave technology is
now generating a series of countervailing arguments about instrument preferences, measurement techniques, and data interpretation,
even though quantitative data is becoming accessible. One can only
hope that the transition period to routine techniques will be short,
because national needs to develop ocean resources will not wait.

For engineering purposes, information on sea-state has highest
priority. Significant wave height (H_s), design wave height for
50 and 100-year storms, wave spectra for length and direction,
probabilistic assessment of design wave heights are all critical
to wave-force analysis. Of lesser importance are data on currents:
magnitude, direction, and distribution with depth. In ice-infested
areas (sea-ice, icebergs), dynamics of ice movement must be predicted
to reduce the chance of structural failure. Because waves and
currents interact with the sea bottom, evaluations of seafloor insta-
bility require these data. Most critical is the need for improvement
in the quality and quantity of climatological data, wave-climate
data in particular.

Conclusions

It seems appropriate to quote from a recent report prepared
by the Panel on Practical Applications of Space Systems for the
Assembly of Engineering, National Research Council 18/: "The statement from the 1967-68 study report that 'satellite technology has not
yet had significant impact on oceanography' is still true. This

does not imply that progress is entirely lacking. Much has been done to develop new instruments and to improve existing ones, *but this work has not progressed to an operational system. Aircraft were used full time to develop these instruments, but at present, no buoy or ship program is specifically operated to provide the sea truth data needed to calibrate spacecraft instruments and to aid in interpretation of signals from these instruments.* All too often, forecasted winds, non-real-time wave measurements, and a few laser profiles have been used for the calibration of the space sensors".

User requirements have been honed to a sharp point and there is no need to keep refining them *ad infinitum*. There is little contention that these requirements are universal, rather than varying from nation to nation. There can be no contention that in the development phase of a viable combination of oceanographic, ocean engineering, and space technology programs, costs initially will be high and will decrease gradually as satellite systems approach operational status. Therefore, international cooperation in the definition of tasks, and in constructing global monitoring systems must be assured, founded on shared responsibilities and shared expenses. Oceanography has largely been an international science and in face of the costs of improving our knowledge we must utilize all available resources, irrespective of nationality or origin. Depending on the outcome of the law-of-the-sea negotiations, ocean engineering could graduate to the level of international cooperation; the memberships of existing ocean mining consortia are indicators of this trend. Efforts by the (International) Engineering Committee on Oceanic Resources (ECOR) to coordinate varied ocean engineering activities around the world reflects this as well.

In the next 10-15 years, the best chance of success in achieving our objectives is the Seasat type of program. No assurance exists at this time that Seasat-A will be followed by Seasat-B and other satellites.

It behooves us to examine our options and decide on the course of events *now*, before the chances for an integrated program, designed and funded to assure continuity, disappear altogether. It requires international participation and commitments for periods longer perhaps than one is accustomed to, but the benefits are well known and worthwhile.

References

1/   Cresswell, G.R., "The French-Australian satellite buoy
            experiment". Australian Meteor. Mag., v. 21, 1973,
            p. 1-17.

2/   National Academy of Sciences-National Research Council, 1959,
         Oceanography 1960 to 1970, vol. 1, Introduction and
         Summary of Recommendation; NAS/NRC, Washington, D.C.

3/   NACOA, "Engineering in the ocean". Report by the National
         Advisory Committee of Oceans and Atmosphere, Washington,
         D.C., 1974.

4/   Econ, Inc., "Offshore oil and natural gas industry, case
         study and generalization". Seasat economic assessment,
         vol. III, Report 75-125-3B, Princeton, N.J., 1975.

5/   Harris, D.L., "Characteristics of wave records in the coastal
         zone". In R.E. Meyer (ed): Waves on Beaches and Result-
         ing Sediment Transport; Academic Press, N.Y., 1972.

6/   Hogben, N., "Environmental prarameters". ECOR Report to the
         4th Joint Oceanographic Assembly, Edinburgh, U.K., 1976.

7/   Huebner, G.L, "The Marine Environment". Ch. 20 in R.G. Reeves,
         A. Anson, and D. Landen (eds.): Manual of Remote Sensing,
         American Society Photogram., Arlington, Va., 1974.

8/   Mathews, R.E., editor, Active Microwave Workshop Report;
         SP-376 National Aeronautical and Space Administration,
         Washington, D.C., 1975.

9/   SPOC, Annual Reports, Spacecraft Oceanography Project,
         U.S. Naval Oceanographic Office, Washington D.C. 1965-
         1969.

10/   NASA, A Survey of space applications; SP-142, Space Appl.
         Programe Office, National Aeronautical and Space
         Administration, Washington, D.C., 1967.

11/   NASA, "The terrestrial environment: Solid earth and ocean
         physics, application of space and astronomic techniques".
         Proc. Conf. sponsored by Mass.Inst. of Technology and
         National Aeronautical and Space Administration, Williams-
         town, Mass., 1969.

12/   Apel, J.R. and Siry, J.W., "The earth and ocean physics
         applications program, ocean dynamics program". In
         Seasat-A Scientific Contributions; National Aeronautical
         and Space Administration, Washington, D.C., 1974.

13/   Cardone, V.J. and Pierson, W.J., "The measurement of winds
         over the ocean from Skylab with application to measuring
         and forecasting typhoone and hurricanes". Proc. 11th
         International Symposium on Space Technology Science,
         Tokyo, Japan. 1975.

14/   Hoge, F.E., "Expected SEASAT-A scientific results". In
        Seasat-A Scientific Contributions; National Aeronautical &
        Space Administration, 1974.

15/   Nagler, R.G., and McCandless, S.W., "Operational oceanographic
        satellites: Potential for oceanography, climatology,
        coastal processes and ice". NASA Office of Applications/
        Jet Propulsion Laboratory, 1975.

16/   TERSSE, "Definition of the total earth resources system for
        the Shuttle era". General Electric Co., Space Div.,
        vol. 1, Valley Forge, Pa. 1974.

17/   NASA, "Outlook for space". Report SP-386 to the NASA Admini-
        strator, Washington, D.C., 1976 .

18/   National Research Council, "Practical applications of space
        systems, supporting paper no. 8: Marine and Maritime
        Uses". Space Appl. Board, Assembly of Eng., National
        Research Council, Washington, D.C., 1974.

THE AVAILABILITY AND USE OF SATELLITE PICTURES

IN RECOGNIZING HAZARDOUS WEATHER

Stanley E. Wasserman
U.S. National Weather Service, Eastern Region Headquarters
Garden City, New York

Introduction

In the last few years tremendous advances have been made in providing real time satellite pictures to meteorologists. Concurrently, meteorologists have made significant advances in their operational use of these pictures. Satellite pictures, especially when used together with other types of observations, such as radar, are now one of the most important aides we have in detecting hazardous weather in its earliest stage of development.

In this paper we will describe the types of satellite picture products available to operational meteorologists in the United States. Then in very general terms we will review what a forecaster looks for as he examines each of the different types of current weather observations for clues that could lead to early recognition of hazardous weather. From this review we will see how satellite pictures fit in with the other types of observations.

Availability of Satellite Picture Products

Meteorologists at forecast offices in the United States receive observations from polar-orbiting satellites and Geostationary Operational Environmental Satellites (GOES).

Polar Orbiting Satellites

The operational polar-orbiting satellites are in a near north/south polar orbit at an elevation of about 1,500 km. These satellites provide operational meteorologists with a 1 km resolution

infrared (IR) image of their local area of interest once every
twelve hours. A 1 km resolution visible picture is also available,
but only once a day. As of September 15, 1976, NOAA-5 is the
operational polar orbiter. The picture-taking time for NOAA-5 is
approximately 9:00 a.m. and 9:00 p.m. local time.

   Pictures from the polar-orbiting satellites are relayed to
operational meteorologists via ground communications from the
Satellite Field Services Stations (SFSS) who in turn receive the
signals from a Central Data Distribution Facility (CDDF) located
in Washington, D.C. The satellite signals are originally received
at one of the command and data acquisition stations located at
Wallops Island, Virginia, Gilmore Creek, Alaska and the SFSS at
San Francisco, California, from where they are relayed by landline
to the CDDF.

   Polar-orbiting pictures are generally not widely used by
operational meteorologists in forecast offices in the eastern
United States because of two disadvantages that are not present
with GOES. These disadvantages are the large interval between
picture-taking time for a given area (12 hours for IR pictures,
24 hours for visible), and the lack of computer-generated grids
superimposed on the pictures. One big advantage of polar-orbiting
satellite pictures is the IR resolution of 1 km as compared to
the GOES IR resolution of 8 km. Another advantage is the coverage
of far northern and far southern latitudes that are out of view of
GOES pictures. GOES, however, does provide adequate coverage of
the United States.

   Because operational meteorologists in the United States do
not generally use pictures from the polar-orbiting satellite, they
will not be included in any further discussions here.

Geostationary Operational Environmental Satellite

   GOES is in a near circular orbit about 36,000 km above the
equator. At this altitude, the satellite and the earth's surface
rotate about the earth's axis in the same direction and in the
same angular distance and time. For these reasons, GOES can be
"parked" above any longitude at the equator. The current operational GOES satellite providing coverage of the eastern United
States and Atlantic Ocean, called GOES-1, is parked at about 75°
W longitude. A second satellite, located at 135°W longitude provides coverage for the western United States and Pacific. (The
latter is an experimental Synchronous Meteorological Satellite,
SMS. SMS-A and B were launched in April 1974 and February 1975.)

   Meteorologists at forecast offices receive the GOES signal
via the same route as they receive the signal from polar-orbiting
satellites. The signal is received at Wallops Island, relayed to

the CDDF, and then transmitted on ground lines to the forecast
office via an SFSS. There are six SFSSs in the United States.
An SFSS located in Washington, D.C., for example, serves fourteen
forecast offices in the eastern United States. Each of these
forecast offices have complete independence in their choice of
satellite product to be received from the SFSS.

There is a wide variety of both IR and visible picture products
available every thirty minutes from GOES. Superimposed on all these
products is a computer-generated grid, providing for easy location
of all features detected by the satellite. The maximum resolution
for IR pictures is 8 km and for visible pictures is 1 km. Images
are received on photographic paper approximately 210 mm by 215 mm.

In the picture distribution system there is a relationship
between resolution of visible pictures and geographical coverage.
To increase the geographical coverage to be included on a given
size paper, the visible resolution is decreased. For IR pictures,
the 8 km resolution is not changed when you vary the geographical
coverage included on a given size paper.

Figure 1, prepared by the U.S. National Environmental Satellite
Service, illustrates the geographical coverage as a function of
resolution for visible pictures. Boxes A, B, and C represent the
area shown on a 210 mm by 215 mm photo using visible resolutions
of, respectively, 1 km (1/2 mi), 2 km (1 mi), and 4 km (2 mi).
8 km (4 mi) IR images are produced for the same aerial coverage
as shown in Figure 1 for visible pictures. Each of the boxes,
A, B. and C on Figure 1 can be independently relocated up, down,
to the left or to the right so as to provide flexibility in the
geographical coverage for any given resolution.

Figures 2 and 3, also prepared by the National Environmental
Satellite Service, show the location of GOES-1 picture boxes (referred to as sectors) over the United States as of July 27, 1976.

Referring to Figures 2 and 3 a meteorologist in the eastern
United States for example, can now request every thirty minutes,
any one of the following sectors:
    DB-5, visible and IR
    KA-5, visible only
    K8-8, visible and IR
    DA-1, visible only
    DA-2, visible only

In addition, if these sectors do not provide the geographical
coverage needed, the required coverage can be made available by
request to the SFSS. Meteorologists located in the western United
States receive different sectors, including those that are developed
from the western satellite mentioned earlier.

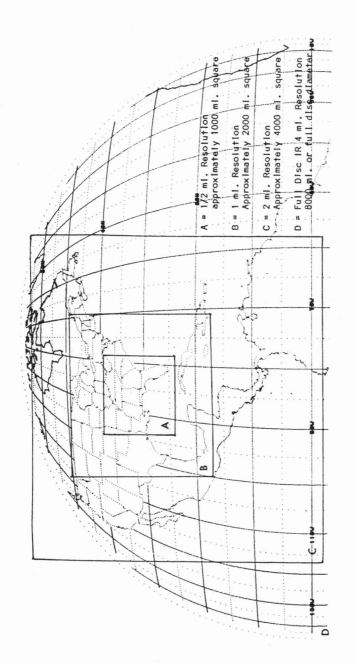

Figure 1.   Resolutions and geographical coverage. (SMS/GOES SUBPOINT 0.0N[1] 75.0W).
See Figure 3 of Wolff and Kesel for a picture of proposed global
coverage by GOES.

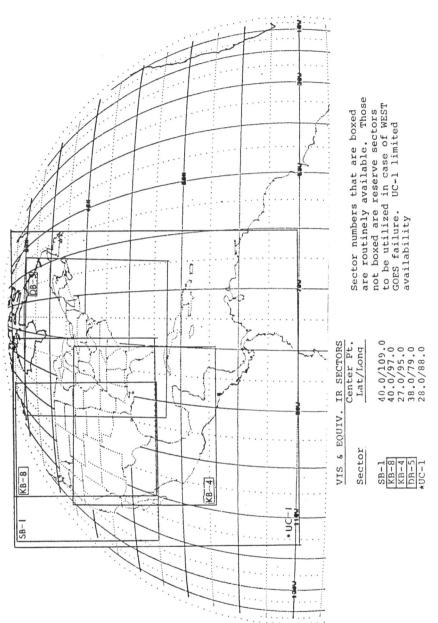

VIS & EQUIV. IR SECTORS

Sector	Center Pt. Lat/Long
SB-1	40.0/109.0
KB-8	40.0/97.0
KB-4	27.0/95.0
DB-5	38.0/79.0
*UC-1	28.0/88.0

Sector numbers that are boxed
are routinely available. Those
not boxed are reserve sectors
to be utilized in case of WEST
GOES failure. UC-1 limited
availability

Figure 2.   GOES-1 (East GOES) 1 and 2 mile resolution sectors. (GOES-1 0.0.N 75.0W).
Sector locations effective July 27, 1976.

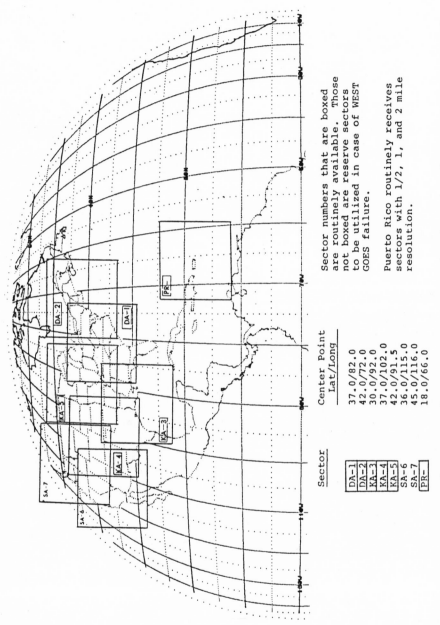

Sector	Center Point Lat/Long
DA-1	37.0/82.0
DA-2	42.0/72.0
KA-3	30.0/92.0
KA-4	37.0/102.0
KA-5	42.0/91.5
SA-6	36.0/115.0
SA-7	45.0/116.0
PR-	18.0/66.0

Sector numbers that are boxed are routinely available. Those not boxed are reserve sectors to be utilized in case of WEST GOES failure.

Puerto Rico routinely receives sectors with 1/2, 1, and 2 mile resolution.

Figure 3. GOES-1 (East GOES) half mile sectors (VIS only). (GOES-1 0.0.N 75.0W).
Sector locations effective July 27, 1976.

It can be seen from what has been described so far, there are many photo products available. For each picture-taking time, a meteorologist can choose one and only one product for reception at his office. Every thirty minutes an updated photo is available and a new choice of product can be made. The system allows for automatic continuous reception of a given picture product, or standard rotation of products, until such time as a change is requested by a telephone call to the SFSS. A rotation of products that has been popular is alternating visible and IR pictures, thus receiving one of each every hour.

A disadvantage that has been experienced with the system is that with only one communication line from the SFSS, a station is unable to receive visible and IR images that were taken at the same time. At best there is a thirty-minute difference in time between a visible and IR image. Furthermore, the inability to receive better than 8 km IR resolution has resulted in a reluctance of meteorologists to request daytime IR pictures. They prefer 1 km resolution visible pictures. The higher resolution is important in early detection of hazardous weather, especially small scale convective activity.

Recently there has been a considerable increase in the types of IR picture products available. Through a process, called enhancement, it is now possible to electronically increase the contrast between features of interest and the background. At the current time, a meteorologist can choose among ten different types of enhancement, depending on what features he wants to highlight. It has indeed become a task in itself for a meteorologist to decide which of the many photo products he should be receiving. Since pictures can only be received in real time (one cannot request something to be retransmitted), the meteorologist must prejudge what product will give the most valuable information for a given situation.

Meteorologists are being trained to understand what photo products best reveal those clues he is looking for in trying to detect the earliest stages of hazardous weather. Also, there is back up support from the SFSS to meteorologists at forecast offices. The SFSS provides forecast offices they serve with satellite information, by either phone or in a teletype message routinely prepared several times daily. Meteorologists at forecast offices, well aware of the better capability of the SFSS, often call the SFSS for consultation.

The SFSS which are staffed twenty-four hours a day every day with highly skilled satellite meteorologists, can and do receive many different types of satellite products at the same time. Some SFSSs also have (and others soon will have) available, in real time, movies automatically prepared from the satellite signals for

projection on a television set. Every thirty minutes the movie
is updated by automatic addition of the latest pictures and dropping
of the earliest picture. Motion presented in the movies provides
information that often cannot be detected in the still pictures.

Forecast offices also have a limited capability to view motion
of cloud fields. This is accomplished by manually registering each
picture on a disk that can rotate at a speed of about 100 RPM. When
viewing the rotating pictures with a synchronous strobe light, motion
becomes apparent. There has been successful application of information gained in using this device. Limitations include the necessity
of receiving the same photo product continuously, and the care and
time required to register the pictures on the disk.

Use of Satellite Pictures in Recognizing Hazardous Weather

Meteorologists receive observations from several different
sources. All of these observations are examined for clues that could
lead to early recognition of hazardous weather. Table 1, for example, presents the different types of observations operationally
available to meteorologists at forecast offices in the eastern
United States. Table 2 presents a list of most of the types of
hazardous weather with which a forecaster in the eastern United
States is concerned. More than one type of hazardous weather can
occur at the same time, as for example with a hurricane or cold
front.

In this section we will present a checklist showing some of
the procedures a meteorologist goes through as he examines the
latest observations for clues that could lead to early detection
of hazardous weather. The procedures listed are not meant to be
all inclusive. Rather, they are presented in general terms to give
the reader a better idea of how satellite picture information fits
in with the other types of observations available. Remember, too,
these observations are being examined by a meteorologist who has
many maps, prepared earlier, showing the current and numerically
predicated state of the atmosphere at many levels. He also has
numerical predictions of what the weather will be in the future.
In this paper, however, we are not talking about predictions beyond
an hour or two. Rather, we are talking about observing what is
happening now, or about to happen. This has been frequently referred
to as "Nowcasting". It will be seen that for some types of hazardous
weather, satellite pictures are not even referred to, but for most
types of hazardous weather satellite information is an important
part of the total information available.

Table 1.
Operationally Available Types of Weather Observations

1. Hourly and special surface observations from network stations.

2. Hourly and special radar observations.

3. Six-hourly synoptic surface observations.

4. Upper-air observations, normally taken twice daily.

5. Coast Guard Station observations.

6. Satellite observations available from the GOES satellite every thirty minutes.

7. Surface observations from cooperating weather observers, such as those that assist in hydrologic, climatological, and marine programs, state police, power companies, off-duty Weather Service employees, etc.

8. Aircraft observations.

9. Ship observations.

10. Telemetered observations from automatic stations.

Table 2.
Types of Hazardous Weather in the Eastern United States

1. Heavy Rain.

2. Flash flooding.

3. Heavy Snow.

4. Freezing precipitation.

5. Poor visibility (Highway, Aviation, Marine).

6. High winds (other than that associated with severe convection activity).

7. Severe thunderstorms and tornadoes (high winds, hail, lightning, heavy showers).

8. Low Cloud ceilings at aviation terminals.

9. Strong low-level wind shear.

10. Coastal or lake flooding.

11. High waves.

12. Aircraft turbulence (clear air or mountain wave).

13. Aircraft icing.

14. River flooding.

15. Frost, freeze, and cold wave.

16. Heat wave.

17. Air pollution potential.

18. Fire weather potential.

19. Beach erosion potential.

1. Heavy Rain Procedures

a. Scan hourly surface observations. Is there a report of moderate
    or greater intensity rain within 100 km of the area of interest?

b. Scan most recent radar observations. Is there indication of
    moderate or greater intensity rain within 100 km of the area of
    interest?

c. Is there a report from cooperative weather observers indicating
    25 mm of rainfall in one hour, 50 mm in six hours, or any amount
    exceeding the 3-hour minimum thought to be required for flash
    flooding?

d. Does information available from satellite pictures indicate
    significant slow moving convective activity within 100 km of
    the area of interest, or upslope flow of moisture originating
    from a marine area?

If the answer to any of these questions is "yes", or if you have
other reason to believe heavy rain is possible, then proceed as
follows:

e. Plot and analyze precipitation amount and intensity reports in
    and near the area of interest at least once every three hours.
    Solicit additional reports for this purpose from cooperative
    observers and remote precipitation gages.

f. Plot hourly any special radar reports. Keep a running account (sum) of hourly estimated precipitation amounts. Coordinate with radar operator. Incorporate radar information into precipitation amount analysis.

g. Solicit pilot reports to determine coverage of convective activity, especially for locations beyond radar range.

h. Keep account of areas of heaviest precipitation as estimated subjectively from satellite pictures. Incorporate satellite information into precipitation amount analysis.

2. Flash Flooding Procedures

a. Know the current guidance concerning minimum amount of precipitation required for flash flooding in area of responsibility.

b. Determine if this minimum amount is being approached or exceeded. (see Item 1, Heavy Rain, for procedure to be followed).

3. Heavy Snow Procedures

a. Scan hourly surface observations. Is there a report of moderate or greater intensity snow within 100 km of area of interest?

b. Scan most recent radar observations. Is there a report of moderate or greater precipitation within 100 km of area of interest, and the precipitation type is known to be snow?

c. Is there a report from cooperative weather observers indicating moderate or greater intensity snowfall, or current rate of snow accumulation equal to or greater than 25 mm per hour?

d. Does information available from satellite pictures indicate significant precipitation within 100 km of area of interest? Do satellite pictures indicate upslope flow of moisture that originated over a marine area?

If the answer to any of these questions is "yes", or if there are other reasons to believe that heavy snow may occur, such as lake or topographical effects then proceed as follows:

e. Plot and analyze snowfall amount and intensity reports at least once every three hours. Estimate snow accumulation from duration and intensity reports as follows:

Snow Intensity	Average Snowfall Rate
Light (S-)	5 mm/hour
Moderate (S)	25 mm/hour
Heavy (S+)	40 mm/hour

f. Plot hourly any special radar information. Keep a running
account (sum) of hourly estimated precipitation. Coordinate
with radar operator. Subjectively incorporate radar information into snowfall amount analysis.

g. Keep account of areas of heaviest snow as estimated subjectively from satellite pictures. Incorporate into snowfall
analysis.

4. Freezing Precipitation Procedures

a. Scan hourly surface observations for reports of freezing
precipitation within or approaching area of interest.

b. Scan hourly surface and radar observations, and satellite
pictures for indications of liquid precipitation within or
approaching that part of the area of interest in which the
surface temperature is near or below freezing, or expected
to fall below freezing.

5. Poor Visibility Procedures

a. Scan hourly surface observations and three-hourly Coast Guard
reports within or near area of interest for reports of visibility below, or decreasing to near, the value at which visibility becomes a hazard to highway, aviation, or marine travel.

b. Receive reports, from established network of cooperative
observers, of visibility below the threshold critical value
determined to be a hazard to highway, aviation, or marine
travel.

c. Examine satellite pictures for indications of fog and stratus
over the area of responsibility. Be concerned about the advec-
tion of thick fog into a critical area.

6. High Winds (other than those associated with severe convective
activity)

a. Scan hourly surface observations within or near area of responsibility for surface wind speed reports above or approaching
the threshold value considered a hazard.

b. Where surface weather maps show tight pressure gradients in
unstable areas, examine hourly surface reports for indications
of further increase in the pressure gradient.

c. Receive reports from established network of cooperative
observers, including tower reports, for wind speed above a
predetermined critical value.

d. Examine satellite pictures for movement and change in intensity of a storm that may produce high winds.

7. Severe Thunderstorm or Tornadoes

a. Scan hourly surface observations. Is there a report of thunderstorms within 100 km of area of interest?

b. Scan hourly radar observations. Is there a report of strong or greater intensity echoes within or immediately adjacent to area of interest? Are there any other radar indications of possible severe weather, e.g. strong intensity, rapid growth, height of tops, line echo wave pattern (LEWP)?

c. Is there a report from cooperative weather observers indicating severe convective activity (high winds, hail, etc.)?

d. Is there a pilot report indicating high and rapidly growing tops?

e. Do satellite pictures contain indications of significant convective activity within 100 km of area of interest, or a triggering mechanism approaching an unstable area?

f. Is the latest observed lifted index ≤ 0 as determined from a nearby sounding?

If the answer to any of these questions is "yes", or if there is other reason to believe that severe convective activity may occur then proceed as follows:

g. Solicit report of severe weather activity from cooperative observers and pilots.

h. Examine satellite pictures for development and movement of convective activity. Coordinate with SFSS.

i. Plot hourly any special radar reports. Keep account of intensities. Coordinate with radar operator.

j. Plot and analyze hourly weather observations to describe what is happening on a small scale. Incorporate radar and satellite information in analysis.

k. Using all information available, keep current on the path, location and rate of movement of severe activity.

8. Low Cloud Ceilings At Aviation Terminals

   a. Scan hourly and special surface observations. Is there a report
      of ceilings that indicate an amendment should be considered to
      the aviation forecast for any of the terminals in your area of
      interest?

   b. Examine satellite pictures for development and change in cover-
      age of low clouds.

9. Strong Low-Level Wind Shear

Vertical wind shear causes changes in air speed and direction which
can be critical if corrective action is not taken by an aware pilot,
especially during approach or takeoff.

   a. Does surface and low-level upper-air observations, or pilot
      reports, available from within and near area of interest
      indicate a sharp vertical or horizontal change in wind speed
      or direction?

   b. Do surface observations or satellite pictures indicate the
      presence of mountain wave clouds, or boundaries of sharp
      discontinuities in the wind flow?

   c. If the answer to either of these questions is "yes", then soli-
      cit pilot reports and, if available, wind reports remoted from
      towers. Determine if low-level wind shear conditions are appro-
      aching or exceeding limites considered hazardous for aircraft
      operations.

10. Coastal or Lake Flooding

   a. Be familiar with where flooding would occur and the degree
      of flooding for the full range of possible water surface heights.

   b. Keep informed of predicted astronomical tide conditions. Take
      note of, and post, dates and times of unusually high (Spring)
      astronomical tides and the serious flooding conditions that
      may result if storm surges are superimposed on the high tide.

   c. Be aware "if, when and where" a report of observed deviation
      from astronomical predicted tide height would produce at least
      minor flooding if the deviation were superimposed on the next
      astronomical high tide.

   d. Arrange for cooperative observers to report when minor coastal
      or lake flooding is beginning or soon will begin if the water
      continues to rise.

   e. Scan Coast Guard Station reports and ship reports near the coast
      or on the lakes for indications of strong winds with an onshore

component, and high wave conditions, both favorable for producing storm surge. The larger the duration of these conditions,
the greater the threat of storm surge.

f. Examine satellite pictures for indications of movement and
change in intensity of storm with which flooding may be associated.

11. High Coastal or Lake Waves

a. Scan ship and Coast Guard Station observations for reports of
high waves.

b. Have cooperative observers report when they note wave heights
approaching or exceeding some predetermined critical value.

c. Examine satellite pictures and radar observations for intense
convective cells over marine areas. Examine satellite pictures
for indications of movement and change in intensity of storm
that may cause high waves.

12. Aircraft Turbulence (clean air, or mountain wave)

Receive pilot reports of aircraft turbulence. (Also, see Item 9.)

13. Aircraft Icing

a. Receive reports of aircraft icing.

b. Examine satellite pictures for movement of clouds associated
with aircraft icing reports.

14. River Flooding

a. Receive reports from remote gages, cooperative observers, and
hydrologists to determine the current stages of rivers in your
area of interest.

b. From river and rainfall statements know the forecast stages
for the rivers, based on past precipitation.

c. Know what the forecast stages for the rivers would be if given
amounts of additional precipitation occur. For this information
see Statements, routinely issued, and special hydrologic guidance issued in critical situations.

d. Keep abreast of the amount of precipitation that is falling by
following the procedures suggested in Item 1, Heavy Rain.

e. Know the effects of melting snow and ice-jam breakups.

15. Frost, Freeze and Cold Wave

a. For the cold season, be aware of the spatial variations of
   minimum temperature (cold spots) in the area of interest.
   This variation is a function of radiation conditions and the
   general synoptic situations. Topography, vegetation (or lack
   of) and proximity to heat and cold sources are important considerations.

b. Be aware of the current effects of frost, freeze and cold wave
   on agriculture, property, and human activity. Receive reports
   from cooperative observers in temperature sensitive areas.

c. When the temperature falls below 5°C and damage or discomfort
   would occur from a frost, freeze or cold wave, maintain a close
   watch for further temperature drop and be alert to the need for
   public advisories.

d. Examine satellite pictures for nighttime cloud-free areas that
   are favorable for radiational cooling. Monitor enhanced I.R.
   pictures for clues to falling temperatures in cloud-free areas.

16. Heat Wave

Is the combination of high temperature and high humidity causing
human discomfort? If yes, and these conditions are expected to
continue, with stagnant circulation, then consider if heat wave
advisories are desirable.

17. Air Pollution Potential

a. Be aware of the seasons and locations, within your area of
   interest, in which air pollution has been or can be a problem.
   For these locations and seasons, if light winds and stable
   atmospheric conditions persist. Consider the advisability of
   issuing air stagnation advisories. (Refer to latest upper-air
   observations and, if available, tower observations.).

b. If the information is available, be aware of the air quality
   and its trend.

18. Fire Weather Potential

If forests in the area of interest have been without rain for several
days, the daily average relative humidity drops to below 25%, and
wind speeds exceed 7 mps there is probably a high fire weather
potential.

19. Beach Erosion Potential

a. Be aware of beaches that are vulnerable to erosion.

b. Receive reports from cooperative observers when they detect erosion.

c. Be concerned about beach erosion if the following setup has existed for at least eighteen hours:

(1) A fetch of at least 500 km directed along the coastline, with an onshore surface wind component of between 0° and 40° (shore to the right, or counterclockwise, from direction of fetch).

(2) A pressure gradient of at least 4 mb./200 km occurring somewhere along the fetch during the setup time.

(3) High astronomical (Spring) tides will be occurring.

d. Examine satellite pictures for indications of movement and change in intensity of a storm that could cause beach erosion.

ACCURATE 24-HOUR WEATHER FORECASTS:

AN IMPENDING SCIENTIFIC BREAKTHROUGH

P.M. Wolff
P.G. Kesel

Ocean Data Systems, Inc.
Monterey, California U.S.A.

Abstract

Efforts over the last two decades in satellite observing techniques and computer solutions of the atmospheric equations are
believed to make possible accurate one-or two-day predictions of
the atmosphere-ocean system in the next five to ten years. Several independent areas of activity are identified which will combine to enhance man's knowledge of his environment. The volume
of satellite observations expected and the required computer
capabilities are defined in some detail.

Introduction

It is always difficult to predict the increase in scientific
knowledge in a particular area as long as the new developments are
in the realm of theoretical research. However, when successful
theoretical investigations of a problem reach a certain point, the
application of engineering technology is often useful to solve the
problem. At this point it is then possible to determine the future
course of the development and define the components that are necessary in order to produce a useful system. The development of
predictive meteorology has reached such a point.

The applications of computers to the solution of the meteorological equations and the applications of remote sensing to the
problem of observing the atmosphere and the ocean surface have been
progressing over the last twenty years. For some time, improvements in the short-range forecasts were difficult for the layman to
detect. Now these developments have reached a point where discern-

ible trends (or particular events) in each of four different areas
show the promise of combining to produce an increase in weather
prediction accuracy of almost incalculable impact. A threshold
will be reached and the additional attainable gains in accuracy
will make almost mandatory the use of these improved predictions
in any aspect of economic behavior in which the environment is
an important factor.

The purpose of this paper is to define and to examine each
of these areas necessary for the better predictions with particular
emphasis on the use of satellite observing systems. In addition,
the computer capacity and characteristics necessary to solve the
predictive equations will be defined.

These four elements are:

1. Certain international programs which will act as a
catalyst to produce action in the next few years.

2. Improvements in satellite systems for observing and
communication.

3. Improved and more complete model physics.

4. Advanced computer systems of the necessary speed and
power.

Each of these developments will be discussed in the following
sections.

International Programs

The international programs which are exciting developments
are these:

1. The First Global GARP Experiment scheduled for 1978-1979.

2. The establishment of the European Centre for Medium-Range
Weather Forecasting (ECMWF).

3. The availability of high-speed point-to-point communications
circuits to transmit weather observations.

The First Global GARP Experiment will stimulate the collection
of conventional and new types of observations of the ocean and the
atmosphere on a global basis. Member states of the World Meteorological Organization (WMO) will increase their expenditures on
environmental observations before and during this period. Sufficient data will be collected to provide realistic data bases

against which to check model performance.

The European Centre has been established as a cooperative effort among European nations in order to make useable forecasts in the four to seven day range. This development is important for the following reasons:

1. It collects a group of outstanding scientists in one place with a single purpose and adequate computer power.

2. In order to make a forecast accurate for four days, the accuracy for the first day will have to be of the order of 75-80% compared to 40-50% available now.

3. The initial data needs of this group should have a good influence on WMO activities.

Weather communications have historically been conducted over the Global Telecommunications Service (GTS) of the WMO. These slow, unreliable circuits are in the process of being replaced by point-to-point links operating at computer or satellite circuit speeds with the resultant increase in capacity, speed and reduced error rates.

Improvements in Satellite Systems

Because of limited space in this paper, it is proposed to discuss the improvements in satellite observation programs mainly by defining the unique approach and capabilities of the NASA efforts which will culminate in the launch of SEASAT A in April 1978. This satellite effort has the following attributes which are important for the reasons stated:

1. The capabilities were developed in response to the requirements of a users group which came into existence six years before the satellite launch.

2. The development of the satellite capabilities has been an engineering effort directed by personnel from NASA and Jet Propulsion Laboratory with a long history of competent successful space programs.

3. The satellite uses active instruments in space such as compressed pulse altimeter and a synthetic aperture radar.

4. Physical variables on earth are measured through computer algorithms which combine data received from more than one sensor simultaneously.

5. Although committees and teams have participated in the
planning and execution of the program, there is a single focal
point of responsibility for each instrument, each interpretation algorithm and each user experiment.

Table 1.   Changes expected in real time data between 1975 and 1980.

REAL-TIME DATA BASE

YEAR	REPORTS DAILY
1975	
Surface Land Reports	18,000
Ship Reports	2,600
Upper-Air Soundings	1,200
Aircraft Reports	1,900
Bathythermograph Reports	150
(Satellite) Upper-Level Wind Vectors	150
(Satellite) Temperature Profiles	1,200
1980* Satellite-Sourced Measurements	
Temperature/Humidity Profiles	20,000
Marine-Wind Vectors	300,000
Upper-Level Wind Vectors	15,000
Spectral Sea-State Reports	2,000
Sea Surface Temperatures	50,000

*In addition to 1975 earth-sourced measurements shown.

Nagler and McCandless (1975) describe the capabilities of
this satellite in detail.

Figures 1 and 2 illustrate current data coverage in the Northern and Southern Hemispheres. It is immediately apparent that
SEASAT A observational capability is concentrated in areas which
have the fewest observations now.

Figure 3 shows the coverage to be achieved by the GOES satellite
system- These satellites have a great potential for transmission
and retransmission of data from earth-based platforms which are
not being utilized.

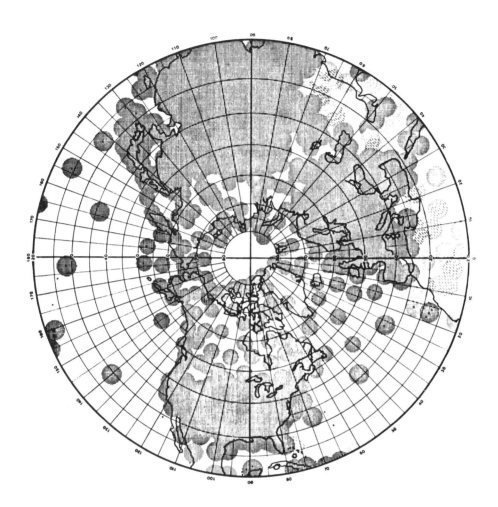

Figure 1.   Surface and upper air observation network-
            Northern Hemisphere (from GARP Pub. No. 11).

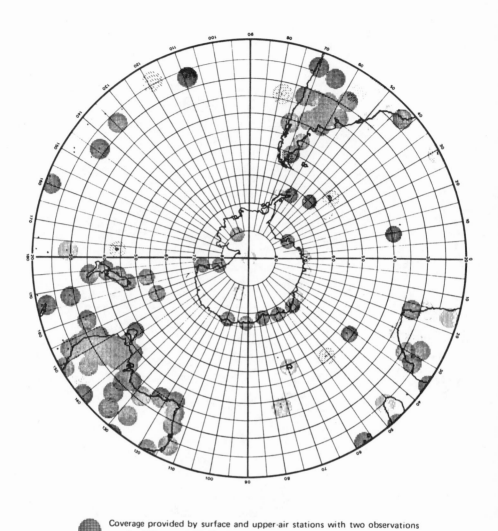

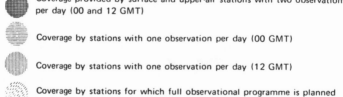

Coverage provided by surface and upper-air stations with two observations
per day (00 and 12 GMT)

Coverage by stations with one observation per day (00 GMT)

Coverage by stations with one observation per day (12 GMT)

Coverage by stations for which full observational programme is planned
by 1975

Figure 2.   Surface and upper air observation network-
            Southern Hemisphere (from GARP Pub. No. 11).

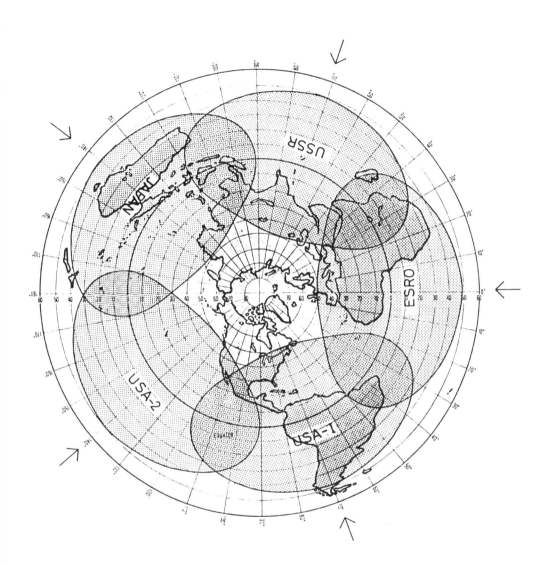

Figure 3.   Proposed GOES satellite coverage
            (from GARP Pub. No. 11).

Definition of the Model Physics Problem

Most activities, whether civilian or military, require adequate
environmental information -- preferably forecasts. Adequacy may be
defined in terms of accuracy, timeliness, coverage, scope and rele-
vancy.

A typical environmental "cycle" for producing such information
contains many steps -- and each takes time.

Table 2.   Elements of environmental cycle.

OBSERVATION

DATA COLLECTION

AUTOMATIC DATA PROCESSING

ANALYSIS OF INITIAL STATE

PREDICTION

DERIVED PRODUCTS

DELIVERY

DISPLAY

INTERPRETATION

Clearly, both automation and computerization are needed to
get such perishable information delivered in the allowable time.

The function of mathematical/physical (computer) models is to
describe the state of the atmosphere in a manner dictated by user
requirements, yet consistent with primary solution constraints
(the allowable time, available computer resources, problem characteristics, the data base). Given these primary constraints, it is
then possible to develop a set of secondary solution constraints
(grid resolution, model type, numerics and physics) which impact
on predictability. In this section, the authors discuss and define
the numerical weather prediction (NWP) problem, and provide an
assessment of computer resources needed to execute any one of a
hierarchy of models of varying resolution.

Before numerical prediction, weather forecasting had no consistent skill, although individuals might occasionally make forecasts
of considerable accuracy. Since 1950, the record of accuracy of
30-day forecasts has been maintained by the National Meteorological
Center of NOAA. Their skill score record has been normalized and
converted to a scale of zero to 100 with averaging from homogenous
periods. The results of these somewhat subjective operations are
shown in Table 3.

Table 3. National Meteorological Center of NOAA's record of
          accuracy of 30 day forecasts. Scale: 0 = No Skill,
          100 = Perfect Forecast

| YEAR | PERCENT | | COMMENTS |
	SURFACE	500 MBS	
1950-1955	10	10	Manual Procedures; Extrapola
1956-1958	12	15	Simple Computer Model; One Level
1959-1962	14	27	Improved Theory, One Level (Still)
1963-1966	25	30	Three Computational Levels
1967-1970	35	45	Complex (PE) Model; Six Levels
1971-1972	47	58	Improved Numerics, Initialization
1973-1975	40	47	

Table 4. A possible record of expected accuracy in the future if
          all the events discussed in this paper materialize.
          Scale: 0 = No Skill, 100 = Perfect Forecast.

| YEAR | PERCENT | | COMMENTS |
	SURFACE	500 MBS	
1977	52	62	Medium Resolution; Static Initial-ization; Improved Numerics.
1978	58	72	Increased Resolution; Dynamic Initialization.
1979	62	76	Satellite Data Base; 4D Data Assimilation.
1981	65	80	Improved Boundary-Layer Physics.

A. Analysis Models

The purpose of an objective analysis is to depict all of the observed variation of the atmosphere consistent with the primary and secondary solution constraints. In some instances, it may suffice to describe only the larger scales (horizontal wavelengths greater than several hundreds of kilometers) of motions, processes and/or phenomena. In others, it may be necessary to describe the fine-scale structure of the given parameter. If an analysis is to be used to provide initial values for a prediction model, then the mass structure and winds should be coupled in such a way as to facilitate trauma-free predictions of the observed tendencies of the state parameters. If, on the other hand, the analyses are to be used to convey qualitative impressions to a user, less sophisticated techniques may suffice. Analyses should be performed on the same (three-dimensional) grid being used for prediction and, to the extent possible, should exhibit comparable complexity (numerics and physics). Enough resolution should be used to describe the smallest desired/necessary scales of motion, but not so much that the problem becomes unmanageable computationally. Even so, the computer time generally used for analysis tends to be small compared to the requirement for prediction. The exception to this statement is related to the (possible) incorporation of off-time satellite data into analysis structures.

B. Prediction Models

Two classes of models are in common usage: gridded models and spectral models. With gridded models, a continuous atmosphere is replaced by its discrete analogue; i.e., by values of each variable located at regularly-spaced points of a suitable grid lattice. With spectral models, the dependent variables are expanded in a truncated series of linearly-independent, differentiable basis functions (such as the spherical harmonic functions which are used for global prediction). Although spectral models do exhibit some advantages (no "pole" problem; no aliasing; no spatial truncation), the computational disadvantage has prevented/delayed their operational usage until recently. The remainder of this material, therefore, deals with gridded models and the related computer requirement. Samples of grids are shown in Figures 4, 5, and 6.

One additional clarification is needed. Atmospheric flows and processes occur in a spectrum of spatial scales ranging from centimeters to thousands of kilometers. The operational data base is not adequate for describing small scale variations of the basic parameters (wind, temperature, pressure, humidity). As a consequence of this and other constraints, many modelers tend to regard the "weather problem" as consisting of the "larger scales" of the "basic parameters". This is unfortunate, but derives from the inability of dynamic models to generate useful information about

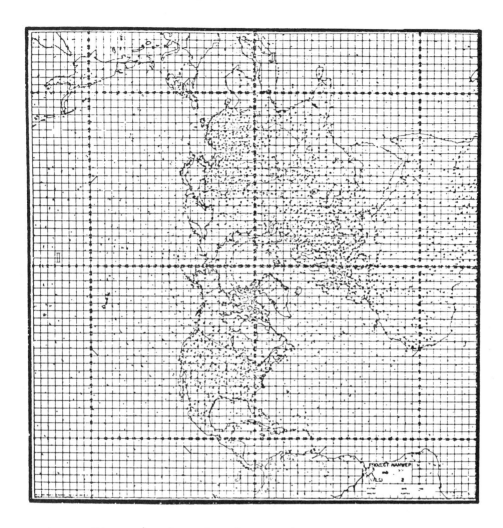

Figure 4.   Example of a prediction model grid.

"sensible weather" such as: visibility, fog, clouds, precipitation,
or ground-level temperatures and winds. What, then, is the model
and what are its outputs?

     The word "model" is used herein to denote a system of predic-
tion equations; all of the physics, physical assumptions and algorithms; all of the numerics, finite-difference methods and computational procedures; the grid lattice; and an appropriate data
base. One seeks a numerical solution to this set of non-linear
partial difference equations in which there are terms representing
physical/dynamical processes and effects which, in the aggregate,
contribute to the forecast. Such processes include: evaporation

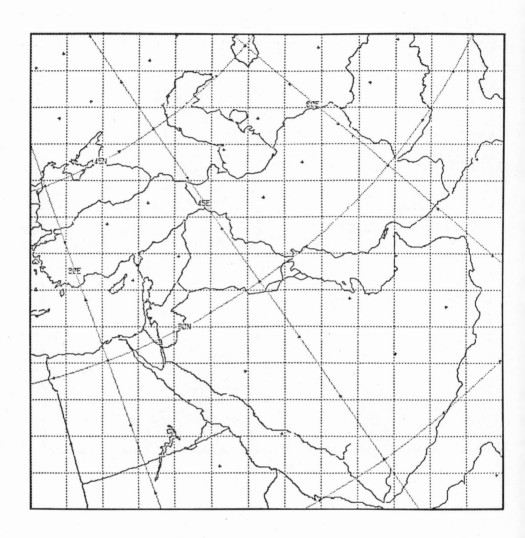

Figure 5.   Example of a prediction model grid.

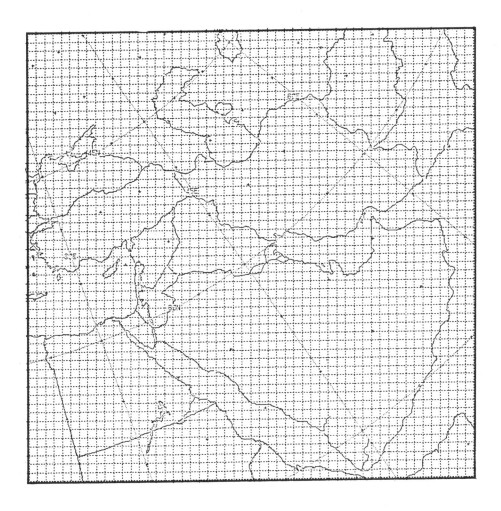

Figure 6. Example of a prediction model grid.

and condensation (synoptic scale and convective); sensible heat exchange (air-ocean and air-land); solar and terrestrial radiation; mountain and frictional effects; land-sea-ice discrimination; three-dimensional transport and divergence (convergence) of heat, momentum and moisture, adiabatic temperature changes, and diffusion. Dry convective adjustment precludes hydrostatic instabilities. Moist convective adjustment (the cumulus parameterization) redistributes heat/moisture in each vertical column, especially in tropical latitudes.

Table 5.   Theoretical and practical limits of predictability peculiar to models of the type discussed in this paper.

TYPICAL SCALE LENGTH (KILOMETERS)	DISTURBANCE TYPE	DETERMINISTIC PREDICTABILITY	
		THEORETICAL	PRACTICAL
10,000	Ultra-Long Waves	10-14 Days	3-6 Days
5,000	Long Baroclinic	3-6 Days	1-3 Days
2,500	Short Baroclinic	1-2 Days	0.5-1.5 Days

50	Cloud Clusters	3-6 Hours	NONE
5	Cloud Cell	1 Hour	NONE

NOTES: (1) Theoretical limits taken from Robinson (1967) and Lorenz (1969). Depend on growth of error energy and the fraction of total energy it represents.

(2) Practical limits estimated from operational experience in 1970-1975 period.

The "model" does not contain an explicit, complex, planetary boundary-layer parameterization. Photochemical effects are not modeled either. The output set consists of arrays of temperatures, wind components, and geopotential-heights at all model levels; of moisture at half of the model levels; precipitation; and sea-level pressures. Sensible weather at the surface is not included.

Finally, the definition of our model must include some of its numerical attributes. First, the method used for performing the time integrations will be explicit. Semi-implicit integration procedures may be used without loss of accuracy. Generally, semi-implicit methods are two to three times faster (depending on memory size and data transfer rates) than explicit methods, but require significantly more memory if one wishes to minimize the input/output (I/O) time. Next, we use second-order approximations to differential operators. Clearly, fourth-order differences may be used to improve the phase speeds of short waves, but stability considerations require a shorter time step which increases computation time by about one-third. Finally, we assume that the model has been suitably initialized. The computer time necessary for this task ranges from (the equivalent of) a few hours to a few days of model integration time. Each of these factors may be used by the reader to modify the tabular estimates provided.

Table 6. Typical model outputs.

Sea-Level Pressure

Marine-Layer Winds

*Sea State

*Sea Surface Temperature

*Atmospheric (Surface to 100,000 Feet)
Heights, Temperatures, Winds and
Relative Humidities

Precipitation

Surface Heat Exchange

*Ocean Thermal Structure

*Ocean Currents

*Water Levels (Storm Surges)

*Objective Fronts

*Not included in sizing estimates.

Sizing the Computer Problem

It is important for the reader to understand that the "weather problem" is, and will probably remain, as big a computational problem as one possesses the resources to address. The computational problem grows in proportion to the processes, the spectrum of scales, and the effects being treated. Even with requisite computer power, other limiting factors will remain operative: the data base; the physics; the numerics. As the problem grows, decision makers must continue to examine the cost-benefits ratio for any proposed weather solution. To achieve the next (model) performance increment may or may not be worth the expense. Potential buyers of large scientific computer systems may settle for something far less than a design to requirements! In this paper, the weather problem is defined in such a way as to exclude the solution elements requiring disproportionate expense/effort, as well as the techniques/procedures in the development stages.

We offer one example of support system capabilities being excluded from the problem definition --- the environmental support for electro-optical (E/O) systems. This requires forecasts of turbidity, (optical) turbulence, ambient effects, absorption, refraction, and wind. Turbidity is based on knowledge of hygrometeors (fog, rain, drizzle, clouds) and aerosols (dust, smoke, salt). Optical turbulence which contributes to (laser) "beam wander" and "scintillation", is based on atmospheric periodicities in the one-second to ten-second range --- which is several orders of magnitude smaller than "meteorological" turbulence. Ambient conditions refer to optical backgrounds affecting signal-to-noise ratios. Refraction arises from spatial gradients of the atmospheric refractive index --- a measure which depends on humidity, temperature, and pressure. Wind is important since it alters the distribution of particulates in the beam path. Thus, to predict optically-relevant E/O parameters is far more difficult/expensive than to predict sensible weather parameters at the surface --- which is far more difficult/expensive than it is to predict the basic parameters with the models described in this paper.

A. Matching the Model and System

The actual design of a prediction model tends to happen in four stages: (1) the combination of a user-driven forecast problem, a data base, and available resources (funding, personnel, allowable time to generate the forecasts) tends to suggest a model (its type, complexity, and resolution); (2) a specification of the computer system (either a "design to requirements" or a "design to costs", or something on that range) is made; (3) a computer system (which satisfies most mandatory requirements and has some desirable features) is procured; and (4) the original attributes of the model are modified to accommodate the new "available" computer resource. In this brief section, the authors tend toward this sort of pragmatism.

Figure 7 contains a schematic of the factors which affect model design along the ultimate path of meeting the needs of the user community. The users tend to define the problem. The main steps in the solution path are related to assembling a data base, processing the data in order to pose the initial state for prediction, and executing a prediction model with suitable numerics and physics as to enhance the predictability of any changes in that initial state.

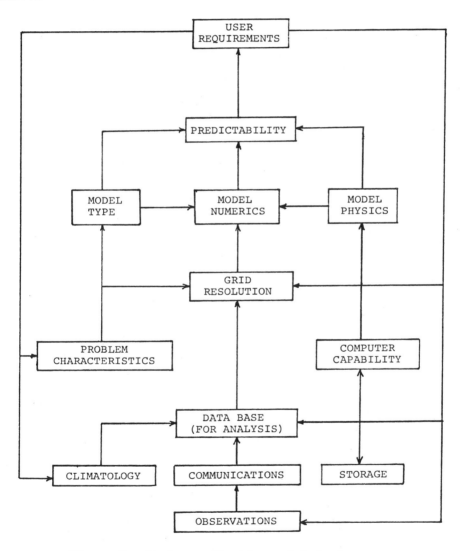

Figure 7. Factors affecting model design.

Because the forecast errors grow rapidly with forecast hour, users generally desire frequent short-range forecasts. Thus, the forecasts must be generated twice each day in a period of one to two hours; that is, the allowable computer time for operational prediction is from 30 to 60 minutes per forecast day!

The factors which impact on the computer requirements for models are: (1) allowable time; (2) the areal extent; (3) the grid resolution; (4) model complexity; and (5) the numerics (order of finite differences; the integration method; computational devices; stability considerations).

Resolution is important to adequate treatment of small scales. (Coarse resolution leads to undermovement of the shorter waves through spatial truncation errors in the finite-difference operators). But, computer power and time determine the number of grid points at which the equations can be solved, as well as the number of calculations allowed per grid point (complexity).

B. Computer Requirements for Models

System characteristics must be considered carefully. These include: the basic cycle time; types and speeds of operations; size/speed of both memory and storage; number and bandwidth of I/O channels; and the type of addressing. These are fixed characteristics. They are critical to the design of efficient program code.

In the seven tables to be referenced in ensuing sections, we will be considering models with the following attributes:

- Global (latitude-longitude) models

- Explicit solution method

- Medium complexity physics (300 calculations per grid point per time step)

- Allowable (wall) time is one hour per forecast day

- Vertical-section data structure

1. Computer Speed

Table 7 shows the number of floating-point calculations per forecast day for a matrix of models with varying resolution. The range is from 10^9 to 10^{11} calculations per forecast day (to produce the "basic" outputs). For more complex models, multiply by a factor of two.

Table 7. Floating-point calculations per forecast day (in billions)

GLOBAL GRID	NUMBER OF LEVELS		
	5	10	15
3°	1.3	2.6	3.9
2°	4.8	9.6	14.4
1°	36.0	72.0	108.0

Since we are allowed one hour per forecast day for the computations, the computer speed (in megaflops) can be readily obtained. Table 8 contains the megaflop rating needed to execute nine typical global models in the allowable time. These range from 0.4 for a 3° - 5 level model to 30.0 for a 1° - 15 level model.

Table 8. Computer performance needed to solve weather problem in allowable time, in millions of floating-point calculations per second (megaflops).

GLOBAL GRID	NUMBER OF LEVELS		
	5	10	15
3°	0.4	0.7	1.1
2°	1.3	2.7	4.0
1°	10.0	20.0	30.0

Table 9 shows estimated computer speeds for several Control Data Corporation systems. (Many of these entries have been verified with weather models.) Note that the CDC 7600 (at 3.0 megaflops) would be appropriate for a 2° - 10 level model. The STAR 100 series would be appropriate for high-resolution models (1° latitude-longitude), as would the CRAY 1 or TI(ASC).

Table 9:   Estimated Computer Speeds

MODEL	MIPS	MEGAFLOPS	
CDC 6400	0.8	0.18	
CDC 6600	2.7	0.6	
CYBER 175	7.2	1.6	
CDC 7600	13.6	3.0	
		SCALAR	VECTOR
STAR 100		1.1 - 1.3	20
STAR 100A		10	20
STAR 100C		20	60

Semi-implicit methods have a significant impact on such assessments -- since they reduce the explicit problem by a factor of two
to three. Vectorization is also an important (additional) consideration. Kesel et al (1976) provide useful information about the
vectorization of atmospheric prediction models.

2. Memory

The required data base for prediction models in often larger
than the available central memory. The techniques/procedures
depend on data structure and management. For explicit gridded
models, the data base should be partitioned in (IK) vertical sections. Using second-order differencing, calculations on the Jth
section require working memory for the following vertical sections:

- Data, Time "t", Section J-1
- Data, Time "t", Section J
- Data, Time "t", Section J+1
- Results, Time "t+1", Section J
- Output Buffer, REsults, Time "t+1", Section J-1
- Input Buffer, Time "t", Section J+2

(For fourth-order differencing, memory for two additional
sections is needed). Additional memory is also needed for temporary use and for code.

A semi-implicit solution involves solving a system of ellip-
tical two-dimensional PDEs encompassing the entire domain. To
minimize I/O leads to a memory-increment requirement for $I*J*K$
words (in addition to the needs of the explicit calculations).
"K" is the number of vertical levels in the model. If one desires
to minimize memory, the implicit calculations can be handled with
an increment of only $6*I*K$ words.

Table 10 shows the fast memory requirement for nine typical
global models. These are gridded, explicit models using vertical-
section data management. The entries range from 72K words for a
3° - 5 level model to 487K words for a 1° - 15 level model. Memory
for buffers and the operating system is not included. This requirement can be met readily with existing scientific computer systems.

Table 10. Fast memory for model variables (thousands of words).

GLOBAL	NUMBER OF LEVELS		
GRID	5	10	15
3°	72	124	176
2°	98	176	253
1°	176	331	487

3. Storage

Since it is assumed that the data base will exceed the size
of fast memory, the mass storage (MS) characteristics become critically important to the efficient utilization of complex architec-
tures. Extended memory is expensive. Most disk devices have adequate capacity (especially if the data base is distributed across
several I/O channels -- each with a device) at low cost. But, the
(effective) data transfer time can be prohibitive unless covered
by computations.

Generally, there are "K" variables which are temporally constant, about "4K" variables of a single time level, and "7K" variables at each of two time levels for each gridpoint in the horizontal domain (total = 12K variables per point). For rotating MS
and physical addressing, it is desirable/necessary to store one
complete copy of all variables on each of two (disk) "systems"
(a system may comprise many I/O channels and devices). One system
holds data for time levels, t and t+1, and the other holds data for
time levels, t and t-1. This facilitates simultaneous read/write
operations. Thus, the total MS requirement (if we neglect storage
for model outputs) would be for 2*(I*J*12K) full-word, floating-
point variables.

Table 11. Mass storage requirements, in millions of words,
           for nine model classes. Clearly, the capacity of
           MS devices is not an issue.

GLOBAL	NUMBER OF LEVELS		
GRID	5	10	15
3°	0.8	1.7	2.6
2°	1.9	3.9	5.8
1°	7.8	15.6	23.3

4. Data Transfer Rates

Using vertical section data storage, and 12*K variables per
horizontal gridpoint, we can calculate the data record/block sizes
for various grids. The number of variables per record is NUMVAR
(= 12*I*K words). Table 12 shows I/O record lengths for various
models.

For calculations proceeding on latitude circles, the requirement is to simultaneously read/write NUMVAR words from/to (rotating)
MS devices. For distributed data bases, these block sizes should
be suitably reduced.

Table 12. I/O record length, in kilowords (NUMVAR).

GRID	NUMBER OF LEVELS		
	5	10	15
3	7.2	14.4	21.6
2	10.8	21.6	32.4
1	21.6	43.2	64.8

Table 13 shows the sustained data transfer rates necessary
to keep the CP busy. The amount of CP time for calculations at
a given latitude circle is shown. We have assumed a 64-bit word
length. Clearly, multiple data paths (distributed data base) are
needed for high-resolution models.

In a virtual memory system, this becomes a requirement on
the implicit I/O rather than on the explicit I/O. To obtain the
corresponding rates for a semi-implicit model, reduce the rates
by a factor of two to three. Even so, the MS capability may be
a critical constraint on model design.

Table 13. Sustained data transfer rates, in megabits per second.

GLOBAL GRID	CP TIME PER LATITUDE CIRCLE (Secs)	NUMBER OF LEVELS		
		5	10	15
3	0.500	2	4	6
2	0.2083	7	14	21
1	0.0555	50	100	150

References

GARP Publication Number 11, "The First GARP Global Experiment-
    Objectives and Plans". WMO, 1973.

Kesel, P.G., Wellck, R.E., McHugh, R.A. and Barkai, D., "The
    STAR Computer and Numerical Weather Prediction: Adaptation
    and Vectorization Assessment of the UCLA General Circulation
    Model, with an Examination of Model and System Constraints
    for Efficient Utilization". ODSI Technical Report under BAO
    9530-139204, PO#1 to Control Data Corporation, Monterey, Cal-
    ifornia, May, 1976.

Lorenz, Edward N., "Studies of Atmospheric Predictability". Final
    Report AFCRL-69-0119 under Contract AF19 (628)-5826, 1969,
    142 pp.

Nagler, R.G., and McCandless, S.W., Jr., "Operational Oceanographic
    Satellites: Potentials for Oceanography, Climatology, Coastal
    Processes, and Ice". Jet Propulsion Laboratory of California
    Institute of Technology, September, 1975, 12 pp.

Robinson, G.D., "Some Current Projects For Global Meteorological
    Observation and Experiment". Quarterly Journal of Royal
    Meteorological Society, 93, pp. 409-418, 1967.

LIST OF CONTRIBUTORS

Co-Directors

Donald J. Clough
Professor of Engineering
University of Waterloo
Waterloo, Ontario, Canada

Lawrence W. Morley
Director-General
Canada Centre for Remote Sensing
Department of Energy, Mines
and Resources
Ottawa, Ontario, Canada

Organizing Committee

S. Galli de Paratesi
Joint Research Centre, Commission
of the European Communities
Ispra Establishment
21020 Ispra, Varese
Italy

B.P. Miller
Vice President Space Systems
ECON Incorporated
Nine Hundred State Road
Princeton, New Jersey, U.S.A.
08540

Gerd Hildebrandt
Professor and Director
Department of Photogrammetry
and Photo Interpretation
Institute of Forestry
and Forest Management
D-7800 Freiburg i.Br.
West Germany

Eric S. Posmentier
Institute Marine and
Atmospheric Sciences
City University New York
Wave Hill/Glyndor Building
675 West 252 Street
Bronx, New York, U.S.A.
10471

Working Group Participants

Dr. Peter W. Anderson
Director
Marine Protection Program
U.S. Environmental
Protection Agency, Region II
Edison, New Jersey, U.S.A.

Dr. Wolfgang Baier
Head, Agmet Section
Chemistry and Biology
Research Institute
Agriculture Canada
Ottawa, Ontario, Canada

Dr. Eric C. Barrett
Lecturer in Climatology
and Meteorology
Department of Geography
University of Bristol
Bristol, U.K., BS8 2SS

Dr. Charles Buffalano
Chief, Applications Assessment
Office (Code 100)
Goddard Space Flight Centre
National Aeronautics and
Space Administration
Greenbelt, Maryland,   20771
U.S.A.

Professor Vincent J. Cardone
Institute Marine and
Atmospheric Sciences
City University New York
Wave Hill/Glyndor Building
Bronx, New York, U.S.A.

Donald J. Clough
Professor of Engineering
University of Waterloo
Waterloo, Ontario
Canada
N2L 3G1

Dr. Robert N. Colwell
Professor of Forestry, and
Associate Director of
Space Sciences Laboratory
University of California
Berkeley, California, U.S.A.

Dr. John M. DeNoyer
Director, EROS Program
U.S.G.S., Department of Interior
1925 Newton Square East
Reston, Virginia, 22090
U.S.A.

Dr. J.L. D'Hoore, Professor
Fakulteit Der
Landbouwwetenschappen
Laboratorium Voor Bodemgenese
En Bodemgeografie
Centrum Voor Tropische Bodemstudie
de Croylaan 42
3030 Heverlee, Belgie

Dr. S. Galli de Paratesi
Joint Research Centre, Commission
of the European Communities
Ispra Establishment
21020 Ispra, Varese
Italy

James J. Gehrig
Professional Staff Member
Committee on Aeronautics
and Space Sciences
U.S. Senate, Senate Office Bldg.
Washington, D.C., U.S.A.

Hervé Guichard, Manager
Groupement pour le
Développement de la
Télédetection Aérospatiale
18 Avénue Edouard Belin
31055 Toulouse, CEDEX, France

Hans Heyman
National Intelligence Officer
Economics
Room 73-E7
Central Intelligence Agency
Washington, D.C., U.S.A.

Dr. Gerd Hildebrandt
Professor and Director
Department of Photogrammetry
and Photo Interpretation
Institute of Forestry and
Forest Management
D 7800 Freiburg, i.Br.
West Germany

Dr. J.A. Howard
Senior Officer
Remote Sensing Unit
Food and Agriculture
Organization of the U.N.
Via Delle Terme di Cavacalla
Rome, Italy

Professor A.D. Kirwan, Jr.
Department of Oceanography
Texas A & M University
College Station
Texas, 77843
U.S.A.

Dr. Philip A. Lapp
Philip A. Lapp Limited
Suite 302
14 Hazelton Avenue
Toronto, Ontario
Canada

Dr. A.K. McQuillan*
Research Scientist
Canada Centre for Remote Sensing
Department of Energy,
Mines and Resources
Ottawa, Ontario, Canada

B.P. Miller
Vice President
Space Systems
ECON Incorporated
Nine Hundred State Road
Princeton, New Jersey
U.S.A.

Dr. Lawrence W. Morley
Director-General
Canada Centre for Remote Sensing
Department of Energy,
Mines and Resources
Ottawa, Ontario
Canada

* - recalled because of family illness.

Dr. Richard Mühlfeld
Head of Photo and Astrogeology
Bundesanstalt für
Geowissenschaften
und Rohstoffe
3 Hannover 51, West Germany

Dr. Archibald B. Park, Manager
Earth Resources Applications
General Electric Co.
Space Division
5030 Herzel Place
Beltsville, Maryland, U.S.A.

Professor Eric S. Posmentier
Institute Marine and
Atmospheric Sciences
City University New York
Wave Hill/Glyndor Building
Bronx, New York, U.S.A.

Professor R.A.G. Savigear
Department of Geography
University of Reading
2 Early Gate
Whiteknights
Reading, U.K., RG6 2AU

Dr. Sigfrid Schneider
Professor & Scientific Director
Bundesforschungsanstalt für
Landeskunde und Raumordnung
53 Bonn-Bad Godesberg
West Germany

Dr. Charles Sheffield
Vice President
Earth Satellite Corporation
7222 47th Street (Chevy Chase)
Washington, D.C.
U.S.A.

Dr. Robert E. Stevenson
Department of the Navy
Office of Naval Research
Branch Office Pasadena
Scientific Liaison Officer
University of California
La Jolla, California, U.S.A.

Dr. Paul G. Teleki
U.S. Geological Survey - MS 915
United States Department
of the Interior
12201 Sunrise Valley Drive
Reston, Virginia
U.S.A.

Stanley E. Wasserman
U.S. Department of Commerce
NOAA, National Weather Service
Garden City, New York, U.S.A.

Dr. Paul Wolff
Vice President
Ocean Data Systems Inc.
Monterey, California, U.S.A.

NATO Observers

Professor Dr. Ing. E. Pestel
Lehrstuhl A. für Mechanik
Technische Hochschule Hannover
3 Hannover, West Germany

Dr. B.A. Bayraktar
Scientific Affairs Division
North Atlantic Treaty Organization
1110 Bruxelles, Belgium